THE CHAOS COOKBOOK

The Chaos Cookbook

a practical programming guide

Joe Pritchard

An imprint of Butterworth-Heinemann Ltd
Linacre House, Jordan Hill, Oxford OX2 8DP

 PART OF REED INTERNATIONAL BOOKS

OXFORD   LONDON   BOSTON   MUNICH   NEW DELHI
SINGAPORE   SYDNEY   TOKYO   TORONTO   WELLINGTON

First published 1992

© Joe Pritchard 1992

All rights reserved. No part of this publication may be reproduced in any
material form (including photocopying or storing in any medium by electronic
means and whether or not transiently or incidentally to some other use of this
publication) without the written permission of the copyright holder except in
accordance with the provisions of the Copyright, Designs and Patents Act 1988
or under the terms of a licence issued by the Copyright Licensing Agency Ltd,
90 Tottenham Court Road, London W1P 9HE, England. Applications for the
copyright holder's written permission to reproduce any part of this publication
should be addressed to the publishers.

NOTICE
The author and the publisher have used their best efforts to prepare the book, including the
computer examples contained in it. The computer examples have been tested. The author
and the publisher make no warranty, implicit or explicit, about the documentation. The
author and the publisher will not be liable under any circumstances for any direct or indirect
damages arising from any use, direct or indirect of the documentation or the computer
examples contained in this book.

TRADEMARKS | REGISTERED TRADEMARKS
Computer hardware and software brand names and company names mentioned in this book
are protected by their respective trademarks and are acknowledged.

This book was typeset and designed by M. A. McKenzie, is set in Donald E. Knuth's
Computer Modern Roman 10pt typeface and was reproduced from camera-ready copy
generated by TeX and LaTeX.

Printed and bound in Great Britain

ISBN 0-7506 0304-6

Library of Congress Cataloging-in-Publication Data
Available from the publishers

British Library Cataloguing in Publication Data
Available from the publishers

1 2 3 4 5 96 95 94 93 92

Contents

Preface		ix
Introduction		1
	The computer	2
	Graphics	2
	Memory	4
	Processing power	4
	Disc drives	6
	Interfaces	7
	Programming languages	8
	Numerical resolution	10
	Numerical accuracy	11
	Numerical overflow	11
	Types of computer	11
	Processing of screen images	12
	Saving video memory	12
	Getting a hard copy	15
	Data export	17
	Language versions used in this book	17
1	**What is chaos?**	19
2	**Iterative functions**	29
	The square root function	31
	The square function	31
	The function $2x(x-1)$	32
	The logistic equation	32
	Other functions	40
	Plotting the attractor	40
	Iterating two-dimensional functions	41
	BBC BASIC listings	45
	Turbo Pascal listings	50

3 Differential equations — 59
- Analysis — 60
- Graphical solution — 61
- A predator–prey model — 65
- Nomenclature of phase diagrams — 67
- The Runge–Kutta method — 70
- Dependence on starting conditions — 71
- Stability — 71
- Forced non-linear oscillators — 72
- More complicated systems — 74
- BBC BASIC listings — 75
- Turbo Pascal listings — 79

4 The Lorenz equations — 91
- First investigations — 94
- Feedback—the common factor — 97
- The Butterfly Effect — 97
- Further analysis of the equations — 99
- Phase portraits of the Lorenz equations — 100
- The strange attractor — 103
- BBC BASIC listings — 105
- Turbo Pascal listings — 100

5 Strange attractors — 119
- The Hénon attractor — 119
- Fractal dimensions and strange attractors — 122
- Other Hénon attractors — 123
- Strange attractors from differential equations — 125
- Other strange attractors — 128
- Strange attractors from real data — 131
- How can strange attractors be found? — 132
- BBC BASIC listings — 134
- Turbo Pascal listings — 137

6 The fractal link — 143
- The Koch curve — 144
- Recursion — 145
- Turtle Graphics — 147
- Fractal Brownian motion — 150
- Cantor set — 151
- Peano — 153
- Dragon and C curves — 155
- Sierpinski carpet — 155
- Drawing with the L-language — 156

CONTENTS   vii

    Fractals in the real world 159
    Fluid flow 161
    Catalysts and enzymes 162
    BBC BASIC listings 164
    Turbo Pascal listings 171

7 The Mandelbrot set   **185**
    Complex numbers 186
    The Mandelbrot set 190
    Nomenclature 193
    Interesting areas 196
    Colour selection and aesthetics 197
    Time-saving steps 198
    Calculation types 199
    Different starting z values 201
    Mandelbroids 202
    Three-dimensional representations 204
    Periodicity in the Mandelbrot set 204
    BBC BASIC listings 207
    Turbo Pascal listings 215

8 Julia sets   **229**
    Trajectories for the Julia set 231
    Plotting the Julia set 231
    Iteration number 232
    The value of c 233
    Periodicity and link to the Mandelbrot set 234
    Newton's method 237
    Quaternions 240
    Inverted Julia sets 241
    Some final comments 242
    BBC BASIC listings 243
    Turbo Pascal listings 249

9 Other fractal systems   **261**
    Fractal landscapes 261
    Mid-point displacement 262
    Fault line modelling 262
    Fractal plants and trees 264
    Iterated function systems 264
    The Collage Theorem 271
    BBC BASIC listings 276
    Turbo Pascal listings 281

10 Cellular automata — 289

- Dimensions in automata — 290
- One-dimensional cellular automata — 290
- The rule set — 292
- Non-totalistic rule sets — 295
- Mutation — 295
- Noise — 296
- Thermodynamics, reversibility and entropy — 296
- Simple two-dimensional automata — 297
- Selection in two-dimensional systems — 299
- Cell eat cell — 300
- Time dependent rules — 301
- The game of Life — 301
- Dynamic structures — 305
- Suggestions for further experiments — 305
- Applications of cellular automata — 305
- Modelling — 306
- Pattern design — 306
- BBC BASIC listings — 307
- Turbo Pascal listings — 317

11 Practical chaos — 335

- The Stock Market — 335
- Weather — 337
- The dripping tap — 337
- A chaotic pendulum — 340
- A driven pendulum — 341
- Electronic chaos — 342
- Driven LCR network — 342
- Plotting phase portraits with an oscilloscope — 344
- Driven Oscillator — 345
- Analogue computers — 346

Appendix — 349

Bibliography — 357

- Periodicals — 358
- Other reading — 360
- Software — 360

Index — 363

Preface

The subject of chaos and fractals has come into popular prominence in the last few years due in the main to the fact that the fractals can create stunning graphical images on computers. Indeed, many magazines have carried programs to generate what is probably the hallmark of the field, the Mandelbrot set. However, beyond the aesthetic pleasure of these images the science of chaos has implications for us all, in areas as diverse as next weeks weather, Jupiter's famous red spot and heart attacks and was actually born thirty years ago! Chaos is likely to cause as much change in the way that scientists think of the world as did the replacement of Newton's classical mechanics with the quantumn theory. Despite this, the basic principles are breathtakingly simple, and anyone with a home computer can get involved and conceivably make a contribution to this still young science.

I hope that this book will provide food for thought for computer hobbyists, students and anyone interested in the subject, and may provide a useful starting point before going in to the more specialist works in the field. The idea of this book is to provide a 'cookery book' for various chaotic systems, and, like all recipes, experimentation will often give new and unexpected results. So, use these programs as starting points for your own work.

Personally, my interest in this area goes back about seven years to when I came across Koch snowflakes for the first time on a small computer. After a while, I encountered the Mandelbrot set, the Lorenz equations and other aspects of the world of chaos, and began to write programs to generate the images seen in books and magazines. And that's where the problems started; I could find nothing to bridge the gap between the pretty pictures and the mathematics. This book, I hope, will provide a practical guide to bridge that gap, and to introduce the subject of chaos and fractals, using small computers as a laboratory for exploring the beautiful and fascinating world on the boundary of order and chaos.

No book is created without a team effort; in this case I'd like to particularly mention Andrew Parr and Peter Dixon, my publisher, for their encouragement and support on this project.

Finally, special thanks must go to my wife, Nicky, for support for what has probably appeared to be a rather odd interest of mine, and to all those people who've given me advice and inspiration during the preparation of this book. Thanks folks!

I would like to dedicate this book to the memory of
Julia Helen Crick,
a dearly loved and greatly missed friend.

JOE PRITCHARD
Sheffield
1991

Introduction

This book is about chaos, a relatively new science, in which the laboratory equipment consists of computer software, the results are often graphical displays of stunning complexity and the subjects of the experiments are numbers themselves. Indeed, many people have already suggested that the term 'Experimental Mathematics' be applied to the whole area of chaos theory and fractals. In this section of the book, I want to look briefly at how this book can be used, and at some of the techniques that we'll be employing.

I suppose that before going any further we should bring in a definition; unfortunately, chaos doesn't lend itself to rapid definitions, so I'll be giving a fuller one later in the book. However, for the time being let's just say that a chaotic system is a deterministic system with great sensitivity to initial conditions. In other words, the behaviour of such a system is predictable by the use of suitable mathematical equations *but varies greatly depending upon what the starting conditions are.* Thus a model of the world's weather on a computer (Chapter 4, The Lorenz equations) can give staggeringly different results when the value of an input parameter is varied by, say, 1 part in 1000. Intuition tells us that small changes in input should give proportional changes at the output; chaos shows us that this isn't necessarily the case.

This book is aimed at anyone with access to a computer running Turbo Pascal or BBC BASIC, although the programs in the book are presented in such a way that they can be duplicated in virtually any language you care to choose, provided that the language has the means of drawing graphical images on a computer display. The programs in the book are designed as starting points for further experimentation, and I'll be giving hints of further things to try in each chapter. It's hoped that the listings will give you a starting point for your own programs.

Although the subject is based in maths, this book is designed as a 'how to do it' book, so I've kept the mathematical content to a minimum. For those of you wishing to follow up some of the mathematics, I've included a substantial list of further reading and computer software. In addition to the computer-based work, I've included in Chapter 11 guidelines for some experimental work in chaotic systems that you might find in the real world. As to actually using this book, the newcomer might like to start at Chapter 1 and work on through; those with some knowledge of chaos might prefer to get right on with the programs featured. Finally, if there's enough interest, who knows; there might be a book called *Advanced Chaos* ...

The computer

To get the best out of this book you need a computer of some sort. I have concentrated on the IBM PC and clones and the BBC Model B. In order to save space, from here on I'll refer to the BBC Model B as the BBC Micro, and to the IBM PC and its clones as a PC. The main points about any computer used for chaos experiments are as follows.

Graphics

The computer needs some medium-to-high-resolution graphics capability to display the images generated. If you have a PC, then the only problem you'll encounter is if you've got a machine fitted with a text-only display adapter. Most graphics systems fitted to computers are colour, but this doesn't matter too much; very attractive monochrome images can be created.

More important than colour is the resolution of the graphics display. This is the number of individually addressable points available on the screen. I would say that a minimum requirement for the work featured in this book is about 200 horizontal by 200 vertical points (*pixels*) (200 × 200). You may find that some authors suggest a higher resolution graphics screen than you have on your machine; if this is the case, then my advice would be to try out the programs on your computer unless the author says the program definitely *won't* work. After all, experimentation is the spice of chaos!

Even if your machine cannot support graphics, some chaos work can be done with a text-only display, by printing out the values rather than graphing them. An example of this is given in Chapter 4 when the Lorenz equations are discussed.

The BBC and IBM PC are two machines capable of supporting good graphics. Here follows a brief description of the graphics modes available on the BBC and the common graphics standards for the PC.

BBC graphics modes

On the BBC Microcomputer, there is a trade-off between screen resolution and memory available for programs. Table 1 shows some common BBC

GRAPHICS

graphics modes. Modes 0 to 2 take up 20k of memory, and this leaves only around 5k or so for programs when the BBC has a disc drive fitted. Modes 4 and 5 take up 10k of memory.

Mode	Colours	Resolution
0	2 colours	640×256 pixels
1	4 colours	320×256 pixels
2	16 colours	160×256 pixels
4	2 colours	320×256 pixels
5	4 colours	160×256 pixels

Table 1. BBC Micro screen modes.

The other BBC screen modes are not really suitable for use in most of the programs in this book, but can be pressed into service for things like the Life program in Chapter 10.

IBM graphics adapters

PCs don't have a range of built-in graphics modes like the BBC; instead, they can be equipped with an add-in card and suitable display to suit the requirements of the user. Once upon a time, all PCs came with *no* display

Adapter	Resolution	Colours	BIOS Mode
CGA	40×25 and 80×25	16 colour text	
	320×200 pixels	4 colours	4 and 5
	640×200 pixels	2 colours	6
EGA	as CGA plus:		
	320×200 pixels	16 colours	13
	640×200 pixels	16 colours	14
	640×350 pixels	16 colours	15
VGA	as EGA plus:		
	640×480 pixels	2 colours	17
	640×480 pixels	16 colours	18
	320×200 pixels	256 colours	19

Table 2. Common IBM PC screen modes.

system at all; a shock to those of us who were more used to home computers! The most common graphics modes supported on IBM machines are shown in Table 2.

The VGA adapter is the latest available on most PC systems. The BIOS mode referred to above is a reference to how the IBM PC operating system BIOS (the program built into a computer that looks after such things as keyboards and displays) turns a particular graphics mode on and off. If you're interested in this sort of thing, I suggest you look at the listings in the Appendix for examples of use of BIOS calls, and look at some of the books listed in the Bibliography. Although it looks like VGA owners are spoilt for choice, it should be pointed out that not all programming languages and pieces of software can actually use all of the modes available unless direct BIOS calls are used.

Memory

The memory of a computer is the amount of space available in it for the storage of computer programs and other data needed for the correct running of a computer program. On BBC Microcomputers, the amount of memory available for the computer program depends upon the resolution of graphics used. On PCs, the amount of memory available is usually 512–640k (1k = 1024 bytes = 1024 characters) irrespective of the graphics mode in use. Some PCs are equipped with more memory than this, but it isn't usually much use to chaos experimenters unless some special programming techniques are employed, so I won't discuss it here. In the topics covered in this book, memory is used for:

1. Storing the program instructions needed.

2. Storing any data needed for the program to run properly.

3. Storing data used by the program or produced by the program.

In general terms, the larger the amount of memory available the quicker the program will run and the less likely the programs are to 'fall over' and fail to work properly.

Processing power

The processing power of a computer depends principally upon the type of central processing unit the computer uses. This is the chip at the heart of the computer that does all the work.

The BBC Microcomputer uses a 6502 chip, quite an old-timer in the microprocessor world. It's also not a terribly powerful chip, and users of the BBC Micro who need a lot of clout behind their programs tend to use second processors—add-on boxes that contain a further CPU that does the

PROCESSING POWER

actual processing in programs, leaving the BBC's own 6502 to look after just the keyboard and screen.

Life is more complicated with the IBM PC. There are three CPU chip families in wide circulation.

8088/8086: These chips are quite old, but are still to be found in many machines, such as the Amstrad PC1512, Toshiba 1200 and many other low-cost clones. Machines using these chips are often called PC machines, as the original IBM PC used one of these processors.

80286: This is a more powerful processor, which also can handle more computer memory than the 8086/8088 processor. Machines using this processor are often called AT clones.

80386/80386SX: These processors are very powerful indeed, and PC clones using these machines represent the most powerful machines in common use. The SX chip is a slightly cut-down version of the 80386, but for our purposes the SX chip is quite adequate.

As well as the processor type, the speed at which it operates is also quite important. For example, the BBC 6502 trundles along at 1MHz (1 million steps per second). This isn't the same as 1 million calculations every second. The 8086/8088 processors usually operate in the 4MHz to 8MHz range, 80286s up to 12MHz and 386SX/386 processors anywhere between 15 and 33MHz. You can get 20MHz 80286 chips which will outpace some of the slower 80386SX chips, but do not offer some of the more subtle advantages of the 80386SX. However, for chaos and fractals work, these chips are fine. Some machines are equipped with a 'Turbo' mode, in which a machine that normally runs at, say, 4MHz can, at the push of a button, run at 7 or 8MHz. The reason for having two modes is that some software requires that the computer be running at an official speed—in this case, the 4MHz speed—and may not function correctly if the higher speed is used.

A useful adjunct to the CPU for any computer that's doing a lot of numerical work is a special chip called a *maths coprocessor*. In the usual run of things, the CPU has to do all the arithmetic, and while it's doing this it can't do anything else. Mathematical operations are not usually the forte of computer chips; they're fine at doing simple arithmetic, but to do complex mathematics they have to execute programs of instructions and this takes time. A coprocessor is designed to do the complicated mathematical operations under the control of the CPU, and it will then pass the results back to the CPU when the calculations are completed. Such a device will make vast improvements in speed to the overall running of mathematically oriented programs, provided that the language used to write the programs can support the maths coprocessor. There is no dedicated maths coprocessor available for the 6502 chip, but for the others, the processor number

needed is obtained by replacing the 6 in the part name with a 7. So the
80286 requires an 80287 maths coprocessor.

Disc drives

The computer disc drive system is simply used for rapid storage and retrieval of data. It's not essential for chaos work; you can use cassette tapes for program and data storage on the BBC Micro, but it makes life easier! Programs can be retrieved more quickly from disc than from tape. On the PCs, there isn't an effective tape system available, and all clones need at least one disc drive, as described below. Furthermore, on PCs many computer languages use the disc as temporary storage whilst you are writing programs, so the disc drive effectively becomes an extension of your computer's memory.

BBC disc systems

There are two main systems on the BBC Micro. These are the 40-track disc, capable of holding 100,000 characters of data (100k) and the 80-track disc, which holds 200k. Figure 1 shows a typical *floppy* disc. The disc fits into a disc *drive* which allows data to be read from and written to the disc as a series of magnetic imprints on the disc surface. The discs are reusable and if you need to use another disc, you simply take the current disc out of the drive, put it in a storage bag, and stick in another one.

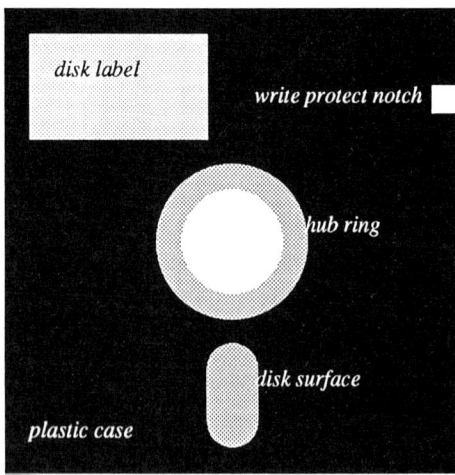

Figure 1. A typical floppy disc.

Floppy discs come in two physical sizes (5.25" and 3.5") and the amount of data that can be stored on them depends upon how the computer *formats* the disc to receive information.

PC disc systems

On PCs there are a variety of floppy disc formats to deal with:

3.5" There are two types of 3.5" disc, called *low density* and *high density*. Low density discs can hold about 720k of data, and high density discs hold about 1400k of data.

5.25" Again, high and low density discs are available, with low density discs holding 360k of data and high density discs holding 1200k of data.

Hard discs

All PC clones will come with at least one floppy disc drive. Some machines have two, but a more common configuration is to have one floppy disc drive and one hard disc. A hard disc is a *non-removable* storage medium; that is, when it fills up you have to erase some old data before you can save new material on to the disc. It's not easy to remove the disc and stick a new one in, but the disc is reusable, and many users move old data from their hard disc on to floppy discs, and then erase the old data from the hard disc to make space for new material. Hard discs can hold large amounts of data—anywhere between 20 million and several hundred million bytes—and are also much faster than floppy discs. A hard disc is well worth the extra cost as it usually makes programming easier and faster and is particularly valuable for storing and recovering screen image files or other large amounts of data.

Interfaces

Computer interfaces are the ways in which the computer talks to other devices in the outside world. Most computers come with at least an interface to allow the computer to send information to a printer to produce hardcopy results from programs. This is usually what is known as a *Centronics* interface, which is a standard means of communication between computers and printers. A further means of communication between computers and peripheral devices is via an *RS232* interface, which allows the computer to communicate with plotters, which can draw graphs in many colours, or allows data transfer between two different computers, even if they are of different makes. If you have an additional piece of equipment called a *modem*, you can communicate with other computers over the telephone line via the RS232 interface.

If you go on to experiment with chaotic systems in the real world, rather than various computer models, then you will require additional interfaces on your computer to convert things happening in the real world into electrical signals that the computer can understand. The BBC Micro is equipped with some of these interfaces, in the form of a *User Port* and *Analogue to Digital Converter* (ADC). The User Port allows simple electrical on or off signals to be read by the computer, whilst the ADC allows continuously varying voltages to be recorded on a computer. Quantities such as light level or temperature can thus be recorded on a computer via a suitable electronic circuit to convert, say, light levels into voltages.

Programming languages

Without suitable programs, computers are simply hi-tech door stops. These programs—sequences of instructions for the computer to follow to get a particular result— are written in particular computer languages. The CPU of a computer only understands one language. This is called machine code, and machine code programs are incredibly fast because the CPU can look at instructions in a machine code program and know exactly what to do without any further processing. However, machine code programs are unintelligible to humans; after all, a machine code program looks like a long list of numbers. Languages such as BASIC or Pascal, called *high level languages*, make more sense to people but cannot be executed directly by the CPU. The programs in these two languages need further processing before they can be executed by the CPU of the computer. There are two ways in which this can be done—*interpretation* and *compilation*.

Interpreted languages

In interpreted languages, like the BASIC on the BBC Micro or the popular GWBASIC for the IBM PC, the set of instructions that is written in the language is examined by a language interpreter and a series of machine code instructions are executed for each instruction in the high level language program. Thus a single high level instruction may give rise to several thousand machine code instructions, which are pre-programmed in to the language interpreter. If we had a simple loop in a program, like:

```
FOR I=1 TO 10
PRINT "HELLO WORLD"
NEXT I
```

then the statement *PRINT "HELLO WORLD"* would be reinterpreted ten times! This, combined with the fact that the machine code statements

PROGRAMMING LANGUAGES

executed when this line is interpreted may not be as efficient as they could be, leads to interpreted languages being a little slow.

Compiled languages

In a compiled language, such as Turbo Pascal or Turbo BASIC on the IBM PC, the set of instructions making up the high level program is examined just *once* by a language *compiler*, which ultimately produces a machine code program for direct execution by the processor of the computer. This program can then be run on other machines that *don't* have the language compiler, and it also runs very quickly. Although the resultant machine code program might not be as efficient as it would be if it were written from the start in machine code, it's still much faster (often ten to twenty times) than a similar program in an interpreted language because the conversion of high level code into machine code instructions is done *once* only, at compilation time, rather than every time the high level instruction is executed.

Which languages are best?

Provided that you can access graphics facilities from a particular language, any language can be used to do experiments in chaos and fractals. My choice of Pascal and BASIC for this book was based on the following criteria.

Availability: It's likely that most readers would have access to one or both of these languages.

Algorithm Description: The algorithm of a computer program is the method used to describe how the task the program performs is done. These two languages are quite well suited to describing how things are done, if good programming practice is followed, so writing the programs in these languages allows programmers in other languages to see easily how the algorithm works.

Familiarity: I know how to program in these languages, so, why not use them. In addition, the languages are widely available and fairly cheap.

Ease of handling of floating point numbers.

However, there's nothing to stop users of other languages writing versions of the software listed in this book in those languages. Assembler language (a human version of machine code) has the advantage of speed but can entail the programmer in writing code for handling floating point arithmetic. In addition, writing assembler code is often a more long-winded process that writing programs in higher level languages. The C language has proved to be quite popular for some chaos experimenters, as it offers

Type	Range			Significant digits	Bytes
real	2.9E-39	to	1.7E+38	11–12	6
int	-32768	to	32767	5	2
long	-2147483648	to	2147483647	10	4

Table 3. Numerical resolution in Turbo Pascal

the advantages of assembler language (fast, efficient programs) with some of the ease of use of higher level languages. However, these two languages fall down when writing a book like this one in that the algorithms used in programs are not as clear as when using BASIC or Pascal; often further documentation of the code is needed before programmers can translate C or Assembler over to other languages.

Some programmers combine the best of both worlds by using assembler code for relatively simple, time-consuming tasks and integrate these routines with the main body of the program in a high level language. It's not a good idea to try and write floating point routines in assembler yourself; the ones provided with most computer languages are quite well written and to improve on them would take a lot of effort.

On the whole, therefore, which language you use is very much a personal choice. However, there are some points to take into account when considering a language, and to be borne in mind when using the language in chaos work. Many of the details about a language will be found in the programming guide for that language, but do bear these points in mind.

Numerical resolution

Computers store numbers, be they integers or floating point numbers, in a finite number of bytes of memory. For example, a computer may store integers in two bytes; this would allow positive numbers between 0 and 65535 to be stored. Increasing the number of bytes allocated to each integer allows a wider range of integers to be stored. The situation is the same with floating point numbers. For example, Turbo Pascal has the numerical resolution shown in Table 3.

The significant digits part is important, especially for the real number. Consider the number 1.6666666666666666666; clearly, if we have to fit this in to eleven significant digits, we'll end up with 1.6666666667, and we have a rounding error in the last position. With maths coprocessors, a resolution of up to twenty significant digits is possible, but then at that point we'll still have some rounding errors if we try and specify a higher resolution than the language can support. With integers, it's often possible to write routines that allow integer arithmetic to be done on integers with several hundred significant digits if required, but extending the floating point resolution by writing your own routines is *not* to be undertaken lightly!

The significance of this to chaos programs lies in the simple definition we gave above; a chaotic system has extreme sensitivity to initial conditions. If the resolution of our system is approached, we lose accuracy and results obtained may actually be *artifacts* caused by this loss of accuracy. These artifacts may give rise to interesting effects themselves, but we need to be aware of their existence when operating at the extremes of numerical resolution.

Numerical accuracy

There's also the possibility of errors of accuracy in programming in some versions of languages. The classical error is something like

$$1 + 1 = 1.999999999$$

This is a rounding error in the arithmetic functions in the language, and when it occurs we can often fudge things to put it right. However, it can still cause us problems. Further accuracy problems might be caused with trigonometrical functions, such as sin and cos, as there are a variety of methods by which values can be computed. A further example is the popular method of arriving at a square root of a number by raising it to the power of a half; this will often give a different result to using the square root function of the same language. Now, the answers obtained will be correct for most practical purposes, but if were to use these values as seeds for some chaotic system, the difference in the fourth or fifth place after the decimal point might cause problems and create artifacts.

Numerical overflow

The largest or smallest number that can be represented in a computer language depends upon the number of bytes used for storage of numbers, whether it's a real number or integer, and so on. It's very easy in chaos work, especially when dealing with things like *iterative functions* (Chapter 2) to generate numbers that overflow the computer's ability to cope. This usually generates a fatal error causing execution of the program to stop.

Types of computer

Although I'll be providing listings for only the BBC Micro and the IBM PC in this book, there is nothing to stop you using other computers, if they have the minimum requirements in terms of graphics mentioned above. However, some computers may not have the processing power needed to complete graphics displays as quickly as the Beeb or the PC. Indeed, some displays take many hours to create on even powerful 386SX computers, and literally *days* to produce on the BBC micro. However, there's nothing to stop you trying out the listings with any computer you've got, but be aware that you'll have to change the listings slightly in many cases to suit the computer used. For the record, the computers used in the writing of this book were as follows:

Packard Bell 386SX: A 386SX-based PC with high-resolution VGA graphics VGA, supporting a 40Mb hard disc and all PC floppy formats. Programs in Turbo Pascal and GWBASIC.

Amstrad PC1512: An 8086-based PC clone, 512k of memory and lower resolution graphics (CGA) than the 386SX. Programs in Turbo Pascal and GWBASIC.

BBC Model B: A 6502-based machine, 32k of memory and low-to-medium-resolution graphics. Programs in BBC BASIC.

Processing of screen images

As already mentioned, many experiments in chaos and fractals generate graphical images. Combine this with the fact that these images often take a long time to assemble and it is not surprising that you need to think about ways of preserving graphical information so that the images can be *quickly* recreated at any time in the future without having to rerun the program that generated them in the first place. Alternatively, you may wish to send images to other people, store them for archive purposes or use them as the basis of other graphics images after further processing. Before starting with any work, it would be useful to examine briefly the various methods of doing this.

Saving video memory

The video memory is the section of the computer's memory that stores the information that makes up the screen image. The easiest way to store an image is thus to simply save the video memory holding the image to disc. Saving it to tape is probably *not* a good idea, as video images tend to be big!

On the BBC Microcomputer, the **SAVE* option from the BBC operating system can be used to save the area of memory used by the graphics

SAVING VIDEO MEMORY

mode in current use. The memory taken up by each screen mode can be calculated from information shown in the BBC User Guide or Advanced User Guide. With the PC, this isn't as straightforward and usually some facilities offered by the language need to be used. For example, in the Turbo Pascal library in the Appendix, I use the *PutImage* and *GetImage* functions to store and recall graphics images from disc.

Although this method of storing images is the simplest, it has some disadvantages. These are as follows:

1. The images are very large. For example, they can be up to twenty kilobytes long on the BBC Microcomputer and a few hundred kilobytes long on PC VGA systems. Thus a few images can take up a large amount of disc space.

2. These images cannot be loaded in to other software packages than those that can handle screen memory images. This isn't too big a problem on BBC and other small computers, as the method of storing screen images as memory dumps is fairly widespread, but on the PC most graphics packages use a more efficient method of storing images. If the images are stored for use in programs that you've written and distributed, though, the mode of storage is up to you.

Using a standard file format for storage

One option adopted by some PC programmers is to write functions in their own software to save screen images in a format that is a standard, such as the PCX file format used by PC Paintbrush and other packages. Popular standards include:

PCX: This is the standard format used by PC Paintbrush, and the file format has been made available by the software house concerned, ZSoft, as an aid to programmers wishing to develop software compatible with this format. PCX is supported by PC Paintbrush, Microsoft Paint, Aldus Pagemaker, WordPerfect 5, Word 5.0 and a variety of other packages.

GIF: The GIF standard was designed by CompuServe, a computer bulletin board provider. It allows great compression of data but is not supported by vast numbers of programs. There are shareware converters available, and the format is becoming more popular.

TIFF: This format is used by many of the applications listed under PCX, with the addition of Logitech's Paintshow graphics software.

The advantage of writing files in these applications is that they can be read back immediately by other applications that use the same format.

For example, I have saved an image of the Mandelbrot set as a PCX file using a screen grabber, loaded it in to WordPerfect and then printed the whole thing out as a combination of text and graphics. The disadvantage is that programming some of these standards, even when the information is available about how the format is put together, can be a little difficult.

Using a screen grabber

A happy medium is to use a screen-grabbing utility, such as FRIEZE, CATCH (Paintshow) or GRAB (WordPerfect). These are resident programs on the PC which can be invoked by a couple of keystrokes and allow you to save the current screen display, *however it was created*, to a file in one of the standard formats. These applications usually come with a graphics package or other software capable of processing graphics and will support at least the graphics format used by that application. Using such software offers the best of all possible worlds; someone else has done the hard work for you, and the output from your programs can be captured, when the software is installed, without you having to modify any existing software.

Run length coding

This is a method of screen compression and storage which is fairly simple to implement, if a little cumbersome in some cases. The idea is simply to store graphics information as sequences of bytes representing the colour of a pixel and the number of subsequent pixels that are the same colour. As soon as a change in the colour is found then the process is repeated for the next colour and run of bytes. As an example, if we started with pixel 0 in the top left corner of the screen, and said that pixel numbers increased from left to right and from top to bottom of the screen, then a simple RLC system could start with the colour of pixel 0 in 1 byte, followed by 2 bytes holding the number of pixels following that have that colour. If a run of more than 65535 pixels of the same colour was obtained, then the fourth byte would have the same colour value as the first one, and the following two bytes would have the number of pixels of that colour still remaining. RLC files are thus built up of clusters of bytes, depending upon how many bytes are used to store the run length. The RLC method of screen compression is most effective when the image consists of larger areas of the same colour, and gives least compression for highly detailed images.

Once a screen image has been filed in some way, it can be recoloured, stretched, shrunk and generally messed about with using whatever graphics facilities you've got. Then it can be printed to give hard copy.

Getting a hard copy

To transfer the screen image on to paper can be quite tricky. The easiest way is to output the data to a printer.

Printer dumps

The PC comes with a function called *GRAPHICS* that transfers the screen data to a printer in some cases, but for much of the time it will not work with high-resolution graphics images. On the BBC Micro, you will certainly have to write or buy a screen dump utility. Alternatively, most graphics programs offer a way of printing out your work, so why not use that if you've got it, by screen grabbing the image, loading it in to the graphics package, then printing it from there.

Screen dump utilities are generally written for one specific printer or a selected few. If you've got an odd printer, you may need to search around a little for a suitable utility. Use of plotters or colour printers gives the same problems; this is why using a graphics package rather than a screen dump utility is often a good idea, as these packages usually support a wide range of output devices.

Screen photography

An alternative to making a printout is to take a photograph from the screen. This is fairly simple providing that some simple rules are followed. The camera used should be some sort of SLR, with a facility for tripod mounting and a timed exposure of several seconds. This is because the photograph will be taken over a period of a few seconds to avoid any risk of dark bands on the screen caused by the shutter of the camera operating half way through the screen being drawn. Any shutter speed of less than 1 second might not be a good idea, especially with the relatively slow film I use. As to film, I tend to use ISO 50 or ISO 100 colour transparency film, as this gives the most versatile finished result. For example, I can put the resulting slide in a projector and show it like that, have it made into a print or even take it to my local printers who'll put it into a colour photocopier and provide me with an A4 colour photocopy of the slide. Once you find a film you're happy with, stick with it. For the record, the slides used in this book were taken on Fuji film with a Canon EOS 600. The steps involved are as follows.

1. Load the camera with the film, then go outside and take a couple of shots of anything that's about, provided that it's a good, contrasty subject. The reason for this is to prevent the automatic equipment that cuts film up into frames deciding that the middle of your first

exposure, which was a little bit dark, is the start of the first frame. The first couple of shots give a reference point for the equipment to prevent this happening.

2. Set up the computer in a darkened room, the darker the better, so as to avoid any unwanted reflections or glare on the computer screen. I tend to take screen shots at night, curtains drawn, lights off. It's a good idea to set up the camera *before* turning all the lights out!

3. Set up the camera with a 50mm lens set to $f5.6$ or $f8$ on a tripod so that the field of view is totally occupied by the computer screen. Ensure that an imaginary line drawn through the lens to the screen would strike the screen at ninety degrees. Turn off any automatic focus the camera has, and focus the camera on the screen. To facilitate this, I have a small BASIC program which, on running, splits the screen into four quarters that are coloured black and white to give an effect like four squares of a chess board. I focus on the intersection of these. Don't select a smaller aperture as the resultant shallow depth of field may prevent the edges of the monitor screen, which in most cases are slightly further away than the centre of the screen, from being in focus.

4. Once focusing is done, load up the image that you want to photograph and adjust the brightness and contrast controls to give a clear image. If your camera has through-the-lens metering, then setting exposure is fairly straightforward. Set the exposure as indicated when focusing the camera on an average part of the image—not total black or total white. If you use a separate light meter, take a reading from an average area of the image. The exposure time will probably be in the range 0.5 to a few seconds. Set the exposure time. If you can't get a time in this range, adjust the brightness of the monitor or open the aperture of the camera. If it's too bright use a neutral density filter.

5. Take the exposure. In order to reduce shake, use a remote cable. If your camera doesn't have the exposure time needed as a setting, use the Bulb setting and a stop watch.

6. Take two *bracketing* exposures, at 0.75 times the indicated time and 1.25 times the indicated time. You may wish to have more bracketing shots at wider time settings. The only drawback with slide film is that good results do depend rather a lot on getting the exposure correct. After a few films, you'll get the hang of things and will probably be able to dispense with bracketing shots.

Expect the first couple of films to be a bit variable in quality, but after that you should have no problems at all. I've found that a photographic

DATA EXPORT

record of my better results is rather useful, and certainly less trouble to show others than having to get the computer set up!

Data export

Although we've examined the different ways of transferring screen images between programs, some programs produce data in a numeric form, such as the differential equations shown in Chapter 3. It's occasionally useful to be able to output such data for processing via another program. On the BBC, there doesn't appear to be a widely accepted standard file transfer format except for ASCII, where information is placed in a file as plain text, numbers being represented as a series of digits. However, on the PC there are a couple of standard ways of passing textual or numeric data between programs. These include:

Comma Separated Value files are files of data with each record of data separated from another by a new line and each piece of data within a record separated by a comma. For example, a series of data points and times might be passed as:

Time	Data	
900,	1234	*Record #1*
901,	1236	*Record #2*
902,	1345	*Record #3*

Fixed Fields where data is sent as a series of strings separated by new lines. The same data as used above would be passed as:

09001234	*Record #1*
09011236	*Record #2*
09021345	*Record #3*

and the software receiving the data would be told to expect the data part of the record to start at character 5.

Language versions used in this book

I've used the following languages in this book. For the Pascal listings I used Turbo Pascal version 4; earlier versions do not support *Units* but will still run the programs with some changes to take the lack of Units into account, typically by including external files with the *$I* compiler directive. In addition, some of the graphics commands, such as *PutPixel*, have different names. See the *Extend* unit in the Appendix for additional functions.

For the BBC BASIC programs I used version 1.2 on a rather venerable BBC Model B. The BASIC listings will also run with the various BBC BASICs available on the Master and Electron computers, as well as versions of BBC BASIC that are available for the PC.

Chapter 1

What is chaos?

In the 17th century, the mathematician Sir Isaac Newton developed three simple rules of motion which set the picture for how the world would be viewed for the next 300 years. Even today, Newton's laws work in all situations except for extreme conditions that are better described by quantum physics and relativity. Indeed, we say that a system is *Newtonian* if it can be described by these laws. The three laws were as follows:

1. An object with no forces acting on it will remain at rest or, if moving, will move in a straight line.

2. The rate at which the speed of an object increases is proportional to the force acting upon that object.

3. To every action there is a further action, equal in magnitude but opposite in direction.

These laws were powerful ideas, and they allowed Newton to explain the motions of the planets, the collisions of objects and the speed at which a dropped ball would hit the ground. They also were interesting from a mathematical point of view, as in deriving and using them Newton had used several new techniques. Firstly, he had used graphs to illustrate algebraic formulas. For example, by drawing a chart like that in Figure 1-1 he was able to show that the speed of an object was equal to the tangent of the line at a particular point. In addition, he showed that such a graph could also show the distance travelled by a moving object up to a certain moment in time as the *area* under the line up to that point. This combination of algebra and geometry lead to the development of *Calculus*, a method of using mathematical equations to describe the rates of change of physical quantities such as speed.

This was very clever stuff for the scientists of the day, and over the next couple of hundred years the principles that Newton had applied to

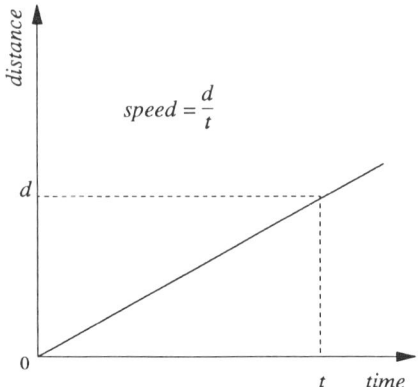

Figure 1-1. Linear graph of speed *versus* time.

dynamics (the science of moving objects, forces, etc.) were applied to other areas of scientific endeavour, such as heat, sound and eventually electricity. At the heart of all these developments was the calculus, which we'll look at in greater detail in Chapter 3. The ultimate expression of this *deterministic* view of the universe was expressed by Laplace, who envisaged what he called a *Vast Intellect* that was able to know the position of and forces acting upon every particle in the universe. If these parameters were known, he argued, at a given moment in time the Vast Intellect could predict the future behaviour of all those particles.

There were a few problems, though. For example, Newton's laws could not explain the outcome of a single, solid ball hitting two other balls *simultaneously*. How would the energy of the first ball be distributed to the two others? If we think about it, Laplace's Vast Intellect would fail at this point as in the universe there are bound to be simultaneous collisions of particles and at each simultaneous collision we lose some information.

In the 19th century, the French mathematician Poincaré applied various techniques to examining the problem of the stability of the solar system. He also came up with some results that compromised the purely deterministic view of things. A further complication was caused by the method people had used when applying mathematics to the various problems that had been modelled using calculus. This was to treat everything as a linear system, usually by talking about small changes in the system and applying the results obtained for small changes to make predictions as to what would happen under conditions of greater change. Had the systems under consideration been linear, then this sort of extrapolation would have been valid. However, the real world isn't like that. For example, a pendulum is a very simple mechanical system, in which, if we displace it from its resting position, it will swing to and fro until friction causes it to come to a stop.

The time taken for each swing when we displace the pendulum by a small amount is the same, but if we change the size of the displacement from the rest position, the time taken for the swing changes by a small amount. The change in period of the pendulum is so small that it tended to be ignored, and this was the case with a variety of systems studied. In some cases this was due to the variation being small, and in other cases it was due to the fact that the variation caused by going away from the linear regions of behaviour was so hard to explain using the existing mathematical models that it was quietly ignored. One useful byproduct of looking at these systems, though, was a technique of representing what was happening in a system as a series of points on a graph. For example, if we look at a (well-behaved) pendulum, we can describe the pendulum at a particular instant in time in terms of its position away from the centre point and the velocity at which it's going. If it's travelling away from the centre, we can say it has a positive velocity, and if it's travelling back toward the centre we can say it's got a negative velocity. We can treat the distance to right and left of the pendulum centre point in the same way. Now, we draw a set of axes, as in Figure 1-2, with the x-axis representing distance and the y-axis representing speed. The intersection of the two axes represents the situation when the velocity of the pendulum is zero and it isn't displaced from the centre point, i.e. it's at rest.

This is called the *phase space* of the pendulum, and is basically a means of representing the status of a physical system at precise instants in time. In Figure 1-2b, we've drawn a series of points that define a typical pendulum's travels, from when it was first displaced to when it finally came to rest due to friction. Each point is a snapshot of the system at that instant in time, and you can clearly see the system spiralling down to rest after friction has dissipated all the energy that we gave the pendulum when we first swung it. This is a classic *dissipative* system, in which energy is lost with time and the system eventually comes to a rest position. The rest position is called an *attractor*, and in this system, there's just the one attractor.

If we were to build a gadget that gave the pendulum a push at the end of each swing that just made up for the energy lost to friction, we'd have an oscillating system. This could be represented by the phase space diagram shown in Figure 1-2c, a closed loop. This is something that's quite important about these phase space diagrams; if you get a point on a phase diagram that's in the same place as another point, then the subsequent points will also be repeats of what has gone before, i.e. a loop will be formed. A closed loop in a phase diagram indicates periodic motion.

The pendulum that we're talking about has two variables to consider; distance from the rest point and velocity. These are called *degrees of freedom*, and when we draw a phase diagram, we have to have as many axes as we have degrees of freedom to accurately represent the phase space of the

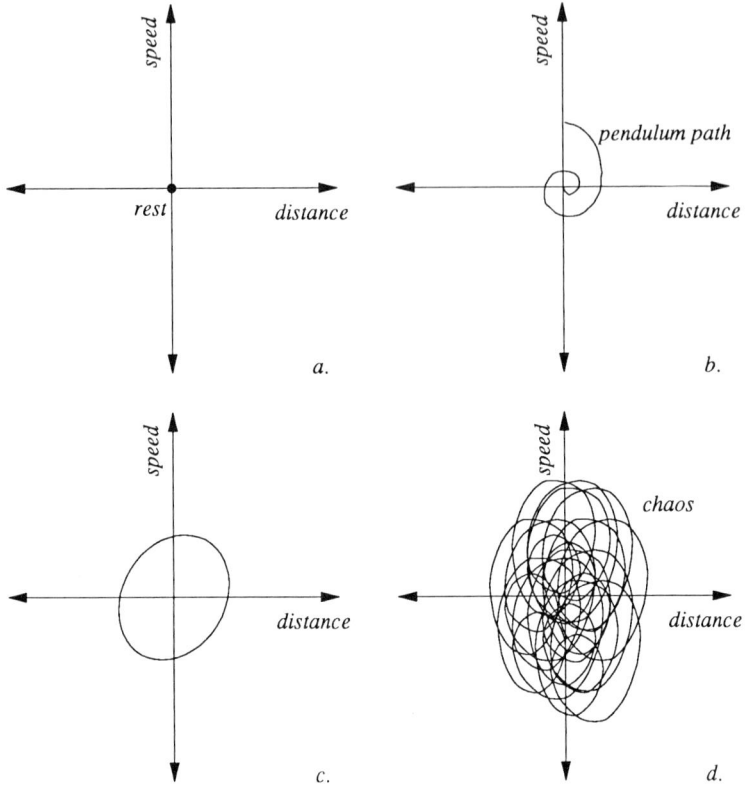

Figure 1-2. Phase diagrams for the motion of a pendulum.

system being examined. So, for a system that has three things that vary with time, we'd need to represent the phase space on a three-dimensional chart, like that shown in Figure 1-3. Here you can see that there are places where from one viewpoint or another we might appear to have a periodic motion (Figure 1-3a), but an examination of the full picture of the phase space indicates that the points on the diagram are actually displaced from each other in phase space and so never coincide and so don't lead to a periodic motion. Phase diagrams can thus be valuable tools for examining dynamic systems—they show us things that might otherwise be hidden within the dynamics of the system. The phase diagram can be examined in a slightly different way; if you were to stick an imaginary piece of plastic through the lines of the phase diagram at 90 degrees to their direction of travel (Figure 1-3b) then you will see the situation shown in Figure 1-3d. The tracks through phase space show as dots, and if we were to observe

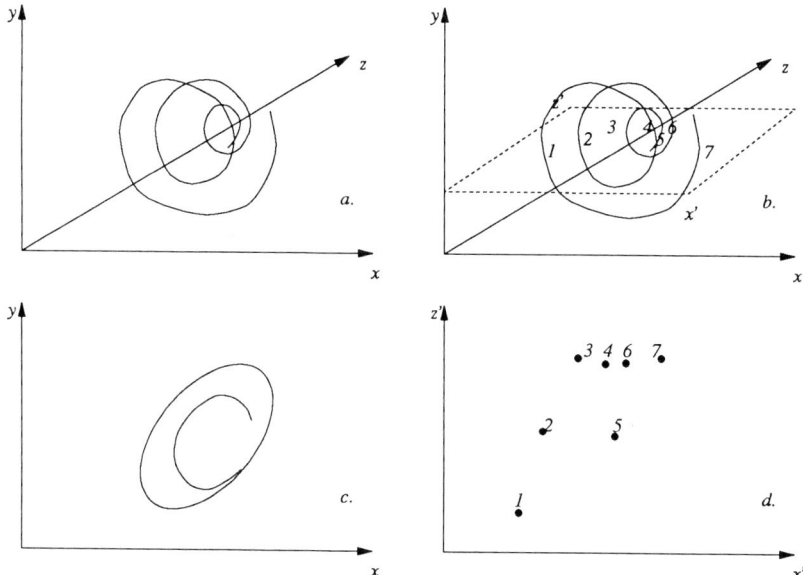

Figure 1-3. A three-dimensional phase diagram.

this section as the phase diagram was being drawn by the system then periodicity would be indicated by two dots being at the same place on the section. This section is called a *Poincaré section*.

If we had a 'funny' pendulum, like that shown in Chapter 11, then the phase diagram would be something like that shown in Figure 1-2d. Here, there's no periodicity apparent from the phase diagram. Although we've concentrated on a pendulum here, the same principles are applicable to other systems as well. For example, a fairly simple system was described by Libchaber and Maurer in their 1982 paper *A Rayleigh Benard Experiment: Helium in a small box* in which a tiny cell was set up to hold a small amount of liquid helium shown in Figure 1-4. When a fluid is heated, convection currents set in and the warm fluid rises and cooler fluid is displaced. In this experiment, the cell was designed so that the convection currents set up two rolls of fluid when a tiny temperature difference was set up between the top and bottom of the cell. With a small temperature difference, the two rolls that are set up are quite stable, but as the temperature is increased, something peculiar happens; a 'wobble' is set up in each roll which runs along the length of the roll of fluid. This wobble has an effect on the temperature at the top of the roll recorded by the temperature sensor, and the situation shown in Figure 1-5 can be recorded, and a phase space diagram can be constructed from the speed at which the wobble moves

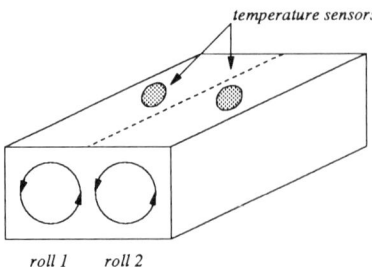

Figure 1-4. The Libchaber-Maurer Helium experiment.

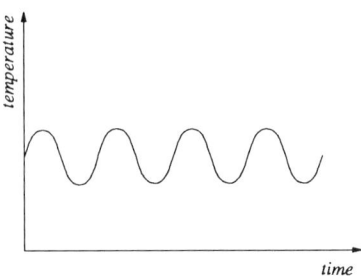

Figure 1-5. The Libchaber-Maurer Helium experiment 'wobble'.

along the fluid roll and the speed at which the wobble moves from side to side. In addition, a similar diagram can be constructed by plotting the temperature recorded from one temperature sensor against the temperature recorded from the other sensor.

Both these diagrams show a single loop, indicating that there is periodicity. Now, if we were to vary the temperature across the cell very slightly, and wait until things settle down again, something rather odd has happened; the system has settled down into a state where the phase space diagram has a double loop. This indicates that some instability has been added to the wobble of the fluid roll, and if we plot the frequency of the ripple of the roll and the amount of energy in each frequency component of the roll's movement, we end up with a *frequency spectrum* as shown in Figure 1-6. A further minute temperature increase leads to a further splitting in the phase diagram, and the production of more peaks in the frequency spectrum shown in Figure 1-7. This splitting is called *period doubling*, and with further temperature increases further period doubling occurs until a point is reached when the motion of the fluid rolls becomes turbulent and unpredictable—the onset of chaos. Each of the splits that occurs is called a *bifurcation*, and the transition of a system from predictability to chaos via a series of bifurcations is very common.

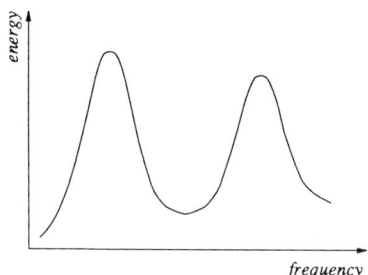

Figure 1-6. Frequency spectrum for the Libchaber-Maurer Helium experiment.

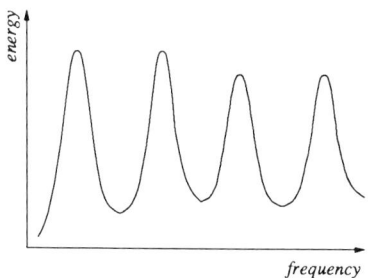

Figure 1-7. Period doubling in the Libchaber-Maurer Helium experiment.

A further example of how a regular system descends into mathematical chaos is given by water flowing through a tap. This motion of a fluid can, in simple circumstances, be predicted by a set of equations called the Navier Stokes equations. However, these only apply under restricted conditions with low levels of flow. You might start off with a regular series of drops, with the same time between each drop falling. Increasing the flow rate will eventually lead to period doubling, where you get drops falling with two times between drops; for example, one drop might be separated by one second, the next by two seconds, and then the next following in one second, then two seconds, and so on. Eventually, this will lead to a sudden transition to turbulent flow of water through the tap—again, chaos via period doubling.

The odd thing about this is that although the system is different, the onset of chaotic behaviour follows the same general pattern, that of gradual change followed by a sudden transition to some chaotic behaviour. A further example is afforded by the *logistic equation* which is used to model population growth. In this equation, a seed value, x, is used to generate a value y and the value of y is plotted on a graph against the corresponding

26   CHAPTER 1. WHAT IS CHAOS?

x value. The calculated y value is then used to generate the next y value, and so on. This process is called *iteration* and I'll look at this in greater detail in Chapter 2, but for now examine Figure 1-8a. If we vary the value

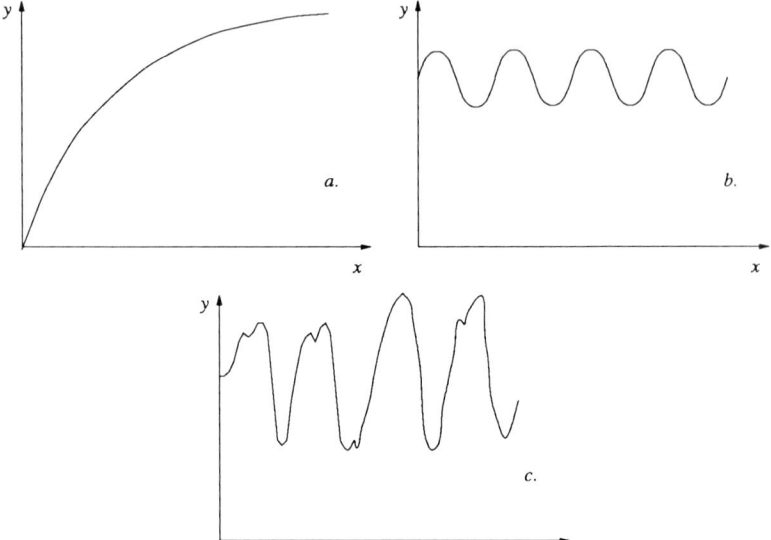

Figure 1-8. Varying values of k *showing transition into chaos.*

of k, we get a graph like that in Figure 1-8b. Varying it further gives us the situation shown in Figure 1-8c, where after k exceeds a particular value chaos sets in with the value of y being unpredictable. If we change the way in which the graph is drawn, to that shown in Figure 1-9, then we plot the ultimate values of y arrived at for the different k values after the iterations have settled down to a steady state. In technical terms, this is called waiting for the transients to die away. The graph obtained is another way of describing the bifurcations experienced in the helium experiment mentioned above. In fact, this diagram is called a *bifurcation diagram* or a *Feigenbaum diagram* and is useful in two ways:

1. It indicates the conditions under which iterative systems go chaotic—here, for example, at a value between 3 and 4 for k.

2. It allows us to make some interesting general observations about chaotic systems, as we'll see in Chapter 2. The most obvious is that within the chaotic region, there are regions of apparent order.

Iterative mathematical systems are very useful for chaos research because they provide an easily managed and apparently simple system that

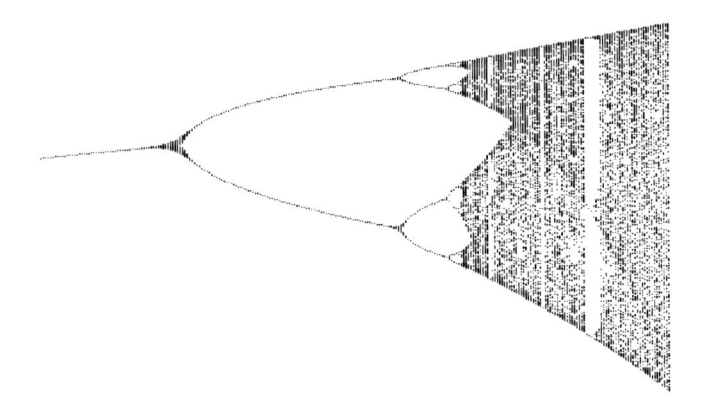

Figure 1-9. An alternative view of the results of the logistic equation.

gives rise to some very complex behaviour. Chaos can also emerge from mathematical models using calculus. It was in such a system that chaos was first examined in the early 1960s by Edward Lorenz, who described a simple mathematical model of a weather system that was made up of three linked differential equations, i.e. equations that showed the rates of change of such parameters as temperature and wind speed. We'll look at these *Lorenz equations* in some detail in Chapter 4, but what Lorenz found was very interesting, and can be summarised as follows:

1. Some equations, which most mathematicians would have considered to be fairly straightforward, gave rise to some surprisingly complex behaviour.

2. The behaviour of this system of equations was very dependent upon the initial conditions of the mathematical model.

This was effectively an exercise in fluid dynamics, and we've already seen in the situations described above that this can give rise to chaos. When the work was originally done, meteorologists were of the opinion that given sufficient computing power you could construct a mathematical model of the Earth's weather that could then be used to give accurate forecasts. Lorenz's model showed that small differences in initial conditions for a model weather system could give large differences in the resulting weather in the model. Now, in the forecasting model, these initial conditions would be the measured pressure, temperature, windspeed, etc. at weather stations that were feeding data in to the model. You can easily imagine that these readings might be error-prone. Prior to this work, it was expected that the small errors thus encountered would be cancelled out, but Lorenz's work showed that this was not so—it also put the end to any dreams the weathermen had of getting a 100% correct computer model, as even the tiniest of errors in the data fed in could lead to vast differences in the forecasted weather, due to the high sensitivity of the system to changes in the initial data.

This high level of sensitivity of the system to initial conditions gave rise to a very evocative description of chaos in weather systems—the *Butterfly Effect*—in which, theoretically, the flap of a butterfly's wing could influence the course of a typhoon on the other side of the world. The idea behind this is that the disturbances in the atmosphere caused by the fluttering of the butterfly would not, as you might expect, simply die away but would have subtle effects which would extend over a wider and wider area.

Chapter 2

Iterative functions

How do we get a laboratory specimen of chaos to play with? The simplest way to experiment with mathematical chaos is to use something called an *iterative function*. The steps involved are as follows:

Select a function

This could be something very simple like $y = 2x$. This particular function, though, wouldn't give rise to chaos. For a function to exhibit chaotic behaviour, the function used has to be *non-linear*. A few seconds thought about $y = 2x$ will show us that if we take a series of x values, and work out the corresponding y values, we'd get a graph something like that shown in Figure 2-1a, a rather boring and predictable straight line. This is a typical

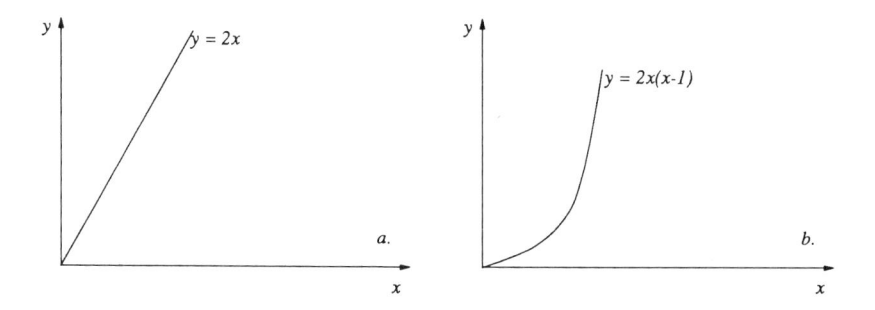

Figure 2-1. Graphs of linear and non-linear functions.

linear equation. However, if we look at something like $y = 2x(x-1)$, then the graph of plotting y against x for different values of x is much more interesting, as shown in Figure 2-1b. A linear equation gives rise to a straight line, and won't give you any chaotic behaviour. A non-linear equation will give rise to a curved line and *may* exhibit chaotic tendencies.

Iterate it

In the examples given above of linear and non-linear equations, we simply took a value of x, applied the equation, and plotted the value of y that resulted on the graph. Iterating the equation is the process of starting with a value of x, evaluating the function to get y, and then using the value of y as the next value of x to put into the function. This is a process known as *mathematical feedback*, and is analogous to the phenomenon noted in rock concerts where a guitarist can generate a howl from his instrument by playing too close to the loudspeakers. This feedback process is common to all chaotic systems. Because we are using the output of the function as its next input, we write these iterative functions in a different way to that we've seen above. Instead of:

$$y = 2x(x-1)$$

we write the equation as:

$$x_n = 2x_{n-1}(x_{n-1} - 1)$$

where n refers to the current value being calculated and $n-1$ refers to the result of the last iteration. In general terms, we can write this as:

$$x_n = f(x_{n-1})$$

where f stands for the function that's being applied to the last value of x. When we iterate functions like this, we actually plot the value of x calculated against the number of iterations of the function. When we write the function in this general form, it becomes clear that there are two important parameters of *any* iterated function, irrespective of what the actual function is. These are the initial value of x, and the number of iterations. We're now in a position to do some practical programming and experience our first dose of chaos. Listing 2-1 in BASIC and Pascal shows a program that allows us to plot the results of iterating a simple function, and to vary the initial value of x. As you can see from looking at the listing, there are three functions in the program. To use one of these, remove the { } from around your chosen function in the Pascal program and remove the *REM* statement from the BASIC version of the program. In each of these cases, you can specify the number of iterations to be carried out and the result is graphed. The results obtained are quite different in each case.

The square root function

If you iterate the square root function (sqrt), with any start value you care to choose, the graph will settle down to a straight line in fairly short order. In fact, if we were to print the value of the function out rather than graph it, the value would be 1. You can try this on a pocket calculator if you like, and you'll get a set of results like:

$$9 \to 3 \to 1.732 \to 1.316 \to 1.147 \to$$
$$1.071 \to 1.034 \to 1.017 \to 1.008 \to 1.004 \to$$
$$1.002 \to 1.001 \to 1.000$$

This function can be described as having a *stable orbit*, in that we can predict the final state of the function if we iterate it enough to get to a steady state. The orbit refers to the series of values that have been derived from a specific initial value. Another property of functions having a stable orbit is that a slight change in initial value will give a similarly slight change in the sequence of values obtained during iteration.

$$9.2 \to 3.016 \to 1.736 \to 1.317 \to 1.148 \to$$
$$1.071 \to 1.035 \to 1.017 \to 1.008 \to 1.004 \to$$
$$1.002 \to 1.001 \to 1.000$$

Here, a change of 0.2 in the initial value is soon lost and the pattern from the iteration sequence is almost identical to that obtained from the first value. If you start with an initial value of between 0 and 1, then the function will iterate to 1 (well, should do; in practice with most calculators or computers you get it to 0.9999999999 ... due to rounding errors in the programming used!). A value of 0 will iterate to 0 immediately.

The square function

Squaring a number gives us a different type of stable orbit; here, any value greater than 1 entered as initial seed will, after a few iterations, 'blow up' as the value it attains exceeds the numerical capability of the calculator or computer used. A seed of 1 will iterate to 1, and a seed of between 0 and 1 will iterate to 0. Again, the function gives rise to stable orbits, even if one of them is infinity!

Stable and unstable orbits

The two functions we've iterated so far are stable; for each one we could draw up a simple set of rules that will indicate the final steady state of the function for a given range of starting values. In addition, we know that changes in initial values will give predictable changes in the values of the iterated sequence. When we graph these functions, we're creating something that's analogous to the phase space diagrams that we saw in Chapter 1. In fact, we can say that the final resting states arrived at by the functions are the attractors for those functions. So far, because these functions have ultimately settled down to a single value, we can say that these functions have *point attractors*. Now, what about unstable orbits? This is the situation where for extremely close initial values the outcome is quite different. We've already encountered an example of this when we looked at the sqr function; an initial value of around 1 will go to 0 if only slightly less than 1 and infinity if marginally greater than 1. If we could remove rounding errors, then a tiny error in setting up the initial value of 1 would lead to a vast difference in the attractor that the function eventually settled down on. This is the only area of potential instability in the sqr function; but what about other functions?

The function $2x(x-1)$

This function is fun! Try values of, say 1.002 and 1.003 for x with 50 iterations. You will get similar, but totally different graphs drawn on the screen. A difference of just 1 part in 1000 changes the outcome completely! If you try larger numbers of iterations, you'll also see that there's no apparent point attractor. Whereas the unstable orbits were for a small range of initial values in the case of the sqr function, here the reverse is the case with the stable orbits being outnumbered by the unstable ones! The patterns obtained are similar, though, and an examination of the graphs drawn will indicate that there are areas within the graph that are similar but not quite the same. Figure 2-2 shows two screen prints that illustrate this.

You might care to amend this program to print the values calculated to a printer or on the screen using *Writeln* in Pascal or *PRINT* in BASIC. This can give you a different perspective on the data being generated. You might also like to experiment with different functions to those given above. In fact, that's just what we'll do now.

The logistic equation

The logistic equation is one of those things that people look at, say 'Yes, I understand that', then go off and look at something else. Fortunately for chaos science, a few people did take a second look at this equation and found it to be very interesting indeed. The logistic equation provides a simple

THE LOGISTIC EQUATION

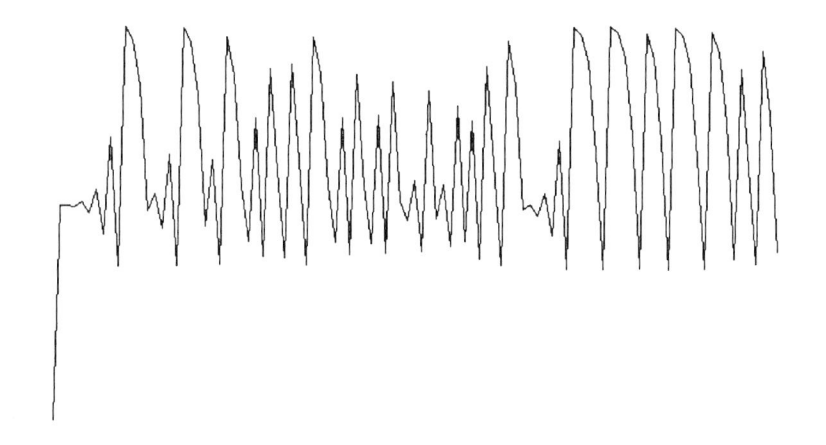

Figure 2-2a. The function $2x(x-1)$ with a particular set of initial conditions.

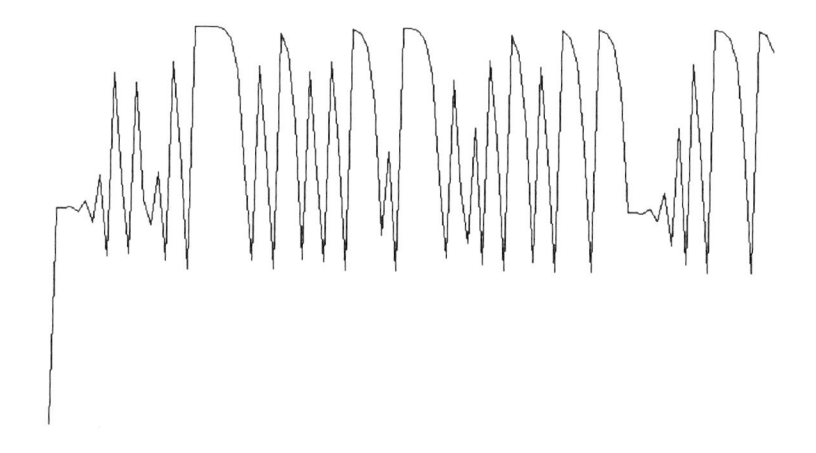

Figure 2-2b. The function $2x(x-1)$ with a different set of initial conditions.

model of population growth for a system where there are no predators. It can be written as:

$$x_{n+1} = kx_n(1 - x_n)$$

where x is the population of whatever we're discussing— fish, fowl, small furry beasties—and k is a constant that represents the rate of growth of the population. The part of the expression reading $(1 - x)$ is effectively a stabilising factor; without it the equation would lead to an ever increasing growth. This doesn't happen in natural population systems. With it, for a given value of k the system might be expected to reach some sort of equilibrium condition, like a population does when the birth rate is just enough to replenish losses in the population due to death or migration.

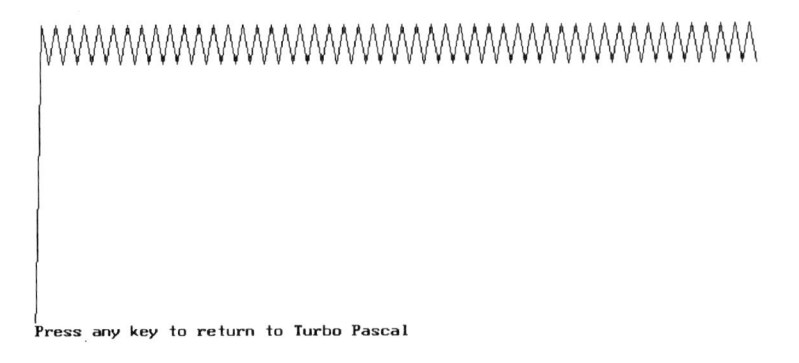

Figure 2-3a. Graph of logistic equation showing period 2.

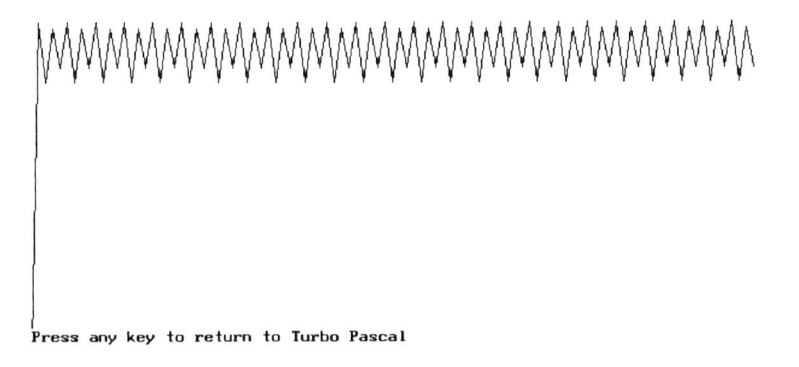

Figure 2-3b. Graph of logistic equation showing period 4.

This equation was explored by ecologists and biologists who used it to model population dynamics and found that by experimenting with the value of k you could change the level at which equilibrium was reached. This we might expect, and we could even reasonably predict that the higher the value of k the higher the equilibrium value will be. This was true—to a point, as we'll now see.

THE LOGISTIC EQUATION 35

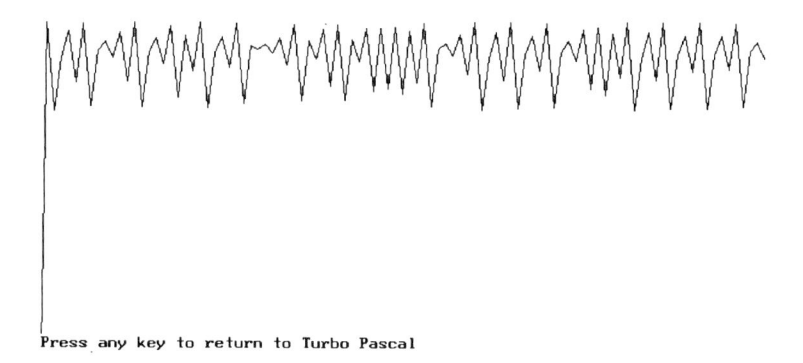

Figure 2-3c. Graph of logistic equation showing chaos.

Listing 2-2 shows a simple program for exploring this equation in two ways; one is to draw a graph and the other is to print to the screen a list of the values generated. You can enter a start value of x, and a value for k. Note that the initial x value should be between 0 and 1; a value of greater than 1 will lead to the equation quickly 'blowing up' as the final value exceeds the limits of numerical resolution of the computer language. In terms of population, the result of each iteration can be viewed as a measure of the population, 0 being nothing alive and 1 being the maximum population that can be supported.

If you start by entering a value of 2 for k, and a few different x values, you'll find that after a while the equation settles down to an equilibrium value of 0.5. The number of steps needed to settle down to this value depends on the initial x value. Similarly, starting with $k = 2.5$ and a number of x values, an equilibrium is reached at 0.6. If you now try $k = 3$, then an interesting effect is noted; the equation settles down after a considerable time into a single value, but while getting there oscillates between two separate values. For example, an initial x of 0.1 will give rise to values of 0.68... and 0.65... which eventually merge together to one intermediate value. Use the program to draw the graph for this. Clearly, something odd is happening; let's carry on varying the value of k, but now let's go up in smaller increments.

If you try $k = 3.25$ and $x = 0.5$ then you'll get a graph like that shown in Figure 2-3a, where the two values are 0.4952 and 0.8124. This time, the oscillation is constant, and the system flicks between these two values forever. If this equation were referring to a population of some sort, the real world equivalent of this would be a population oscillating between two values. This system is said to have a *periodicity* of 2, or is said to be *period* 2.

Starting with $k = 3.5$ gives rise to period 4, where the final equilibrium state oscillates between four discrete values—in the case of a starting x of 0.5 these values are 0.8749, 0.3828, 0.8269 and 0.5008. The graph of this system is shown in Figure 2-3b.

Life gets really interesting if we let $k = 3.75$; here, we get a totally irregular looking graph, as shown in Figure 2-3c. The system has decayed into chaos, and in the real world this would indicate a population that varied wildly from year to year. By looking at individual values of k in this way, we can get pictures of how the equation evaluates to different values with each iteration, but we can't see, for example, the precise value of k at which the transition from period 2 to period 4 takes place, or at what value of k chaotic behaviour is first noted. To do this, we need to have a means of plotting the final values reached by the logistic equation for a particular value of k. The steps involved in this are as follows:

1. Pick a value of x to act as the start value of x in each case—say we pick $x = 0.5$.

2. Select a value by which we'll increment k; this might, for example, be 0.1, or any other value, depending upon whether we're interested in looking at the fine detail of the resultant graph, in which a smaller increment would be needed, or in the big picture.

3. Select a start and end range for values of k. After the work above, we might consider a range of 2.5 to 4.

4. At the start k value, iterate the logistic equation a number of times to ensure that it's settled down into its final state. This 'settling down' is a common requirement when investigating dynamic systems, so try 100 times for the iterations needed to get rid of the transients, then plot or print the final equilibrium value(s) for a further 100 or so iterations.

The program shown in Listing 2-3 does this, and draws a graph of the attractors of the logistic equation for each value of k in the specified range.

The program is easy to use, and a good set of parameters to start with is a k range from 2.7 to 4.3, a step size of 0.005, y scale of 300 and y offset of 30. This gives us a graph like that shown in Figure 2-4, where the bifurcations and the eventual chaotic behaviour are easily seen. This type of graph is known as a *bifurcation plot* or a *Feigenbaum diagram*, after Mitchell Feigenbaum, who put a lot of work in to this topic. As you can see, chaos is reached by a series of bifurcations, where the number of values assumed by the logistic equation doubles. What is quite fascinating about this diagram is that even within the chaotic area of the diagram, there are areas of regularity, as seen in Figure 2-4b. This diagram introduces some

THE LOGISTIC EQUATION

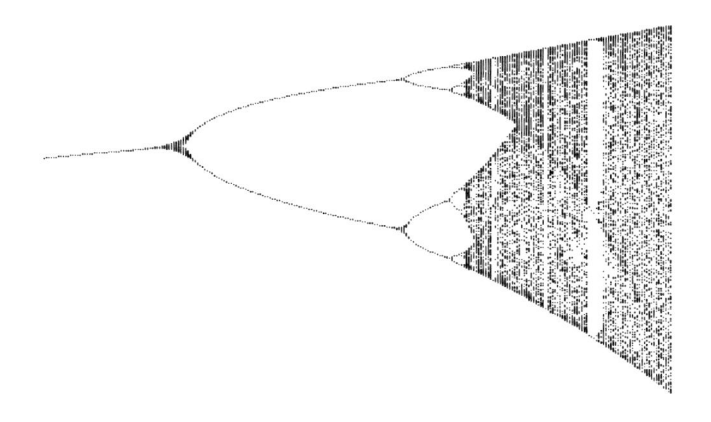

Figure 2-4a. Bifurcation diagram for the logistic equation.

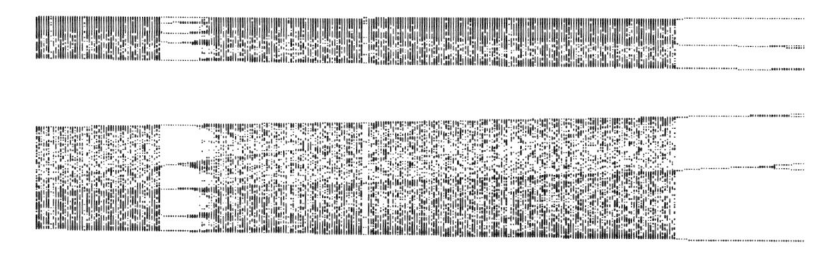

Figure 2-4b. Different bifurcation diagram for the logistic equation.

new concepts, and also shows ideas we met in Chapter 1 in a different light. We can split the diagram into three parts.

Order

This section of the diagram is what we have up to the first bifurcation point, and the line drawn to that point represents a point attractor for the equation for those values of k.

Period doubling cascade

From the first bifurcation to the onset of chaos, we have a repeated series of period doublings, usually known as a period doubling cascade. In this section of the graph, we no longer have a point attractor but we have a period attractor. Within this section of the graph, the bifurcations get closer and closer together with increasing values of k until at some point the system breaks down into chaotic behaviour. For the logistic equation, the onset of chaos occurs at a value of about 3.57. The areas within the period doubling cascade are often called n–cycle areas; for example, the area where we have two possible values of x for a particular k value can be called 2–cycle, and that area where we have four values of x for a given k value is called 4–cycle.

Chaotic behaviour

From the onset of chaos, we have a seemingly random jumble of dots, in which are embedded areas of regularity. A closer examination of these areas reveals that they too consist of regions of order, period doubling cascades and finally chaos again. This concept of *self-similarity* will turn up again in Chapter 6 when we examine fractals. Within the chaotic region, we no longer have a period attractor but a *strange attractor* which consists of all the values given by the equation for a specific value of k.

You can explore the bifurcation diagram in greater (though more tedious!) detail by using the program in Listing 2-4. Rather than plotting a graph, this program simply lists the values of x to the screen. On the Pascal version, use the *Pause* key to stop the scrolling; on the BBC BASIC version hit any key except *Escape* or *Break*. In each case, hit *Enter* to start things rolling again. This program allows you to evaluate the value of k for which a period doubling takes place, as you can see when the number of discrete values returned from the iterations of the equation changes. As well as the period doubling of 1 value, to 2, to 4, etc., you can also see areas where a periodicity of 3 is encountered. It's been proved mathematically and by numerical experiment that if you get period 3 (i.e. a repeating sequence of three similar values) showing up in a system, then chaos will soon follow. In the bifurcation diagram for the logistic equation, a period 3 area starts at values of k around 3.83 and continues until about 3.84, where the period doubles to give a period 6 region, before chaos reasserts itself. There are a couple of points to note when looking at the numbers produced by the program in Listing 2-4.

The first point is that the values arrived at by the equation do gradually change between bifurcation points. This can be a little bit confusing on first examination, but you soon see when genuine bifurcations have happened. You might find it easier to start by displaying the numbers to, say, two or three decimal places at first until you get used to the results. (However, let the calculations be done to the *full* resolution of your computer.)

The second thing to note is that you will almost certainly see detail that you didn't get from the graphs that were drawn. In fact, one of the more interesting features of a bifurcation diagram was discovered by Mitchell

THE LOGISTIC EQUATION

Feigenbaum by doing the sums on a calculator and then writing the results down; whilst doing this he had time to think about the numbers and came up with an interesting relationship. In general, the value of:

$$(k_n - k_{n-1})/(k_{n+1} - k_n)$$

where k_n is the value of k giving a bifurcation point, is a constant with a value to three decimal places of 4.669. This is called the *Feigenbaum constant* and is not found only in the bifurcation diagram of the logistic equation; in any system where bifurcations or period doublings are observed, the Feigenbaum constant turns up. You might like to have a shot at calculating this constant from the programs we've explored so far. It's not easy, though, but here's what to do.

1. Take the value of three sequential bifurcation points. For example, for the logistic equation we might get the first three points as:

n	k
1	2.998
2	3.441
3	3.540

2. Start with $n = 2$ and substitute the values in the equation given above. This will give us:

   $$(3.441 - 2.998)/(3.540 - 3.441)$$

3. The value given by this rough calculation is 4.7. Don't expect to get spot-on values without doing some quite considerable work, though!

Not surprisingly, there is an easier way to do this; we let the computer do the work. Listing 2-4 shows a simple program to calculate the values of k at which bifurcation points occur. Although the program may look a bit complicated, it is not too difficult to understand. All that is done is to take a start value for k and iterate the equation until a point at which all transients have been suppressed is reached. Now, take the value of x reached and iterate again b times, where b is the number of values we anticipate at that value of k. If we find that the value of x arrived at is the same as the one we started with, then we've not yet reached the next bifurcation point. If, on the other hand, we don't get back to roughly the same value, then we've passed through a bifurcation point. We then increase the value of k by a tiny amount until the next bifurcation point is encountered.

This program can be used for any function to work out bifurcation points, provided that you don't mind putting up with the delays involved when running this program; it can be a bit slow! The only points to notice with the program are as follows:

1. The value by which *k* in the program is incremented controls the speed of running of the program—the smaller this value is the longer the time will be but the better the resolution will be.

2. The value by which *x* and *xtemp* are allowed to differ before it's assumed that a bifurcation point has been reached should be as small as is possible; however, if it's too small you may find that rounding errors in the arithmetic cause problems.

Other functions

It's not only the logistic equation that gives a bifurcation diagram. Other functions do as well, but the diagram is slightly different and the value of *k* at which chaos begins is also different. For example, Figure 2-5 shows the

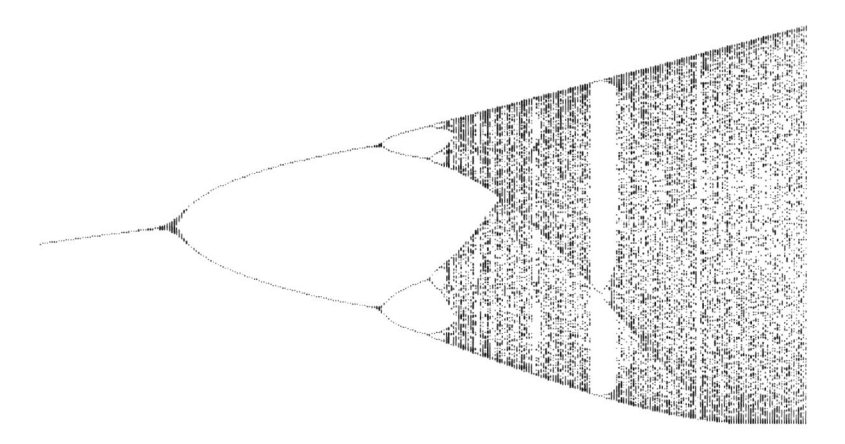

Figure 2-5. Bifurcation diagram for $x_n = kx_{n-1}(1 - \sin(x_{n-1}))$.

bifurcation diagram generated by the function:

$$x_n = kx_{n-1}(1 - \sin(x_{n-1}))$$

for the range of $k = 3$ to $k = 12$ with a step size for Listing 2-3 of 0.01. To reproduce this, simply replace the two occurrences of the logistic equation in the program with the above function.

Plotting the attractor

By varying values of *k*, we've seen how the logistic equation exhibits point, period and strange attractors, as shown in the different parts of Figure 2-4.

ITERATING TWO-DIMENSIONAL FUNCTIONS

Here, each dot on the graph represents a point on the attractor for the value of k being examined. However, there's another way in which we can show the behaviour of an equation as we alter k, and that is as follows:

1. Select a start value of x, say 0.5.

2. For a specified value of k, iterate the equation a few times to allow the transients to die out.

3. For a given number of iterations, say 200, repeat the following steps.

    a. Store the current value of the equation being iterated in a variable, called x_1.

    b. Iterate the equation, using x_1 as the start value, and call the resulting value of the function x.

    c. Plot the point (x_1, x) on the screen of the computer, and repeat from stage a until all 200 iterations are completed.

    d. Increment the value of k by a small amount, reinitialise x to a starting value and repeat the sequence again.

The program in Listing 2-5 does this, and produces images like that shown in Figure 2-6.

This example was produced by entering a start k of 3, a finish k of 4 and a step size of 0.01. Try a y scale of 200 and an offset of 30. The actual evolution of the shape drawn is very informative; watch it as it is drawn, and you can spot where the bifurcation points occur. The parabolic part of the attractor only appears once when we are in the chaotic region of the bifurcation diagram. This method of drawing an attractor is quite useful, and we'll see it used again in Chapter 5.

Iterating two-dimensional functions

The logistic equation is an example of a *one-dimensional* system. There is just the one variable, x, which is used in the equation. Other systems of equations represent two-dimensional systems, and can be written, in general terms, as:

$$x = f_0(x, y) \qquad y = f_1(x, y)$$

In this system of equations, the two functions, f_0 and f_1, can involve both x and y. This makes matters more complicated than we've seen with the logistic equation, but some very interesting patterns can be generated from these functions.

One of the best known sets of equations is called *Martin's mappings*, after the discoverer. They were popularised in *Scientific American* under the name 'Hopalong', which describes how these functions draw points on the computer screen when a suitable program is written. Some typical Martin's mappings are:

$$\begin{aligned} x &= y + \sin(x) \\ y &= a - x \end{aligned}$$

$$\begin{aligned} x &= y + \text{sgn}(x)(bx + c) \\ y &= a - x \end{aligned}$$

and they are of the general form:

$$\begin{aligned} x &= y + f(x) \\ y &= a - x \end{aligned}$$

In each case, the values of x and y are calculated, plotted on a graph and then the functions are iterated again using the x and y values. The values of x and y depend upon the values of constants in the functions, such as a, b and c and the start values of x and y.

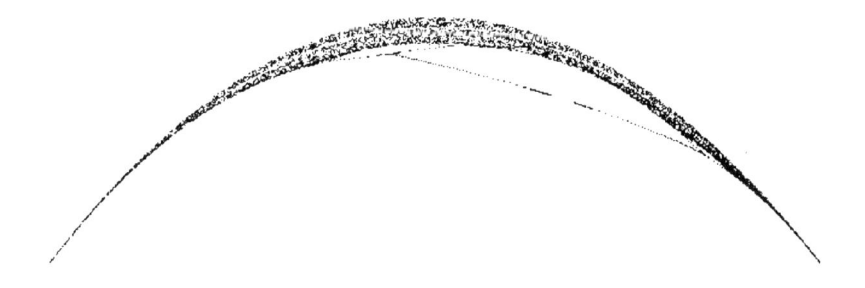

Figure 2-6. Typical output from the Martin's mapping program in Listing 2-6.

Listing 2-6 is a program for generating Martin's mappings, and typical output from this program is shown in Figure 2-7. The default figures given will generate this image, and you can play around with these values to change the shape generated. Again, it's valuable to look at the shape as it is being generated, as the order in which points are plotted is quite interesting. Each point plotted is on the attractor of the system of equations, so we're again seeing a representation of the x and y values that are possible from the particular system of equations. The pattern generated can be prettied up by the use of colour changes every couple of hundred iterations.

Listing 2-7 shows a program to do this with a different mapping, which gives rather different results to the mapping in Listing 2-6.

ITERATING TWO-DIMENSIONAL FUNCTIONS

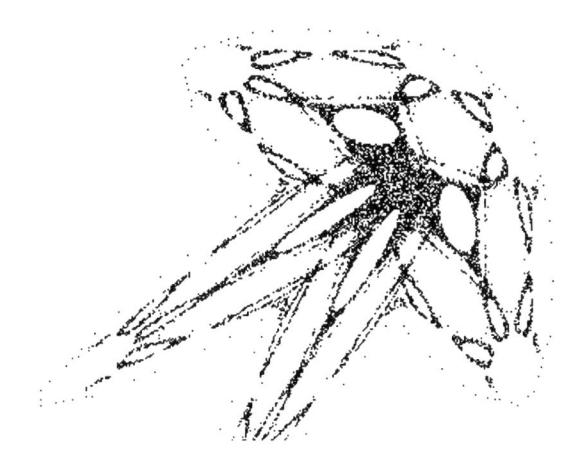

Figure 2-7. A rather beautiful example of a Martin's mapping.

There's nothing to stop you having three-dimensional systems, in which we might have three equations describing the behaviour of three different parameters, x, y and z, with time. However, I'll leave discussion of these until later in the book.

There are other systems that are well worth exploring using the tools that we have examined in this chapter. To summarise, we've seen a variety of ways of analysing dynamic systems. These are:

1. Plot the equations directly.

2. Plot the bifurcation diagram of the system.

3. Calculate when bifurcation points occur in the system, and so work out the Feigenbaum constant.

4. Plot the attractor of the system.

To experiment further, you could modify the parameters given in the equations described in this chapter, or write your own functions and place them in the programs provided, replacing the existing functions. A couple of points to remember are:

1. Don't forget to use the new equation in *all* places in the program where the function is called; this is particularly important where the equation is iterated to get a stable value, as in Listing 2-3.

2. When using new equations, keep the initial values of x and/or y and the values of any constants fairly small until you get a feel for how the system is going to behave. Many dynamic systems 'blow up' very quickly, and will generate a numeric overflow error from your computer.

BBC BASIC listings

Listing 2-1

```
  5 REM Logistic Equation
 10 REM First of all, clear screen and get the initial X value
 15 CLS
 16 INPUT "Initial value of x: ",x
 30 INPUT "Maximum Number of Iterations: ",MaxIteration
 50 MODE4:MaxY=1000:MaxX=1000:XShift=INT(MaxX/MaxIteration):MOVE0,0
 60 FOR IterationNumber=1 TO MaxIteration
 70   REM Here is SQUARE number line.
 80   REM y=x*x
 82   REM Here is the SQUARE ROOT line....
 83   REM y=SQR(x)
 85   REM Here is the complex function line....
 87   y=2*x*(x-1)
 90   REM Now calculate the value that will be plotted on the screen.  This
      is the
100   REM integer portion of the value from the function multiplied by 100 to
      bring
110   REM it in to a sensible range for plotting.
120   ploty=INT(MaxY/2)+INT(100*y)
130   DRAW XShift*IterationNumber,ploty ELSE
140   REM Make the next value of x equal to the value just calculated....
150   x=y
160   REM and round again until all iterations completed.
170 NEXT
180 ch=GET
190 END
```

Listing 2-2

```
  5 REM Logistic Equation
 10 REM First of all, clear screen and get the initial X value
 15 CLS
 16 INPUT "Initial value of x: ",x
 20 INPUT "Initial value of k: ",k
 30 INPUT "Maximum Number of Iterations: ",MaxIteration
 40 INPUT "Draw a graph? ",Graphit$
 50 IF Graphit$="Y"
    MODE4:MaxY=1000:MaxX=1000:XShift=INT(MaxX/MaxIteration):MOVE0,0
 60 FOR IterationNumber=1 TO MaxIteration
 70   REM Here is the Logistic Equation
 80   y=k*x*(1-x)
 90   REM Now calculate the value that will be plotted on the screen.  This
      is the
100   REM integer portion of the value from the function multiplied by 100 to
      bring
110   REM it in to a sensible range for plotting.
120   ploty=INT(MaxY/2)+INT(100*y)
130   IF Graphit$="Y" DRAW XShift*IterationNumber,ploty ELSE PRINT Y:ch=GET
140   REM Make the next value of x equal to the value just calculated....
150   x=y
160   REM and round again until all iterations completed.
170 NEXT
180 ch=GET
190 END
```

46   CHAPTER 2. ITERATIVE FUNCTIONS

Listing 2-3

```
10 REM Draw Bifurcation Diagram
11 REM Try yscale=500, YOffset=100
15 CLS
20 INPUT "Start value of k  ",MinValk
30 INPUT "Finish value of k ",MaxValk
40 INPUT "Step Size         ",StepSize
50 INPUT "Y Scale           ",YScale
60 INPUT "Y Offset          ",YOffset
70 MODE 4
80 i=1
90 k=MinValk
95 REM Start value of k set up
100 REPEAT
110    x=0.5
112    REM Start value of x
113    REM Iterate the equation until the equilibrium conditions have been
       established.
114    REM Keep x value to below 2E6 to prevent problems with the program
       'blowing up'
120    FOR j=1 TO 50
130      IF ABS(x)<2E6 THEN x=x*k*(1-x) ELSE x=2E6
135    NEXT
140    REM Now plot the first 200 values assumed by the equation as it's
       iterated
150    REM further.  For some values of k, of course, all these points will
       fall on
160    REM the same place on the graph.  Again, check for 2E6
170    FOR j=1 TO 200
180      IF ABS(x)<2E6 THEN x=x*k*(1-x) ELSE x=2E6
190      REM Now plot the point on the screen.  The YScale parameter sets the
         'spread' of the vertical points
200      REM and the YOffset adjusts how far up and down the screen the graph
         sits.
210      GCOL0,1
215      PLOT69,i*2,INT(x*YScale)+YOffset
220    NEXT
230    k=k+StepSize
235    REM increment k by stepsize
240    i=i+1
242    REM move to next horizontal graphing position.
250 UNTIL (k>=MaxValk) OR (i*2>=1000)
251 REM repeat until all r values done OR full screen graph drawn.
255 REM Now beep when the graph is drawn, then press any key to go on
260 PRINT CHR$(7)
270 G=GET
280 END
```

Listing 2-4

```
10 REM Draw Bifurcation Diagram
15 CLS
20 INPUT "Start value of k  ",MinValk
30 INPUT "Finish value of k ",MaxValk
40 INPUT "Step Size         ",StepSize
70 MODE 3
80 i=1
90 k=MinValk
95 REM Start value of k set up
100 REPEAT
110    x=0.5
112    REM Start value of x
```

BBC BASIC LISTINGS

```
113     REM Iterate the equation until the equilibrium conditions have been
        established.
114     REM Keep x value to below 2E6 to prevent problems with the program
        'blowing up'
120     FOR j=1 TO 50
130       IF ABS(x)<2E6 THEN x=x*k*(1-x) ELSE x=2E6
135     NEXT
170     FOR j=1 TO 20
180       IF ABS(x)<2E6 THEN x=x*k*(1-x) ELSE x=2E6
200       PRINT k;"      ";x
205       G=INKEY(100)
220     NEXT
230     k=k+StepSize
232     REM Now wait
235     REM increment k by stepsize
240     i=i+1
250 UNTIL (k>=MaxValk)
251 REM repeat until all r values done OR full screen graph drawn.
255 REM Now beep when the graph is drawn, then press any key to go on
260 PRINT CHR$(7)
270 G=GET
280 END
```

Listing 2-5

```
10 REM Draw Logistic Equation Attractor
15 CLS
16 REM Try YScale=500, YOffset=100
20 INPUT "Start value of k  ",MinValk
30 INPUT "Finish value of k ",MaxValk
40 INPUT "Step Size         ",StepSize
50 INPUT "Y Scale           ",YScale
60 INPUT "Y Offset          ",YOffset
70 MODE 4
75 MaxX=1000
80 i=1
90 k=MinValk
95 REM Start value of k set up
100 REPEAT
110   x=0.5
112   REM Start value of x
113   REM Iterate the equation until the equilibrium conditions have been
      established.
114   REM Keep x value to below 2E6 to prevent problems with the program
      'blowing up'
120   FOR j=1 TO 50
130     IF ABS(x)<2E6 THEN x=x*k*(1-x) ELSE x=2E6
135   NEXT
140   REM Now plot the first 200 values assumed by the equation as it's
      iterated
150   REM further. For some values of k, of course, all these points will
      fall on
160   REM the same place on the graph. Again, check for 2E6
170   FOR j=1 TO 200
175     x1=x
180     IF ABS(x)<2E6 THEN x=x*k*(1-x) ELSE x=2E6
190     REM Now plot the point on the screen. The YScale parameter sets the
        'spread' of the vertical points
200     REM and the YOffset adjusts how far up and down the screen the graph
        sits.
210     GCOL0,1
215     PLOT 69,INT(x1*MaxX),INT(x*YScale)+YOffset
220   NEXT
```

48   CHAPTER 2. ITERATIVE FUNCTIONS

```
230    k=k+StepSize
235    REM increment k by stepsize
240    i=i+1
242    REM move to next horizontal graphing position.
250    UNTIL (k>=MaxValk) OR (i*2>=1000)
251    REM repeat until all r values done OR full screen graph drawn.
255    REM Now beep when the graph is drawn, then press any key to go on
260    PRINT CHR$(7)
270    G=GET
280    END
```

Listing 2-6

```
 10 REM Martins Mapping
 20 REM Try p=0.987
 30 REM Try q=1.234
 40 REM Try r=0.654
 50 REM Try x=2.3
 60 REM Try y=2.3
 70 REM Try xs=10
 80 REM Try ys=10
 90 REM Try n=20000
100 REPEAT
110    MODE 6
120    CLS
130    REM  Next section gets input from the user as to what parameters are
       desired
150    PRINTTAB(2,2)"Martins Mappings"
160    INPUT "Constant P ",p
180    INPUT "Constant Q ",q
200    INPUT "Constant R ",r
220    INPUT "Start X ",x
240    INPUT "Start Y ",y
260    INPUT "Iterations ",n
280    INPUT "X Scaling ",xs
300    INPUT "Y Scaling ",ys
320    MODE 5
330    xc=500
340    yc=500
350    CurrentIteration=1
360    PixelColor=1
370    REPEAT
380      REM Next lines evaluate the (n+1) state of x and y.  The equation is
390      REM fairly simple in this case.  If you want to change the equation
400      REM here, see text
410      x1=y-SGN(x)*(q*x+r)
420      y=p-x
430      x=x1
440      REM Now plot the point for this iteration
450      GCOL0,1
460      PLOT69,INT(xc+x1*xs),INT(yc+y*ys)
470      CurrentIteration=CurrentIteration+1
480    UNTIL (INKEY(1)<>-1) OR (CurrentIteration>=n)
490    PRINT CHR$(7)
500    ch=GET
510 UNTIL (ch=ASC("X")) or (ch=ASC("x"))
520 END*EDIT
```

Listing 2-7

```
 10 REM Martins Mapping
 20 REM Try p=0.987
 30 REM Try q=1.234
 40 REM Try r=0.654
 50 REM Try x=2.3
 60 REM Try y=2.3
 70 REM Try xs=10
 80 REM Try ys=10
 90 REM Try n=20000
100 REPEAT
110   MODE 6
120   CLS
130   REM Next section gets input from the user as to what parameters are desired
150   PRINTTAB(2,2)"Martins Mappings"
160   INPUT "Constant P ",p
180   INPUT "Constant Q ",q
200   INPUT "Constant R ",r
220   INPUT "Start X ",x
240   INPUT "Start Y ",y
260   INPUT "Iterations ",n
280   INPUT "X Scaling ",xs
300   INPUT "Y Scaling ",ys
320   MODE 5
330   xc=500
340   yc=500
350   CurrentIteration=1
360   PixelColor=1
370   REPEAT
380     REM Next lines evaluate the (n+1) state of x and y. The equation is
390     REM fairly simple in this case. If you want to change the equation
400     REM here, see text
410     x1=y-SGN(x)*(q*x+r)
420     y=p-x
430     x=x1
440     REM Now plot the point for this iteration in a different colour
450     GCOL0,(PixelColor MOD 4)
460     PLOT69,INT(xc+x1*xs),INT(yc+y*ys)
470     CurrentIteration=CurrentIteration+1
475     PixelColor=PixelColor+1
480   UNTIL (INKEY(1)<>-1) OR (CurrentIteration>=n)
490   PRINT CHR$(7)
500   ch=GET
510 UNTIL (ch=ASC("X")) or (ch=ASC("x"))
```

Turbo Pascal listings

Listing 2-1

```
Program Equations;

{ Listing 2.1 in Chaos Programming Book
  Joe Pritcahrd, July 1990                                        }

Uses Crt,Graph,Extend;

Var            x,y,k              :     Real;
               MaxX,MaxY          :     Integer;
               IterationNumber    :     Integer;
               plotx,ploty        :     Integer;
               MaxIteration       :     Integer;
               XShift             :     Integer;
               Graphit,ch         :     Char;
Begin

{ First of all, clear screen and get the initial X value }

   ClrScr;
   TextColor(Yellow);
   Write('Initial value of x: ');
   Read(x);
   Write('Maximum Number of Iterations: ');
   Readln(MaxIteration);

{ Use a Library Function to choose a graphics mode and palette and then
  get the maximum Y value on the screen. Move to bottom left of screen
  (note - in Turbo Pascal, top left is 0,0. }

   ChoosePalette;
   MaxY:=GetMaxY;
   MaxX:=GetMaxX;
   XShift:=trunc(MaxX/MaxIteration);
   MoveTo(0,MaxY);

   for IterationNumber:=1 to MaxIteration do begin

{ Here is theset of equations -  only include ONE equation
  }

{ SQR - square the value of x

      y:=sqr(x);

}

{ SQRT - take the square root of x

      y:=sqrt(x);
}

{ Use the complex function  }

      y:=2*x*(x-1);

{ Now calculate the value that will be plotted on the screen. This is the
  integer portion of the value from the function multiplied by 100 to bring
  it in to a sensible range for plotting.   }

      ploty:=trunc(MaxY/2)-trunc(100*y);
```

TURBO PASCAL LISTINGS

{ Now use the standard Turbo Pascal Line Drawing function to draw a line
  on the screen from the last point visited to the current point being
  calculated.   }

```
    LineTo(XShift*IterationNumber,ploty);
```

{ Make the next value of x equal to the value just calculated.... }

```
    x:=y;
```

{ and round again until all iterations completed.   }

```
  end;

    Readln;

end.
```

Listing 2-2

```
Program LogisticEquation;

{ Listing 2.2 in Chaos Programming Book
  Joe Pritcahrd, July 1990                                          }

Uses Crt,Graph,Extend;

Var                x,y,k              :      Real;
                   MaxX,MaxY          :      Integer;
                   IterationNumber    :      Integer;
                   plotx,ploty        :      Integer;
                   MaxIteration       :      Integer;
                   XShift             :      Integer;
                   Graphit,ch         :      Char;
Begin

{ First of all, clear screen and get the initial X value }

    ClrScr;
    TextColor(Yellow);
    Write('Initial value of x: ');
    Read(x);
    Write('Initial value of k: ');
    Read(k);
    Write('Maximum Number of Iterations: ');
    Readln(MaxIteration);

    Write('Draw a graph? ');
    Read(GraphIt);

{ Use a Library Function to choose a graphics mode and palette and then
  get the maximum Y value on the screen. Move to bottom left of screen
  (note - in Turbo Pascal, top left is 0,0. }

if (Graphit='Y') then begin
    ChoosePalette;
    MaxY:=GetMaxY;
    MaxX:=GetMaxX;
    XShift:=trunc(MaxX/MaxIteration);
    MoveTo(0,MaxY);
end;
```

CHAPTER 2. ITERATIVE FUNCTIONS

```pascal
        for IterationNumber:=1 to MaxIteration do begin

{ Here is the Logistic Equation                                           }

        y:=k*x*(1-x);

{ Now calculate the value that will be plotted on the screen. This is the
  integer portion of the value from the function multiplied by 100 to bring
  it in to a sensible range for plotting.   }

        ploty:=trunc(MaxY/2)-trunc(100*y);

{ Now use the standard Turbo Pascal Line Drawing function to draw a line
  on the screen from the last point visited to the current point being
  calculated.                                                            }

If (Graphit='Y') then
    LineTo(XShift*IterationNumber,ploty)
else
    begin
      Writeln(y);
      ch:=ReadKey;
    end;

{ Make the next value of x equal to the value just calculated.... }

        x:=y;

{ and round again until all iterations completed.                         }

     ond;

     Readln;

end.
```

Listing 2-3

```pascal
Program Bifurcation;
{ For Chaos Programming Book, Joe Pritchard July 1990
  Allows drawing of Bifurcation diagram for the Logistic Equation   }

Uses Crt,Graph,Extend;

Var

    i,j,yscale,yoffset       :            integer;
    x,k,MaxValk,MinValk      :            real;
    StepSize                 :            real;

{ Get the start parameters in from the user           }

Begin
  ClrScr;
  Write('Start value of k  : ');
  Readln(MinValk);
  Write('Finish value of k : ');
  Readln(MaxValk);
  Write('Step Size         : ');
  Readln(StepSize);
  Write('Y Scale           : ');
  Readln(YScale);
  Write('Y Offset          : ');
```

TURBO PASCAL LISTINGS

```pascal
    Readln(YOffset);

    ChoosePalette;             { Use Extend function to set up graphics mode }
    i:=1;                      { Left hand edge of screen. }
    k:=MinValk;                { Start value of k set up }

    Repeat

      x:=0.5;                  { Start value of x }

{ Iterate the equation until the equilibrium conditions have been established.
  Keep x value to below 2E6 to prevent problems with the program 'blowing up' }

      for j:=1 to 50 do begin
          if abs(x)<2E6 then
             x:=x*k*(1-x)
          else
             x:=2E6;
      end;

{ Now plot the first 200 values assumed by the equation as it's iterated
  further. For some values of k, of course, all these points will fall on
  the same place on the graph. Again, check for 2E6                        }

      for j:=1 to 200 do begin
          if abs(x)<2E6 then
             x:=x*k*(1-x)
          else
             x:=2E6;

{ Now plot the point on the screen. The YScale parameter sets the 'spread' of
  the vertical points
  and the YOffset adjusts how far up and down the screen the graph sits.    }

          putpixel(i*2,(GetMaxY-trunc(x*YScale)-Yoffset),3);
      end;

      k:=k+StepSize;           { increment k by stepsize }
      i:=i+1;                  { move to next horizontal graphing position. }
    until (k>=MaxValk) or (i*2>=GetMaxX);  { repeat until all r values done OR full
                                             screen graph drawn.        }

{ Now beep when the graph is drawn, then press any key to go on }

    Sound(220);
    Delay(250);
    NoSound;
    Readln;
    CloseGraph;

end.
```

Listing 2-4

```pascal
Program PrintBifurcation;
{ For Chaos Programming Book, Joe Pritchard July 1990
  Allows printing of Logistic Equation   }

Uses Crt,Graph,Extend;

Var
```

CHAPTER 2. ITERATIVE FUNCTIONS

```
    i,j                     :            integer;
    x,k,MaxValk,MinValk     :            real;
    StepSize                :            real;

{ Get the start parameters in from the user           }

Begin
  ClrScr;
  Write('Start value of k  : ');
  Readln(MinValk);
  Write('Finish value of k : ');
  Readln(MaxValk);
  Write('Step Size         : ');
  Readln(StepSize);

  i:=1;                       { Left hand edge of screen. }
  k:=MinValk;                 { Start value of k set up }

  Repeat

    x:=0.5;                   { Start value of x }

{ Iterate the equation until the equilibrium conditions have been established.
  Keep x value to below 2E6 to prevent problems with the program 'blowing up' }

    for j:=1 to 50 do begin
        if abs(x)<2E6 then
           x:=x*k*(1-x)
        else
           x:=2E6;
    end;

{ Now print the first 20 values assumed by the equation as it's iterated
  further.  Again, check for 2E6                       }

    for j:=1 to 20 do begin
        if abs(x)<2E6 then
           x:=x*k*(1-x)
        else
           x:=2E6;

      writeln(k,x);
      Delay(100);

    end;

    k:=k+StepSize;         { increment k by stepsize }
    i:=i+1;                { move to next horizontal graphing position. }

    writeln(' ');

  until (k>=MaxValk);   { repeat until all r values done OR full
                                         screen graph drawn.    }

{ Now beep when all done, then press any key to go on }

Sound(220);
Delay(250);
NoSound;
Readln;

end.
```

Listing 2-5

```pascal
Program DrawAttractor;
{ For Chaos Programming Book, Joe Pritchard July 1990
  Allows drawing of Attractor diagram for the Logistic Equation  }

Uses Crt,Graph,Extend;

Var

   i,j,yscale,yoffset       :           integer;
   x,k,MaxValk,MinValk,x1   :           real;
   StepSize                 :           real;

{ Get the start parameters in from the user         }

Begin
  ClrScr;
  Write('Start value of k : ');
  Readln(MinValk);
  Write('Finish value of k : ');
  Readln(MaxValk);
  Write('Step Size       : ');
  Readln(StepSize);
  Write('Y Scale         : ');
  Readln(YScale);
  Write('Y Offset        : ');
  Readln(YOffset);

  ChoosePalette;              { Use Extend function to set up graphics mode }
  i:=1;                       { Left hand edge of screen. }
  k:=MinValk;                 { Start value of k set up }

  Repeat

   x:=0.5;                    { Start value of x }

{ Iterate the equation until the equilibrium conditions have been established.
  Keep x value to below 2E6 to prevent problems with the program 'blowing up' }

     for j:=1 to 50 do begin
         if abs(x)<2E6 then
            x:=x*k*(1-x)
         else
            x:=2E6;
     end;

{ Now plot the first 200 values assumed by the equation as it's iterated
  further. For some values of k, of course, all these points will fall on
  the same place on the graph. Again, check for 2E6                        }

     for j:=1 to 200 do begin
        x1:=x;
        if abs(x)<2E6 then
           x:=x*k*(1-x)
        else
           x:=2E6;

{ Now plot the point on the screen. The YScale parameter sets the 'spread' of
  the vertical points
  and the YOffset adjusts how far up and down the screen the graph sits.    }

        putpixel(trunc(x1*GetMaxX),(GetMaxY-trunc(x*YScale)-Yoffset),3);
     end;
```

56   CHAPTER 2. ITERATIVE FUNCTIONS

```
        k:=k+StepSize;        { increment k by stepsize }
        i:=i+1;               { move to next horizontal graphing position. }

    until (KeyPressed) or (k>MaxValk);  { repeat until all r values done OR full
                                          screen graph drawn.        }

  { Now beep when the graph is drawn, then press any key to go on }

  Sound(220);
  Delay(250);
  NoSound;
  Readln;
  CloseGraph;

  end.
```

Listing 2-6

```
Program MartinMapping;

Uses Crt,Graph,Extend;

Var
    x,x1:                       Real;
    y,xc,yc,xs,ys:              Real;
    r,p,q:                      Real;
    CurrentIteration,n:         LongInt;
    g:                          Integer;
    ch:                         Char;
    xt,yt,qs,rs,ps,is,sx,sy:    String;
    yst,xst,nst:                String;

Begin
  ps:='0.987';
  qs:='1.234';
  rs:='0.654';
  sx:='2.3';
  sy:='2.3';
  xst:='10';
  yst:='10';
  nst:='20000';
  Repeat
   TextBackground(Blue);
   ClrScr;

{ Next section gets input from the user as to what parameters are desired
  for the program run. Uses functions from the EXTEND library.         }

   PrintAt(2,2,'Martins Mappings, Implementation Joe Pritchard, 1989',Yellow);
   PrintAt(2,4,'Constant P:',Green);
   StringEdit(ps,20,20,4,Yellow);
   Val(ps,p,g);
   PrintAt(2,5,'Constant Q:',Green);
   StringEdit(qs,20,20,5,Yellow);
   Val(qs,q,g);
   PrintAt(2,6,'Constant R:',Green);
   StringEdit(rs,20,20,6,Yellow);
   Val(rs,r,g);
   PrintAt(2,7,'Start X:',Green);
   StringEdit(sx,20,20,7,Yellow);
   Val(sx,x,g);
   PrintAt(2,8,'Start Y:',Green);
   StringEdit(sy,20,20,8,Yellow);
```

TURBO PASCAL LISTINGS

```
    Val(sy,y,g);

    PrintAt(2,10,'Iterations:',Green);
    StringEdit(nst,20,20,10,Yellow);
    Val(nst,n,g);
    PrintAt(2,11,'X Scaling:',Green);
    StringEdit(xst,20,20,11,Yellow);
    Val(xst,xs,g);
    PrintAt(2,12,'Y Scaling:',Green);
    StringEdit(yst,20,20,12,Yellow);
    Val(yst,ys,g);
    ChoosePalette;
    xc:=trunc(GetMaxX/2);
    yc:=trunc(GetMaxY/2);

    CurrentIteration:=1;
    PixelColor:=1;

    Repeat
       Begin

{      Next lines evaluate the (n+1) state of x and y. The equation is
       fairly simple in this case. If you want to change the equation
       here, see text                                                }

          x1:=y-sgn(x)*(q*x+r);
          y:=p-x;
          x:=x1;

{         Now plot the point for this iteration                      }

          PutPixel(trunc(xc+x1*xs),trunc(yc+y*ys),1);
          CurrentIteration:=CurrentIteration+1;
       end;
    until (KeyPressed) or (CurrentIteration>=n);
    Sound(200);
    Delay(100);
    NoSound;
    ch:=readkey;
    CloseGraph;
 until (ch='X') or (ch='x');
end.
```

Listing 2-7

```
Program MartinMAp;

Uses Crt,Graph,Extend;

Var
   d,x,x1,u,v:                    Real;
   y,xc,yc,xs,ys:                 Real;
   r,p,q,t:                       Real;
   CurrentIteration,n:            LongInt;
   g:                             Integer;
   ch:                            Char;
   xt,yt,qs,rs,ps,is,sx,sy:       String;
   yst,xst,nst:                   String;

Begin
   ps:='0.987';
   qs:='1.234';
   rs:='0.654';
```

```
  sx:='2.3';
  sy:='2.3';
  xst:='20';
  yst:='20';
  nst:='10000';
  Repeat
   TextBackground(Blue);
   ClrScr;
   PrintAt(2,2,'Martins Mapping, Implementation Joe Pritchard, 1989',Yellow);
   PrintAt(2,4,'Constant P:',Green);
   StringEdit(ps,20,20,4,Yellow);
   Val(ps,p,g);
   PrintAt(2,5,'Constant Q:',Green);
   StringEdit(qs,20,20,5,Yellow);
   Val(qs,q,g);
   PrintAt(2,6,'Constant R:',Green);
   StringEdit(rs,20,20,6,Yellow);
   Val(rs,r,g);
   PrintAt(2,7,'Start X:',Green);
   StringEdit(sx,20,20,7,Yellow);
   Val(sx,x,g);
   PrintAt(2,8,'Start Y:',Green);
   StringEdit(sy,20,20,8,Yellow);
   Val(sy,y,g);

   PrintAt(2,10,'Iterations:',Green);
   StringEdit(nst,20,20,10,Yellow);
   Val(nst,n,g);
   PrintAt(2,11,'X Scaling:',Green);
   StringEdit(xst,20,20,11,Yellow);
   Val(xst,xs,g);
   PrintAt(2,12,'Y Scaling:',Green);
   StringEdit(yst,20,20,12,Yellow);
   Val(yst,ys,g);
   ChoosePalette;
   xc:=trunc(GetMaxX/2);
   yc:=trunc(GetMaxY/2);

   CurrentIteration:=1;
   PixelColor:=1;

   Repeat
      Begin

{ Next two lines evaluate the (r+1) state of x and y. The equation is
  fairly simple in this case.                                        }

         x1:=y-sgn(x)*sqrt(abs(q*x-r));
         y:=p-x;
         x:=x1;

{        Now plot the points and draw each point in a different colour        }

         PutPixel(trunc(xc+x1*xs),trunc(yc+y*ys),Pickpixelcolor(CurrentIteration,100));

         CurrentIteration:=CurrentIteration+1;
      end;
    until (KeyPressed) or (CurrentIteration>=n);
    MoveTo(290,200);
    ch:=readkey;
    CloseGraph;
  until (ch='X') or (ch='x');
end.
```

Chapter 3

Differential equations

We've already examined systems that start life as *differential* equations in Chapter 2. If we start by looking at a non-differential equation, like:

$$total\ cost = unit\ price * number\ purchased$$

then we can quickly see that we're interested only in the final price we have to pay. We're not interested in the rate of change of the price over the next few days, levels of stock of the items we're buying or anything else—just the bottom line of cost. In physics, we might consider the equation:

$$V = I * R$$

where V is the voltage across a circuit, I the current through it and R the electrical resistance of that circuit. These equations are called *algebraic* equations. We can easily rearrange them, for example, to calculate current given resistance and voltage, and stick values for R and I in the equation and calculate V. However, what if we're dealing with dynamic systems, that is, physical systems or sets of equations in which the value of at least one parameter is changing? Here we might be interested in the final values to which the system settles down but we're also interested in the rates of change of the changing parameters of the system. When you've got the rates of change, you can see roughly how a system behaves with time, and so attempt to predict the behaviour of a system given the initial parameters for any point in the future. When we're looking at these systems, we're interested in how a parameter of the system changes from value a at time t to value b at time t_1. The rate of change is given by:

$$(a - b)/(t_1 - t)$$

For example, if a and b were distances of a dropped object away from the hand that let it go, the rate of change would give us the velocity of the object at a specific time. If we say that x is the distance, then we can write down such an equation as:

$$\frac{dx}{dt} = f(t)$$

which tells us that the rate of change of x with time (the $\frac{dx}{dt}$) part of the equation is simply a function of time t. The essential point to remember is that this expression $\frac{dx}{dt}$ is actually referring to a very tiny change in x and an equally small change in t. This is an example of a *differential equation*. Now, if the function were, say,

$$\frac{dx}{dt} = k*t$$

then this indicates that the velocity of the object is simply proportional to the elapsed time. In fact, a few seconds will show that the longer the elapsed time is, the greater the rate of change is—the object is, in fact, accelerating away from its rest position. This is what happens to all dropped objects; they accelerate under the force of gravity. A similar equation lies behind the logistic equation—there we were concerned with how a population grew. We could write that generally as:

$$\frac{dx}{dt} = k*x$$

where x is the population. The rate of change here depends upon the initial population level, x, and a constant, k, which covers other aspects of the real-world system we're modelling. Systems modelled by such equations, such as populations, moving objects, etc. are often called *dynamic systems*.

What if we want to know what the population will be at a specific point in the future? Can we work that out from this equation? Well, as we've already used the equation the answer is 'Yes', but the process of getting from a differential equation like this that describes a system to an algebraic equation into which we can feed numbers and watch what happens is not always easy. In principle, the process of solving these differential equations is that of finding an algebraic function that allows us to stick in a value of t and a start value of x and get the x value at time t out of the equation. The process of getting the function to do this is called *integration*. There are two approaches to integrating a simple differential equation.

Analysis

Here the differential equation is integrated using some basic rules of integration and tables of integrated functions. For example, the simple function $\sin(x)$ can be found in a table of integrals and its corresponding integral substituted in the function being derived from the differential equation.

GRAPHICAL SOLUTION

This process will eventually give us an algebraic equation for the differential equation being integrated, and into this equation we can then plug some numbers and start calculating. Of course, if the equation is changed, you have to reintegrate the whole thing again. This process gives us the most direct solution to the problems of solving differential equations, but is also a rather complicated activity that takes time. In addition, it would be rather nice if we had a more general purpose method of solving differential equations; after all, in chaos we're likely to run in to them fairly frequently!

Graphical solution

Let's take a few steps back. I've got a differential equation, say:

$$\frac{dx}{dt} = 2*t+3$$

that I want to solve. If I wanted to, I could get an analytical solution to this equation as described above. If I did that, I could put numbers into the resulting algebraic equation and calculate the resultant x value for times $t = 1, 2, 3 \ldots n$. If I now draw these points on a graph, then I've got a graphical representation of how the function in question varies with time. Let's take a look at a small part of this graph in Figure 3-1. As you can see,

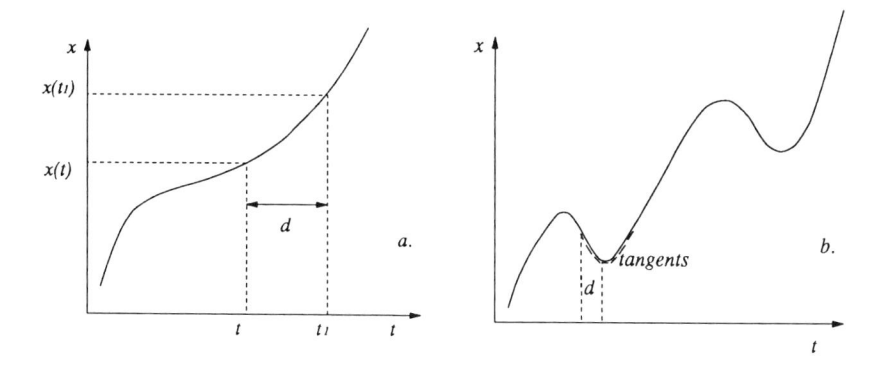

Figure 3-1. Graphs showing $\frac{dx}{dt}$ as tangent to curve.

there are two values of t (t and t_1) and I've marked the corresponding x values x and x_1. Now I've got two values of x corresponding to two different times. I could calculate the rough rate of change of x at this point by:

$$rate\ of\ change = (x_1 - x)/(t_1 - t)$$

Hold on a minute—rate of change? Isn't that what we started with in the differential equation? Yes—the rate of change of x between two closely spaced times on the graph is given by calculating the slope of the graph between the two times. Obviously the slope of a curve isn't too easy to calculate, so what we do instead is to draw a tangent to the curve, as shown in Figure 3-1a. The slope of the tangent can then be calculated and this gives us the approximate rate of change at point t. If we were to draw a series of tangents with small values of d, we could actually approximate the curve of the equation by a series of short, straight lines. Thus, the term $\frac{dx}{dt}$ is actually the slope of the line obtained from plotting a very tiny change in x against a similarly tiny change in the value of t. This is a very useful technique to have available.

Alternatively, we could do a computerised 'join the dots' problem and specify the line as a series of spot $\frac{dx}{dt}$ values, joining the dots to give the appearance of a smooth(ish) curve. Clearly the smaller the value d, the more accurately the resulting line would represent the function. This is true to a great extent, but in practical terms you often find that a too small value of d causes problems. However, we can cross that bridge when we arrive there. This would give us an alternative way of approximating the curve as a series of short, straight lines. From this graph, we could then read off x values corresponding to specific t values, rather than putting the figures directly into the algebraic equation. Thus, the results obtained by plotting a series of x values against t as calculated by using an algebraic equation obtained by integrating the original function, or by plotting a series of rates of change values as obtained from the original differential equation are equivalent. Thus, we can see how a differential equation behaves by simply putting numbers in to the differential equation to obtain approximate rates of change, and plot those on a graph. This technique is called *Euler's method* and is a rather nice way of getting a glimpse at how differential equations behave. The technique can be summarised as follows:

1. Select a function, say $\frac{dx}{dt} = 2 * x * x + 3$.

2. Select a value for d by which t will be incremented for each calculation.

3. Calculate a value called $xinc$, the amount by which x is to be incremented at time t. Calculate this by: $xinc := f() * d$. In this case, it would be $xinc := (2 * x * x + 3) * d$.

4. Add the value of $xinc$ to the current value of x. Plot this value of x against t.

5. Increment t by d. Repeat from 3.

Listing 3-1 shows this algorithm programmed. To experiment with the program, first of all try putting in different values of d after running the program with the value of d given. Figure 3-2 shows the graph that should be obtained from different d values with the equation given in the listing.

GRAPHICAL SOLUTION

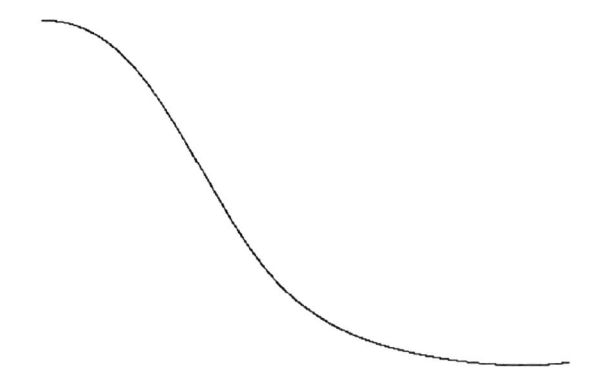

Figure 3-2. Graph obtained from Listing 3-1.

This method of drawing the function of a differential equation works, but it is not as accurate as some other methods. The main difficulty presented by the Euler method is that it is at its most accurate when the value of d is small; however, as you decrease the value of d, there is an increased possibility of rounding errors being made by the mathematics routines of your computer language. The overall accuracy of the Euler method with small values of d will therefore depend upon the language compiler or interpreter that is in use. In addition, with the program shown you should note the following:

1. The larger the value of d used, the greater the chances are of a numerical overflow error being generated, which will most likely crash the program. The value of d that will cause this depends upon the number of iterations of the equation (in the *for* loop) and the actual equation itself.

2. The amount of the function in terms of the range of t covered will depend upon the value of d and the number of times the equation is evaluated. A large value of d will cover a longer period of time on the graph but in relatively low detail, as the time increment, d will no longer be approximating to a smooth transition from t to $t+1$. To cover a long period of time in greater detail, a small value of d and a larger number of iterations are required.

You might like to try putting various functions of your own together and running this routine with them to draw the resulting graph. You simply need to alter the function in the *Equation* function. Also, try experimenting with different starting values of x and t.

With these simple equations we are dealing with situations where the behaviour of the system being modelled depends only upon the change of value of one parameter, x, with time. This is true of many cases, for example, the logistic equation we've already seen, the temperature of a cup of tea as it cools down, and so on.

Although a variety of real-world systems are modelled using equations such as these, there are systems where the behaviour depends not just upon the change of value of one parameter with time, but of the change of two or more parameters with time. We can write such a pair of equations in general terms as:

$$\frac{dx}{dt} = f(x,y) \qquad \frac{dy}{dt} = f_1(x,y)$$

There are two ways in which we can get graphs from these systems: we can plot two lines on a graph, one of x against t and one of y against t. This gives a graph commonly known as a *time history*, *time series* or *time progression* of the system, and for the range of time covered by the graph gives us the values of x and y obtained for different values of t for a particular set of initial x and y values. If we changed the initial conditions, then we could plot a different time progression.

Alternatively, we can use a totally new approach and leave t out of the graphing procedure altogether. The value of y at an instant in time is plotted against the corresponding value of x, the area covered by such a plot being called the *phase plane*.

The graphs obtained from these equations by plotting the x and y values are often called *trajectories* of the system, and they represent the x and y values attained for an equation given a particular set or initial x and y values. The complete collection of trajectories for a range of initial x and y values is called the *phase diagram* or *phase portrait* of the system. We've already encountered the idea of phase space in Chapter 1. These pairs of equations can describe a whole host of real-world situations, from the position of a particle in space to what are called *predator–prey* systems, where one variable represents the population of a predator, such as a large bird, and the other the population of some prey animal, such as an insect or small mammal. The useful thing about these diagrams is that they tell us a great deal about the *long-term* behaviour of a system. For example, with a phase diagram we can tell at a glance whether the system will settle down into periodic behaviour, whether it will exhibit chaotic behaviour, and so on. By examining a phase diagram *while* it is drawn we can also learn a lot about the equations being solved. This type of thing would be next to impossible from a time progression alone, and for this reason phase diagrams are the favoured way of examining systems of differential equations.

A PREDATOR–PREY MODEL

Let's start our work with pairs of equations by considering a simple predator–prey model.

A predator–prey model

We assume that, in the absence of any diseases, predators or any other limits on population growth the prey population in future will be dependent upon its level today, and that the rate of change will also be dependent upon the current population level. So, for the prey we can write:

$$\frac{dx}{dt} = k * x$$

where k is a constant. Now, if we apply the same logic to the predator, who needs the prey for food, in the absence of prey the population will die out. We can write this as:

$$\frac{dy}{dt} = -k_2 * y$$

where k_2 is a constant and the $-$ sign refers to the fact that the population will decrease with time without any prey to eat. Clearly there's some interaction; the prey, if the predators are at all good at hunting, tend to get eaten. We can model this by the expression:

$$k_1 * x * y \quad or \quad k_3 * x * y$$

where k_1 is represents the probability of being hunted and k_3 the probability of hunted prey being killed and x and y are the current population levels of prey and predator, respectively. This will be added to the predator population, as it will increase the numbers of predators if they can eat enough prey to survive and reproduce, whilst it will decrease the prey population. We can now write these equations in full as:

$$\frac{dx}{dt} = k * x - k_1 * x * y$$

$$\frac{dy}{dt} = -k_2 * y + k_3 * x * y$$

These equations are often known as *Volterra's equations* after the mathematician who first explored them. We can apply a modified version of Listing 3-1 to this problem, and the program in Listing 3-2 allows us to graph these equations in two ways:

1. Plot x and y separately against time. This gives us a graph with two lines on it which represent the populations of predator and prey at particular times.

2. Plot y against x. This gives us the phase portrait for the system.

The default values given produce a reasonable set of results, but you can enter any value you like for x, y or any of the four constants. To switch between graph or phase portrait simply enter P or G when asked; to exit the program enter X at the *Graph or Portrait* prompt; a graph will be drawn and on pressing a key the program will finish. After each graph or portrait is drawn a tone will sound; press *Return* to continue.

Once the program is working, you should obtain a default phase portrait that is in the form of a closed loop. The size and shape of the portrait will vary for x and y values and for different values of the four constants. A closed loop like this indicates that the system is periodic for the number of time-increments examined; that is the graph obtained repeats itself. Of course it might be that if you increased the number of points examined (the *mpoints* variable) you might see some non-periodic behaviour at some point in the future of the system being examined. A totally jumbled phase portrait would indicate little, if any, periodicity in the system.

Now, the observant amongst you may have noticed a problem here; if you watch carefully while the phase portrait is being drawn the second pass around the loop isn't quite at the same position as the first pass. Does this indicate that the system isn't periodic for the default k values? Well, no. If you vary the value of d, say to 0.01, then the loop becomes more of a spiral. This is an *artifact* of the Euler method. It doesn't exist in the system of equations but is introduced as an error by the method of calculating the points to graph. After all, the Euler method only approximates solutions to the differential equations involved. The smaller the value of d, the more accurate the solutions will be until rounding errors crop up. Larger values of d, however, cause errors like this and also often lead to the computer coming to a halt with a numerical overflow error. We'll see how this particular problem can have wider implications later in this chapter.

Apart from varying the parameters in the system, you might like to try keeping the values of the k constants the same and varying x and y. This will give you a family of phase portraits, one for each start x and y value. An enhanced method of doing this is to plot all the resultant phase diagrams on the same graph. Listing 3-3 does this, though here the equation used is that that describes the motion of a pendulum (not considering friction or air resistance) as it swings:

$$\frac{dx}{dt} = y$$

$$\frac{dy}{dt} = -k * \sin(x)$$

where x is the angle by which the pendulum is displaced from the vertical (in radians) and y is the velocity of the pendulum. A typical output from this program is shown in Figure 3-3, and you might like to try varying the value of k. Again, the graph shows some inaccuracies due to the Euler

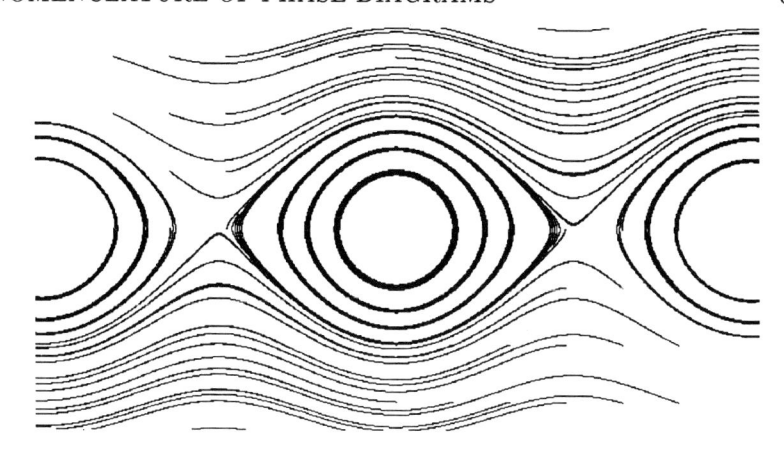

Figure 3-3. A phase diagram for the motion of a pendulum.

method of drawing the portraits. One thing to note is that you may need to alter the value of d (make it smaller) for certain values of k, x and y should numerical overflow occur.

Nomenclature of phase diagrams

Phase portraits are made up of a variety of features, and if you modify Listing 3-3 to use other functions, then you could easily come across some of these aspects of phase diagrams; if you read some of the books listed at the end of this book, then you'll certainly encounter them.

Equilibrium points

Points in the phase diagram where, in the systems of two equations we've looked at so far, $\frac{dx}{dt}$ and $\frac{dy}{dt}$, both equal 0 are called *equilibrium points* or *critical points*. The physical significance of these points is that they indicate situations of no change in the system; for example, the pendulum is stationary at this point. Several of the terms below refer to classifications of equilibrium points.

Trajectory

As we've already seen, the trajectory is the path that a pair of equations takes for a particular initial x and y value. Each point on the trajectory represents a point arrived at from initial x, y values after a length of time has elapsed. The length of time that the trajectory is calculated for (in our listings, the number of points plotted) will give the length of the trajectory; if insufficient points are plotted the trajectory will only be a small part of the full trajectory available from the initial x and y values chosen. One important point to note in these systems of two equations (often called two-dimensional systems) is that two trajectories can't cross with each other. In fact, no matter how many equations you have to describe a system, trajectories can't occupy the same space, although they might get very close to each other. One practical result of this, for example, is that a trajectory starting within a limit cycle (see below) can't escape from the limit cycle. To do this, the trajectory would have to cross the trajectory of the limit cycle and this simply can't happen.

Attractor

On phase diagrams, the attractor is described as a trajectory followed by the equations when any transients have died away. A *period attractor* or *limit cycle* is a simple closed loop, like those shown in Figure 3-3. A further type of attractor is called a *sink* and is just a point somewhere in phase space to which trajectories lead. If we have trajectories that pass very close to each other in phase space, but don't form a limit cycle, and eventually shoot off to give trajectories in different parts of phase space then we have what's called a *strange attractor*. We'll discuss these in greater detail in Chapters 4 and 5.

Nodes

Certain equation pairs give rise to the situation shown in Figure 3-4a, where trajectories lead to or from a particular equilibrium point in the phase diagram. This is an example of where it's important to see *how* the trajectory was drawn, rather than just how it looks when it's completed. If you watch the trajectory being drawn, then if the trajectories leading to a point go from somewhere in the portrait *to* the point, we have a *stable node*. If, on the other hand, the trajectories lead from the node to somewhere else in the phase portrait then we have an *unstable node*.

This fact, that we can only determine the precise nature of a point in the diagram by seeing how it was arrived at, again underlines that we are dealing with a dynamic system. If we don't see the formation of the trajectories, we can easily miss the significance of a point in a phase portrait.

Saddles

The situation illustrated in Figure 3-4b is called a *saddle*. Again, it is a type of equilibrium point.

NOMENCLATURE OF PHASE DIAGRAMS

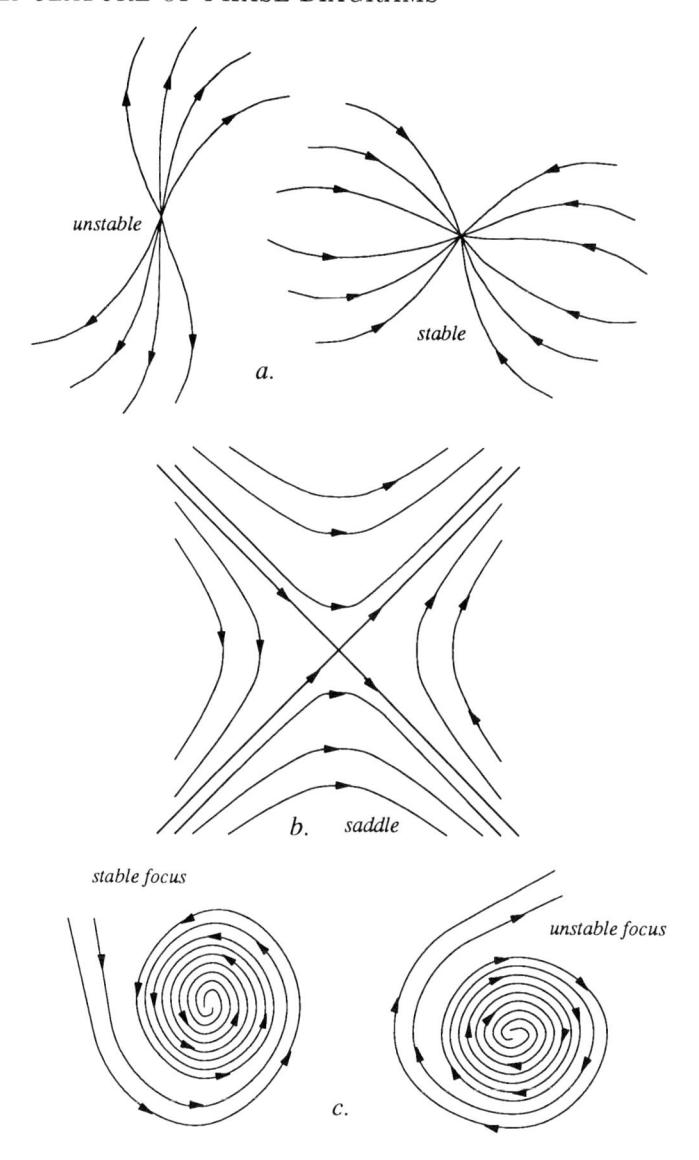

Figure 3-4. Various phase diagrams.

Focuses

If we have a point where trajectories spiral to or from the equilibrium point, we have a focus. Again, we need to see how the trajectory develops to

The Runge–Kutta method

If you experiment with the programs used so far in this chapter to develop phase portraits of different systems of equations, you will soon come to an interesting conclusion; the Euler method of solving the equations isn't always going to give us a true picture of what the phase portrait of a particular system looks like. We've already seen how, for certain values of d, limit cycles don't close up precisely but overlap to give a broader trajectory than is correct. As mentioned before, this is due to the fact that the Euler method only approximates to a solution of the equations being examined. If the approximation is really bad, however, we might end up with a limit cycle looking more like a stable focus, with the trajectory spiralling inwards rather than closing up.

This problem will exist to a greater or lesser extent with all methods of solving differential equations because they are always methods that solve the equations by approximations. However, there are more accurate methods of solving equations than Euler's method. One of the most popular is called the *Runge–Kutta method*, and is simply a more accurate method of finding the value of $xinc$ in the earlier listings. The method consists of the following steps, assuming a function $\frac{dx}{dt} = f(x,t)$ and a step size of d:

$$xinc_1 = f(x,t) * d$$
$$xinc_2 = f(x + xinc_1/2, t + d/2) * d$$
$$xinc_3 = f(x + xinc_2/2, t + d/2) * d$$
$$xinc_4 = f(x + xinc_3, t + d) * d$$
$$x = x + (xinc_1 + 2*xinc_2 + 2*xinc_3 + xinc_4)/6$$

I'm afraid that as far as this book goes, you'll have to take this on trust! However, it does work.

Listing 3-4 shows the Runge–Kutta method used to analyse our pendulum again. The program produces a phase portrait that is more accurate than that produced by the Euler method in that the limit cycles close up on themselves quite smoothly. One thing to note about this method is that, due to the larger number of calculations needed for each x and y increment, the program is slower than the Euler method. However, it's certainly worth using to get higher accuracy. You might like to try altering the value of d and the number of points plotted—you will be able to get d smaller before you lose accuracy. The default number of points given for each trajectory in Listing 3-4 is 5000. This can be increased if necessary to plot more of the trajectory for an initial pair of x, y values.

Dependence on starting conditions

In the phase portrait of the pendulum we saw that for varying initial x and y values a wide range of different trajectories was followed by the equations. This is what we might expect; different starting conditions should lead to different outcomes in terms of the behaviour of the system. What is interesting to look at, though, is the way in which the system behaves for similar initial starting values. To explore this, you might like to run the program listed in Listing 3-5. Here the Runge–Kutta method is used to examine the Pendulum equations for a series of starting values of $x = -2$ and different y values between -1 and -0.8, in 0.02 steps. If you run the program, you'll see that both periodic and aperiodic behaviour is noted, even though we're moving through a narrow range of values. Clearly, at some point the behaviour of the system moves from a limit cycle to non-periodic behaviour. If we were to select initial y values around this point, then for two values of y quite close together we might easily get totally different behaviour—periodic behaviour in one case and non-periodic behaviour in the other case. This, of course, implies the existence of sets of equations which exhibit chaotic behaviour—sensitive dependence upon initial conditions. Looking back over what we've just said about attractors, we can see that this would be most likely to occur when small changes in initial x and y values of the system placed the trajectory of that particular solution to the equations in to a region of the phase portrait occupied by a strange attractor.

Stability

You've probably noticed that the real world is a fairly stable place provided that you leave out the people! If I gently swing a real pendulum, with friction and air resistance, it will settle down to a regular periodic motion followed by the pendulum stopping at an equilibrium point—hanging stationary from its support. If I repeat the operation then the pendulum will still attain the same sort of final equilibrium state, irrespective of whether the initial starting conditions in terms of position of pendulum bob and speed at which I pushed it were the same or not. These variations are cancelled out within the system, and we can quite happily predict the exact motion of a pendulum provided that we don't push it too hard, or in any other way fiddle around with it. This is a typical example of a stable system.

In the world of differential equations, we also have a concept of stability that's applied to equations. If we have a pair of equations with initial values x and y, then they're said to be *stable* if the solution of the equations with very similar values of x and y gives a similar phase portrait. If the phase portrait were different, then we would term the equations *unstable*. The terms of stability and instability used here refer to particular ranges of x and y values; you can have systems of equations which are stable for certain values but that become unstable for certain x, y values.

This concept of stability has implications for us when we're solving equations as we have been doing in this chapter. Small inaccuracies due to the method of solution used will not be important when looking at stable systems of equations, as the very stability of the equations will tend to cancel them out. However, if we're interested in the unstable areas of phase portraits, then we can have serious problems, because inaccurate methods of solving the equations will lead to trajectories being drawn that are not necessarily correct for the initial conditions under consideration; the errors will mount up during the calculations and we could end up in the situation where we don't know whether the phase diagram we've obtained is a true picture of the unstable system under consideration or an artifact of the calculation method used! This is particularly so when the Euler method is being used, and for serious work with differential equations the more long-winded Runge–Kutta method is preferred.

Forced non-linear oscillators

There, I bet that brought you up sharp! Any system that settles down into a limit cycle is exhibiting oscillation. For example, a system of equations may be put forward to describe the behaviour of an electronic circuit known as a negative resistance oscillator—this is a circuit that produces a continuously varying signal:

$$\frac{dx}{dt} = y \quad , \quad \frac{dy}{dt} = k*(1-x*x)*y - x*x*x$$

If you put these equations in the program listings seen already, then you will see a limit cycle obtained. If you want to try this, then set up the program with $k = 0.5$ and $x = y = 1$ as starting conditions. You'll find that the system will eventually settle down into a squarish limit cycle. Now, it's possible to feed signals into such an electronic circuit of a different frequency to the natural frequency of oscillation of the system, and have these oscillations affect the behaviour of the system. The additional signal is called a *forcing signal*, and the resultant circuit is called a *forced oscillator*. We can represent this in our model by changing the equations to read:

$$\frac{dx}{dt} = y$$

$$\frac{dy}{dt} = k*(1-x*x)*y - x*x*x + k_2*\cos(k_3*t)$$

where k_2 and k_3 are constants that refer to the magnitude and frequency of the forcing oscillation, and t is the time elapsed during the analysis of

FORCED NON-LINEAR OSCILLATORS

the system. The cos function is used in this expression to provide the forcing function, because it is a simple function that gives us a sinusoidally varying value with time, as is required to force the system. This pair of equations was first proposed by Ueda and Akamatsu and documented in *Chaos* edited by A. V. Holden. There are, however, a variety of other systems that model forced oscillations. We can model these systems using Euler or Runge–Kutta, but we now have to deal with a phase portrait that includes a time variable as well as x and y. None of the phase portraits examined so far have included time as an explicit part of any function. Actually, it's not really difficult to include; the only thing to remember is that time is incremented by the step size, d, used in solving the equation, and that time is incremented whenever a new value for x and y is calculated.

Listing 3-6 shows how this can be done, and allows us to explore the equations. If you use the default values for the various constants, and then enter 1 for the transient decay time, you will see that before the system achieves a limit cycle there is a period of time in which the trajectory seems to be settling down. This is precisely what is happening; the values of x and y for the system between $t = 0$ and t getting to a couple of hundred are transients that don't reflect the long-term behaviour of the system. It is useful in any system that exhibits transient behaviour to write the program so that the transients are allowed to decay before plotting of x and y values commences. A short examination of the listing will show how this is done. Of course, while the value of t gets to the value specified, no points will be plotted on the screen, and the totally blank screen can be a little disconcerting!

For this system, the more interesting results are obtained for $k < 1$. Start with initial conditions of $x = y = 0$, as in the program. In general, the behaviour of the system falls into the following three categories:

$k_2 = 0$ If $k_2 = 0$, then the forcing term will also evaluate to 0 so it will not affect the system at all. The system will settle down to a limit cycle.

$k_2 > 0$ The value of k_2 is a measure of the magnitude of the forcing oscillation, and so as this increases the effect on the system will increase. $k_2 = 1$, $k_3 = 4$ will, after transient decay, show a limit cycle with the forced oscillation superimposed on it.

$k_2 >> 0$ If k_2 is allowed to get very large, the resultant attractor is very complicated, and is, in fact, a strange attractor. Try values of k_2 above 17 or so to see this, with k_3 values of 4. In order to see the structure of the attractor clearly, it's important to allow the transients to decay by selecting a suitable transient suppression time (at least 200).

More complicated systems

We've encountered systems of differential equations consisting of one or two separate equations. It is quite possible to have systems containing three or four separate equations in three- or four-dimensional systems. In these, strange attractors are more common and the overall situation is more complicated, although we can still examine these systems with the programs we've used in this chapter. The most famous three-dimensional system is called the Lorenz equations, so we'll look at that next.

BBC BASIC listings

Listing 3-1

```
10 REM Euler
20 REM Program to draw a differential equation by the Euler method
30 REM for use in the Chaos book.
40 REM Initialise start values for the system.  ds should be the same as d,
   as it is simply
45 REM a string representation of d for data input.
60 REM Try d=0.001
70 x=0.77
80 t=-1
90 REM Allow user to enter new d values
100 MODE 6
110 CLS
130 INPUT "Value of d: ",d
150 REM Set graphics mode up
160 MODE 4
170 FOR n=1 TO 2000
180   PLOT 69,(n/4*1000*d),(x*150)+200
190   REM Increment value of x is calculated and placed in Xinc.
200   xinc=FNEquation(x,t)*d
210   REM Now update x and t for next pass
220   x=x+xinc
230   t=t+d
240 NEXT
245 END
260 REM put the equation to be graphed into a function
270 REM of its own so that it can be called when needed.
280 DEFFNEquation(xi,ti)=3*(SIN(xi*xi)-COS(ti*ti))
```

Listing 3-2

```
10 REM EulerVolterra
20 REM Program to draw a Volterra equations by the Euler method
30 REM for use in the Chaos book.
80 d=0.001
160 REM Try the following values:    x=10
170 REM y=1
180 REM k=3
190 REM k1=2
200 REM k2=2
210 REM k3=1.5
220 REM  Allow user to enter new values
230 REPEAT
240   MODE6
250   CLS
260   INPUT "X Value: ",x
280   INPUT "Y Value: ",y
300   INPUT "k Value: ",k
320   INPUT "Value of k1",k1
340   INPUT "Value of k2",k2
360   INPUT "Value of k3",k3
380   INPUT "Phase Portrait or Graph?",porg$
390   REM  Set graphics mode up
400   MODE 4
410   IF (porg$="P") OR (porg$="p") THEN mpoints=5000 ELSE mpoints=1000
420   FOR n=1 TO mpoints
430     REM Calculate mpoints points, and graph on screen. Use the porg
        variable
440     REM to see whether a phase portrait or graph is needed.
```

```
450     IF (porg$="P") OR (porg$="p") THEN PLOT69,INT(x*20)+100,INT(y*20)+100
        ELSE GCOL0,1:PLOT69,n,x*20+300:GCOL0,2:PLOT69,n,y*20+600
460     REM Increment value of x is calculated and placed in Xinc.
470     xinc=FNEquationX(x,y)*d
480     yinc=FNEquationY(x,y)*d
490     REM Now update x and t for next pass
500     y=y+yinc
510     x=x+xinc
520   NEXT
530   PRINT CHR$(7)
540   temp=GET
550 UNTIL (porg$="X") or (porg$="x")
560 END
570 DEF FNEquationX(xi,yi)=k*xi-k1*xi*yi
580 DEF FNEquationY(xi,yi)=-k2*yi+k3*xi*yi
```

Listing 3-3

```
10 REM Program EulerPendulum
20 REM Program to draw pendulum equations by the Euler method
30 REM for use in the Chaos book.
60 d=0.005
90 REM Try values of mpoints=5000, k=3
110 REM Allow user to enter new values
120 MODE 6
130 CLS
140 INPUT "Value of k: ",k
160 INPUT "Points for Phase Portrait: ",mpoints
180 REM Set graphics mode up
190 MODE 4
200 REM Look at the curves obtained from x values between -5 and +5
210 REM and y=-5 to +5
220 FOR xn=-5 TO 5
230   FOR yn=-5 TO 5
240     x=xn
250     y=yn
260     FOR n=1 TO mpoints
270       REM Calculate mpoints points, and graph on screen. Again, adjust Y coordinate to
280       REM suit graphics origin.
290       PLOT69,INT(x*50)+500,INT(y*50)+500
300       REM Increment value of x is calculated and placed in Xinc.
310       xinc=FNEquationX(x,y)*d
320       yinc=FNEquationY(x,y)*d
330       REM Now update x and t for next pass
340       y=y+yinc
350       x=x+xinc
360     NEXT
370   NEXT
380 NEXT
390 PRINT CHR$(7)
400 temp=GET
410 END
420 DEF FNEquationX(xi,yi)=yi
430 DEF FNEquationY(xi,yi)=-k*SIN(xi)
```

Listing 3-4

```
10 REM Program RungeKuttaPendulum
20 REM Program to draw pendulum equations by the Runge-Kutta method
30 REM for use in the Chaos book.
80 d=0.005
100 REM Try the following values
110 REM mpoints=5000
120 REM k=3
140 REM Allow user to enter new values
190 MODE 6
220 INPUT "Value of k ",k
230 INPUT "Number of points ",mpoints
290 MODE 5
310 REM Look at the curves obtained from x values between -5 and +5
320 REM and y=-5 to +5
350 FOR xn=-5 TO 5
360   FOR yn=-5 TO 5
380     x=xn
390     y=yn
430     FOR n=1 TO mpoints
450       REM Calculate points and adjust Y coordinate to fit coordinate
          system.
470       PLOT69,x*50+500,y*50+500
490       REM Now start the Runge-Kutta solution, first for x
510       xinc1=FNEquationX(x,y)*d
520       xinc2=FNEquationX(x+xinc1/2,y+d/2)*d
530       xinc3=FNEquationX(x+xinc2/2,y+d)*d
540       xinc4=FNEquationX(x+xinc3,y+d)*d
560       REM Now repeat for the Y Equation
580       yinc1=FNEquationY(x,y)*d
590       yinc2=FNEquationY(x+yinc1/2,y+d/2)*d
600       yinc3=FNEquationY(x+yinc2/2,y+d)*d
610       yinc4=FNEquationY(x+yinc3,y+d)*d
650       x=x+(xinc1+2*xinc2+2*xinc3+xinc4)/6
660       y=y+(yinc1+2*yinc2+2*yinc3+yinc4)/6
700     NEXT
710   NEXT
720 NEXT
740 PRINT CHR$(7)
750 G=GET
760 END
820 DEFFNEquationX(xi,yi)=yi
840 DEFFNEquationY(xi,yi)=-k*SIN(xi)
```

Listing 3-5

```
10 REM Program InitialConditionDependance
30 REM Program to draw pendulum equations by the Runge-Kutta method
40 REM for use in the Chaos book.
220 d=0.005
240 REM Allow user to enter new values
250 REM Try mpoints=5000
280 INPUT "Number of points in Phase Portrait ",mpoints
310 REM Set graphics mode up
330 MODE 5
350 REM Look at the curves obtained from initial x value of -2 and y values
360 REM varying
370 REM from -1 to -1.2 by incrementing y by 0.02 each time
400 y=-1
410 FOR yn=1 TO 10
430   x=-2
440   y=y+0.02
```

78   CHAPTER 3. DIFFERENTIAL EQUATIONS

```
480    FOR n=1 TO mpoints
500    REM plot the pixels, changing colour each time.  Colours may be used
       more
520    GCOL0,yn MOD 4
530    PLOT69,x*50+500,y*50+500
550    REM Now start the Runge-Kutta solution, first FOR x
570    xinc1=FNEquationX(x,y)*d
580    xinc2=FNEquationX(x+xinc1/2,y+d/2)*d
590    xinc3=FNEquationX(x+xinc2/2,y+d)*d
600    xinc4=FNEquationX(x+xinc3,y+d)*d
620    REM Now repeat FOR the Y Equation
640    yinc1=FNEquationY(x,y)*d
650    yinc2=FNEquationY(x+yinc1/2,y+d/2)*d
660    yinc3=FNEquationY(x+yinc2/2,y+d)*d
670    yinc4=FNEquationY(x+yinc3,y+d)*d
700    y=y+(yinc1+2*yinc2+2*yinc3+yinc4)/6
710    x=x+(xinc1+2*xinc2+2*xinc3+xinc4)/6
750  NEXT
760 NEXT
780 PRINT CHR$(7)
790 G=GET
820 END
870 DEFFNEquationX(xi,yi)=yi
890 DEFFNEquationY(xi,yi)=-1*SIN(xi)
```

Listing 3-6

```
10 REM Program RungeKuttaForcedOscillator
20 REM       Program to draw Forced Oscillator equations by the Runge-Kutta
             method
30 REM       for use in the Chaos book.
40 REM       First of all, put the equations to be graphed into a function
50 REM       of its own so that they can be called when needed.
60 REM       Initialise start values for the system.  ?s should be the same as
             ?, as it is simply
70 REM       a string representation of ? for data input.
80 MODE 6
90 d=0.005
100 t=0
110 REM  Allow user to enter new values
120 REM  Try k=0.5,k2=0,k3=2,transient=100
130 INPUT "Value of k ",k
140 INPUT "Value of k2 ",k2
150 INPUT "Value of k3 ",k3
160 INPUT "Time for transient supression ",transient
170 REM  Set graphics mode up
180 MODE 5
190 REM  Initialise x and y values
200 x=1
210 y=1
220 REPEAT
230   REM     Only plot after the transients have decayed
240   IF (t>transient) THEN PLOT69,(x*20)+500,(y*20)+500
250   REM Now start the Runge-Kutta solution, first for x
260   xinc1=FNEquationX(x,y)*d
270   xinc2=FNEquationX(x+xinc1/2,y+d/2)*d
280   xinc3=FNEquationX(x+xinc2/2,y+d)*d
290   xinc4=FNEquationX(x+xinc3,y+d)*d
300   REM Now repeat for the Y Equation
310   yinc1=FNEquationy(x,y)*d
320   yinc2=FNEquationy(x+yinc1/2,y+d/2)*d
330   yinc3=FNEquationy(x+yinc2/2,y+d)*d
340   yinc4=FNEquationy(x+yinc3,y+d)*d
```

```
350   x=x+(xinc1+2*xinc2+2*xinc3+xinc4)/6
360   y=y+(yinc1+2*yinc2+2*yinc3+yinc4)/6
370   REM         Increase value of t
380   t=t+d
400   REM  Repeat until a key is pressed
410 UNTIL INKEY(0)<>-1
420 PRINT CHR$(7)
430 G=GET
440 END
450 DEFFNEquationX(xi,yi)=yi
460 DEFFNEquationy(xi,yi)=k*(1-xi*xi)*yi-xi*xi*xi+k2*COS(k3*t)
```

Turbo Pascal listings

Listing 3-1

```
Program Euler;
{       Program to draw a differential equation by the Euler method
        for use in the Chaos book.
}

Uses Crt,Graph,Extend;

Var

n                   :     Integer;
x,t,d,xinc          :     Real;
ds                  :     String;
xi,ti               :     Real;

{   First of all, put the equation to be graphed into a function
    of its own so that it can be called when needed.            }

Function Equation(xi,ti : real) : real;
Begin
   Equation:=3*(sin(xi*xi)-cos(ti*ti));
end;

Begin

{   Initialise start values for the system.  ds should be the same as d, as it
    is simply
    a string representation of d for data input.
}

   ds:='0.001';
   d:=0.001;
   x:=0.77;
   t:=-1;

{ Allow user to enter new d values }

   TextBackground(Blue);
   ClrScr;
   PrintAt(2,4,'Value of d',Green);
   StringEdit(ds,20,20,4,Yellow);
   Val(ds,d,n);

{ Set graphics mode up }
```

80   CHAPTER 3. DIFFERENTIAL EQUATIONS

```
    ChoosePalette;

    for n:=1 to 2000 do begin

  { Calculate 2000 points, and graph on screen.  The GetMaxY clause is needed
    because the PC graphics screen top left is Y=0 in coordinates, rather than
    the more usual Y=Max.  Y Coordinate         }

         PutPixel(trunc(n/4*1000*d),GetMaxY-(trunc(x*150)+200),1);

  {        Increment value of x is calculated and placed in Xinc.     }

         xinc:=Equation(x,t)*d;

  {        Now update x and t for next pass                           }

         x:=x+xinc;
         t:=t+d;

    end;
end;
```

Listing 3-2

```
Program EulerVolterra;
{       Program to draw a Volterra equations by the Euler method
        for use in the Chaos book.
}

Uses Crt,Graph,Extend;

Var

n,mpoints                  :     Integer;
x,t,d,xinc                 :     Real;
y,yinc,k,k1,k2,k3          :     Real;
xs,ys,ks,k1s,k2s,k3s       :     String;
xi,yi                      :     Real;
porg                       :     String;
temp                       :     Char;

{   First of all, put the equation to be graphed into a function
    of its own so that it can be called when needed.                  }

Function EquationX(xi,yi : real) : real;
Begin
   EquationX:=k*xi-k1*xi*yi;
end;

Function EquationY(xi,yi : real) : real;
Begin
   EquationY:=-k2*yi+k3*xi*yi;
end;

Begin

 {   Initialise start values for the system.  ?s should be the same as ?, as it
     is simply
     a string representation of ? for data input.}

    d  :=0.001;
    xs :='10';
```

TURBO PASCAL LISTINGS 81

```pascal
    ys :='1';
    ks :='3';
    k1s:='2';
    k2s:='2';
    k3s:='1.5';
    porg:='G';
    x:=10;
    y:=1;
    k:=3;
    k1:=2;
    k2:=2;
    k3:=1.5;

{ Allow user to enter new values                              }

Repeat

    TextBackground(Blue);
    ClrScr;
    PrintAt(2,4,'Value of x',Green);
    StringEdit(xs,20,20,4,Yellow);
    Val(xs,x,n);
    PrintAt(2,5,'Value of y',Green);
    StringEdit(ys,20,20,5,Yellow);
    Val(ys,y,n);
    PrintAt(2,6,'Value of k',Green);
    StringEdit(ks,20,20,6,Yellow);
    Val(ks,k,n);
    PrintAt(2,7,'Value of k1',Green);
    StringEdit(k1s,20,20,7,Yellow);
    Val(k1s,k1,n);
    PrintAt(2,8,'Value of k2',Green);
    StringEdit(k2s,20,20,8,Yellow);
    Val(k2s,k2,n);
    PrintAt(2,9,'Value of k3',Green);
    StringEdit(k3s,20,20,9,Yellow);
    Val(k3s,k3,n);
    PrintAt(2,10,'Phase Portrait or Graph? ',Green);
    StringEdit(porg,20,30,10,Yellow);

{ Set graphics mode up }

    ChoosePalette;

if (porg='P') or (porg='p') then
    mpoints:=5000
else
    mpoints:=GetMaxX;

    for n:=1 to mpoints do begin

{ Calculate mpoints points, and graph on screen.  Use the porg variable
  to see whether a phase portrait or graph is needed.  Again correct for
  graphics origin of PC graphics screen                  }

        if (porg='P') or (porg='p') then
            PutPixel(trunc(x*20)+100,GetMaxY-(trunc(y*20)+100),2)

        else begin
            PutPixel(n,trunc(x*20)+100,1);
            PutPixel(n,trunc(y*20)+100,2);
        end;

{       Increment value of x is calculated and placed in Xinc.   }
```

82   CHAPTER 3. DIFFERENTIAL EQUATIONS

```
            xinc:=EquationX(x,y)*d;
            yinc:=EquationY(x,y)*d;
{       Now update x and t for next pass                        }

            y:=y+yinc;
            x:=x+xinc;

      end;
      sound(200);
      Delay(30);
      NoSound;
      temp:=Readkey;
      CloseGraph;
    until (porg='X') or (porg='x');

    end.
```

Listing 3-3

```
Program EulerPendulum;
{       Program to draw pendulum equations by the Euler method
        for use in the Chaos book.
}

Uses Crt,Graph,Extend;

Var

n,mpoints               :       Integer;
x,d,xinc                :       Real;
y,yinc,k                :       Real;
xi,yi                   :       Real;
mpointss,ks             :       String;
temp                    :       Char;
xn,yn                   :       Integer;

{       First of all, put the equations to be graphed into a function
        of its own so that they can be called when needed.              }

Function EquationX(xi,yi : real) : real;
Begin
   EquationX:=yi;
end;

Function EquationY(xi,yi : real) : real;
Begin
   EquationY:=-k*sin(xi);
end;

Begin

{       Initialise start values for the system.  ?s should be the same as ?, as it
        is simply
        a string representation of ? for data input.
}

      d:=0.005;
      ks:='1';
      mpointss:='5000';
      mpoints:=5000;
      k:=3;
```

TURBO PASCAL LISTINGS   83

```pascal
{ Allow user to enter new values
}
   TextBackground(Blue);
   ClrScr;
   PrintAt(2,6,'Value of k',Green);
   StringEdit(ks,20,20,6,Yellow);
   Val(ks,k,n);
   PrintAt(2,7,'Number of points for Phase Portrait? ',Green);
   StringEdit(mpointss,20,40,7,Yellow);

{ Set graphics mode up }

   ChoosePalette;

{ Look at the curves obtained from x values between -5 and +5
  and y=-5 to +5                                              }

for xn:=-5 to 5 do begin
   for yn:=-5 to 5 do begin

      x:=xn;
      y:=yn;

      for n:=1 to mpoints do begin

{ Calculate mpoints points, and graph on screen.  Again, adjust Y coordinate to
  suit graphics origin.              }

         PutPixel(trunc(x*50)+trunc(GetMaxX/2),GetMaxY-(trunc(y*50)+trunc(GetMaxY/2)),2);

{        Increment value of x is calculated and placed in Xinc.       }

         xinc:=EquationX(x,y)*d;
         yinc:=EquationY(x,y)*d;

{        Now update x and t for next pass                             }

         y:=y+yinc;
         x:=x+xinc;

      end;
    end;
   end;

   sound(200);
   Delay(30);
   NoSound;
   temp:=Readkey;
   CloseGraph;

end.
```

Listing 3-4

```
Program RungeKuttaPendulum;
{       Program to draw pendulum equations by the Runge-Kutta method
        for use in the Chaos book.
}

Uses Crt,Graph,Extend;

Var

n,mpoints                :         Integer;
x,d,xinc                 :         Real;
y,yinc,k                 :         Real;
xinc1,xinc2,xinc3,xinc4:           Real;
yinc1,yinc2,yinc3,yinc4:           Real;
xi,yi                    :         Real;
mpointss,ks                        String;
temp                     :         Char;
xn,yn                    :         Integer;

{     First of all, put the equations to be graphed into a function
      of its own so that they can be called when needed.              }

Function EquationX(xi,yi : real) : real;
Begin
   EquationX:=yi;
end;

Function EquationY(xi,yi : real) : real;
Begin
   EquationY:=-k*sin(xi);
end;

Begin
{     Initialise start values for the system.  ?s should be the same as ?, as it
      is simply
      a string representation of ? for data input.

}

   d:=0.005;
   ks:='1';
   mpointss:='5000';
   mpoints:=5000;
   k:=3;

{ Allow user to enter new values
}

   TextBackground(Blue);
   ClrScr;
   PrintAt(2,6,'Value of k',Green);
   StringEdit(ks,20,20,6,Yellow);
   Val(ks,k,n);
   PrintAt(2,7,'Number of points for Phase Portrait? ',Green);
   StringEdit(mpointss,20,40,7,Yellow);
   Val(mpointss,mpoints,n);

{ Set graphics mode up }

   ChoosePalette;

{ Look at the curves obtained from x values between -5 and +5
```

TURBO PASCAL LISTINGS

```
        and y=-5 to +5                                        }
for xn:=-5 to 5 do begin
   for yn:=-5 to 5 do begin

      x:=xn;
      y:=yn;

      for n:=1 to mpoints do begin

{       Calculate points and adjust Y coordinate to fit coordinate system. }

           PutPixel(trunc(x*50)+trunc(GetMaxX/2),GetMaxY-(trunc(y*50)+trunc(GetMaxY/2)),2);

{ Now start the Runge-Kutta solution, first for x }

           xinc1:=EquationX(x,y)*d;
           xinc2:=EquationX(x+xinc1/2,y+d/2)*d;
           xinc3:=EquationX(x+xinc2/2,y+d)*d;
           xinc4:=EquationX(x+xinc3,y+d)*d;

{ Now repeat for the Y Equation                        }

           yinc1:=Equationy(x,y)*d;
           yinc2:=Equationy(x+yinc1/2,y+d/2)*d;
           yinc3:=Equationy(x+yinc2/2,y+d)*d;
           yinc4:=Equationy(x+yinc3,y+d)*d;

           x:=x+(xinc1+2*xinc2+2*xinc3+xinc4)/6;
           y:=y+(yinc1+2*yinc2+2*yinc3+yinc4)/6;

        end;
      end;
   end;

   sound(200);
   Delay(30);
   NoSound;
   temp:=Readkey;
   CloseGraph;

end.
```

Listing 3-5

```
Program InitialConditionDependance;

{       Program to draw pendulum equations by the Runge-Kutta method
        for use in the Chaos book.
}

Uses Crt,Graph,Extend;

Var

n,mpoints                    :       Integer;
x,d,xinc                     :       Real;
y,yinc,k                     :       Real;
xinc1,xinc2,xinc3,xinc4:             Real;
yinc1,yinc2,yinc3,yinc4:             Real;
xi,yi                        :       Real;
mpointss,ks                  :       String;
```

86   CHAPTER 3. DIFFERENTIAL EQUATIONS

```
temp                  :     Char;
xn,yn                 :     Integer;

{    First of all, put the equations to be graphed into a function
     of its own so that they can be called when needed.              }

Function EquationX(xi,yi : real) : real;
Begin
   EquationX:=yi;
end;

Function EquationY(xi,yi : real) : real;
Begin
   EquationY:=-1*sin(xi);
end;

Begin

{    Initialise start values for the system.  ?s should be the same as ?, as it
     is simply
     a string representation of ? for data input.

}

   d:=0.005;
   mpointss:='5000';
   mpoints:=5000;

{ Allow user to enter new values
}

   TextBackground(Blue);
   ClrScr;
   PrintAt(2,7,'Number of points for Phase Portrait? ',Green);
   StringEdit(mpointss,20,40,7,Yellow);
   Val(mpointss,mpoints,n);

{ Set graphics mode up }

   ChoosePalette;

{  Look at the curves obtained from initial x value of -2 and y values varying
   from -1 to -1.2 by incrementing y by 0.02 each time }

   y:=-1;
   for yn:=1 to 10 do begin

      x:=-2;
      y:=y+0.02;

      for n:=1 to mpoints do begin

{      plot the pixels, changing colour each time.  Colours may be used more
       than once, depending upon the graphics mode.                      }

           PutPixel(trunc(x*50)+trunc(GetMaxX/2),GetMaxY-(trunc(y*50)+trunc(GetMaxY/2)),yn);

{ Now start the Runge-Kutta solution, first for x }

           xinc1:=EquationX(x,y)*d;
           xinc2:=EquationX(x+xinc1/2,y+d/2)*d;
           xinc3:=EquationX(x+xinc2/2,y+d)*d;
           xinc4:=EquationX(x+xinc3,y+d)*d;
```

TURBO PASCAL LISTINGS   87

```
{ Now repeat for the Y Equation                    }

        yinc1:=Equationy(x,y)*d;
        yinc2:=Equationy(x+yinc1/2,y+d/2)*d;
        yinc3:=Equationy(x+yinc2/2,y+d)*d;
        yinc4:=Equationy(x+yinc3,y+d)*d;

        y:=y+(yinc1+2*yinc2+2*yinc3+yinc4)/6;
        x:=x+(xinc1+2*xinc2+2*xinc3+xinc4)/6;

      end;
   end;

   sound(200);
   Delay(30);
   NoSound;
   temp:=Readkey;
   CloseGraph;

end.
```

Listing 3-6

```
Program RungeKuttaForcedOscillator;
{       Program to draw Forced Oscillator equations by the Runge-Kutta method
        for use in the Chaos book.
}

Uses Crt,Graph,Extend;

Var

x,d,xinc                 :       Real;
y,yinc,k,k2,t,k3         :       Real;
xinc1,xinc2,xinc3,xinc4:         Real;
yinc1,yinc2,yinc3,yinc4:         Real;
xi,yi                    :       Real;
ks,k2s,k3s,transients    :       String;
temp                     :       Char;
n                        :       Integer;
transient                :       Integer;

{       First of all, put the equations to be graphed into a function
        of its own so that they can be called when needed.             }

Function EquationX(xi,yi : real) : real;
Begin
   EquationX:=yi;
end;

Function EquationY(xi,yi : real) : real;
Begin
   EquationY:=k*(1-xi*xi)*yi-xi*xi*xi+k2*cos(k3*t);
end;

Begin

{       Initialise start values for the system. ?s should be the same as ?, as it
        is simply
        a string representation of ? for data input.
}
```

CHAPTER 3. DIFFERENTIAL EQUATIONS

```
    d:=0.005;
    k:=0.5;
    k2:=0;
    k3:=2;
    transient:=100;
    ks:='0.5';
    k2s:='0';
    k3s:='2';
    transients:='100';
    t:=0;

{ Allow user to enter new values
}

    TextBackground(Blue);
    ClrScr;
    PrintAt(2,4,'Value of k? ',Green);
    StringEdit(ks,20,50,4,Yellow);
    Val(ks,k,n);
    PrintAt(2,5,'Value of k2? ',Green);
    StringEdit(k2s,20,50,5,Yellow);
    Val(k2s,k2,n);
    PrintAt(2,6,'Value of k3? ',Green);
    StringEdit(k3s,20,50,6,Yellow);
    Val(k3s,k3,n);

    PrintAt(2,7,'Time for transient supression? ',Green);
    StringEdit(transients,20,50,7,Yellow);
    Val(transients,transient,n);

{ Set graphics mode up }

    ChoosePalette;

{ Initialise x and y values    }

      x:=1;
      y:=1;

    Repeat

        Begin

{       Only plot after the transients have decayed             }

        if (t>transient) then

            PutPixel(trunc(x*20)+trunc(GetMaxX/2),GetMaxY-(trunc(y*20)+trunc(GetMaxY/2)),2);

{ Now start the Runge-Kutta solution, first for x }

            xinc1:=EquationX(x,y)*d;
            xinc2:=EquationX(x+xinc1/2,y+d/2)*d;
            xinc3:=EquationX(x+xinc2/2,y+d)*d;
            xinc4:=EquationX(x+xinc3,y+d)*d;

{ Now repeat for the Y Equation             }

            yinc1:=EquationY(x,y)*d;
            yinc2:=EquationY(x+yinc1/2,y+d/2)*d;
            yinc3:=EquationY(x+yinc2/2,y+d)*d;
            yinc4:=EquationY(x+yinc3,y+d)*d;

            x:=x+(xinc1+2*xinc2+2*xinc3+xinc4)/6;
            y:=y+(yinc1+2*yinc2+2*yinc3+yinc4)/6;
```

```
{       Increase value of t    }
        t:=t+d;
   end;
{ Repeat until a key is pressed                    }
   until (keyPressed);
   sound(200);
   Delay(30);
   NoSound;
   temp:=Readkey;
   CloseGraph;
end.
```

Chapter 4

The Lorenz equations

The birth date of chaos could be said to be January 1963, when Edward Lorenz sent a revised version of a paper he had been working on, 'Deterministic Nonperiodic Flow', to the American *Journal of the Atmospheric Sciences*. This paper described the work the Lorenz had been doing for the previous three years on a mathematical model of a weather system in his laboratory at the Massachusetts Institute of Technology. The paper did little for the confidence of weather forecasters the world over; it basically pointed out that they were never going to get the weather right through mathematical models.

The weather system of the earth can be considered as an enormous experiment in fluid dynamics, the fluid involved being air. There are turbulences, pressure and temperature differences and a whole host of other things to take into account. Of course, to create a system of equations to model the atmosphere in every small detail would be an enormous task, so Lorenz decided to base his model on just a dozen equations, of which three were to become known as the *Lorenz equations*. These equations attempted to represent convection processes in the atmosphere; that is, the processes involved when air warms, rises, cools and falls again. There are two basic types of convection; the steady, stable type where nothing unpredictable happens and the unstable, less predictable type of convection—turbulence—the sort of thing that gives us problems with the weather. In addition, with this system of equations, there is a steady-state condition in which the temperature difference in the system is not quite enough to set up convection; in this situation we simply have a temperature gradient from top to bottom in the system but no fluid movement.

The Lorenz system of equations is:

$$\frac{dx}{dt} = a*y - a*x$$

$$\frac{dy}{dt} = r*x - y - x*z$$

$$\frac{dz}{dt} = x*y + b*z$$

As you can see, it is a set of three differential equations. To give some background about what the equations represent in the weather system, let's look at what each letter represents.

x This value is proportional to the speed of motion of the fluid due to convection in the system being modelled.

y This is a measure of the temperature difference between the warmer, rising fluid in the system and the cooler, falling fluid. In other words, it's a measure of the horizontal temperature difference in the system.

z In any convective system, there will be a difference in temperature as you move vertically through the system. This is called the vertical temperature profile and z is a measure of this, representing the vertical temperature difference in the system.

The three constants obviously have a great deal of impact on the system. They represent real-world features and so it is usual, when experimenting with the Lorenz equations, to make these three constants positive numbers. Of course, there's nothing to stop you messing around with them after you've got the programs working.

a This is proportional to a value called the *Prandtl number* and is a value based upon physical parameters of the fluid involved, such as its density and how efficiently it conducts heat. In the original work, a value of 10 was assigned to this number.

b This is a measure of the size of the area being represented by the equations. In the original work, a value of 8/3 (2.666...) was assigned to this constant.

r This is the important one, as far as we're concerned, and is the *Rayleigh number* for the system. This is simply a parameter which indicates the point at which convection starts for a particular system. Below this critical value we have steady convection, and above it unstable convection starts up.

Lorenz first looked at these equations from an analytical point of view, and was able to ascertain that the critical value of r for his system, with $a = 10$ and $b = 8/3$, was about 24.74. However, to see what happened after the unstable convection started, Lorenz did what we did in the previous chapter; he reached for his computer. This was a stroke of luck for all concerned, as without some computing power the task of solving the equations would have proved very time consuming.

Once he started looking at the equations graphically, he could see the instabilities in the system. He also noticed that the system of equations had a very sensitive dependence upon the initial x, y and z values for $r > 24.74$ and that the phase portrait for this system was, well, interesting, to say the least. Let's start work on these equations by using a program to draw the time progression of x, y and z for various x, y and z initial values and different r values. Listing 4-1 shows a program to do this.

The program uses the Runge–Kutta method to solve the equations, as this is more accurate than the simpler Euler method, and for these equations we need all the accuracy we can get! There are a couple of points to note in this listing that are worth bearing in mind when investigating other systems of equations.

The DisplayAfter variable

The routine only starts plotting the lines representing x, y and z values after a specified length of time has elapsed in the equation system (that is after a specified number of x, y and z values have been calculated but *not* displayed). This allows us to see the values of the three variables after many thousands of iterations, even though the screen width may only allow us to plot a few hundred iterations at once. As we shall later see, this can be quite helpful.

Plotting three variables

I've used a different colour for each of the three variables, and offset them from $y = 0$ to make them clearer. I've subtracted the value to be plotted from the maximum y value so that an increasing value to a particular parameter leads to the line representing that parameter moving up the screen (in the PC version of the program).

Solving the equations

The Runge–Kutta method is used, but the x, y and z variables are not updated until all calculations of $xinc$, $yinc$ and $zinc$ variables has been carried out. This is because the original values of x, y and z should be available to the three equations, and if we were to calculate the new value for x and use the *new* x value to calculate the new y and z values rather than using the *old* x value, the solution obtained would be incorrect.

First investigations

We can use this program to examine the basic properties of the Lorenz equations. We'll start off with an r value of 28 (the value Lorenz used in the original work) and $x = y = z = 1$. The value of 28 ensures that the system is well within the area of unstable convection, so let's start by looking at the time progression for the first few hundred iterations and the progression after 3000 iterations ($DisplayAfter$=1 and 3000 respectively). Typical plots are shown in Figure 4-1; in each graph the lines, from the top, are z, y and x.

As you can see from Figure 4-1a, which shows the first few hundred iterations, there is an initial hump in each of the three variables, followed by gently undulating behaviour. This looks fairly boring, so let's jump ahead and examine the graph for the few hundred iterations following iteration 3000 (Figure 4-1b). The undulations have now increased in magnitude and have, in some cases, become most irregular. If you repeat this for different time spans, you'll see similar behaviour.

Now, repeat the experiment but with $x = y = 1$ and $z = 1.001$, still with an r value of 28. Before starting the program, common sense would tell us not to expect too much difference between the behaviour of the system with these start conditions and the behaviour of the system with the previous start conditions. Indeed, for the first few hundred iterations (Figure 4-2a) common sense works! In fact, you can lay one trace on top of the other and see no apparent difference. We might be forgiven at this point for expecting the few hundred iterations after iteration 3000 to be similar to Figure 4-1b. However, we'd be totally wrong. Figure 4-2b shows the first few hundred iterations for the system after iteration 3000. We have smooth, undulating lines, very similar to 4-2a, but totally different to Figure 4-1b. Now, the important thing to note is that we've only changed the value of z by 1 part in 1000—a perfect example of sensitive dependence on initial conditions; in other words, we've got a chaotic system.

You might like to repeat this for other starting values, comparing the traces obtained at different iteration numbers. In order to show up where differences occur, I use a sheet of tracing paper over the VDU screen; after the first trace is drawn, I trace on to the paper the graphs shown on the screen. Then, leaving the paper in place, I change the initial conditions slightly and allow the graphs to be redrawn. Using this technique I can spot small differences between the graphs quite easily.

You might like to try repeating this experiment with a smaller value of r, say 10. The system will now accommodate large changes in the initial

FIRST INVESTIGATIONS

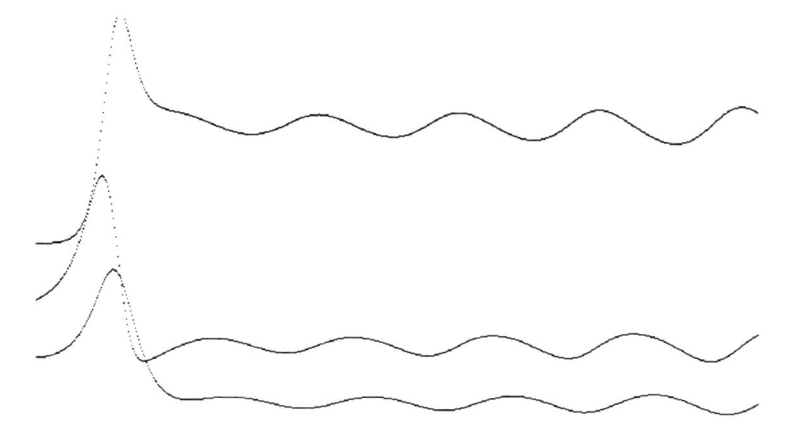

Figure 4-1a. Time progression plot for the Lorenz equations after a few hundred iterations.

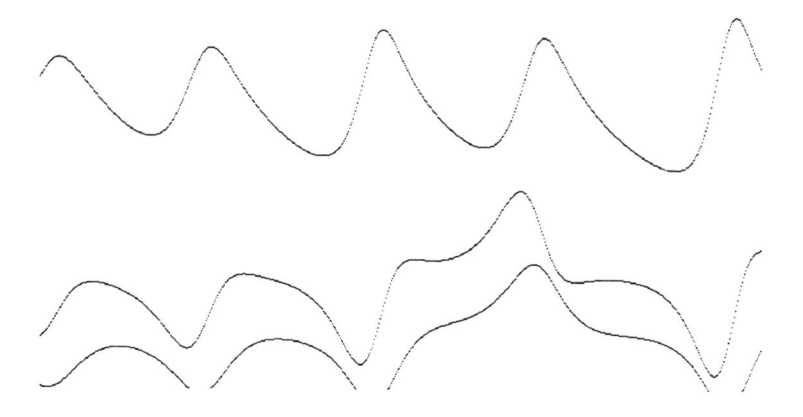

Figure 4-1b. Time progression plot for the Lorenz equations after a few thousand iterations.

values of the three variables without affecting the medium- or long-term behaviour of the system. The discrepancies introduced by different initial conditions very quickly cancel out. This is, after all, what we'd expect, as with values of r as small as 10 the system is modelling a state of steady convection in which we wouldn't expect to find any unstable behaviour. If your computer will allow you to, try very large values of r; these exhibit less chaotic behaviour as well, and it's been demonstrated that for large and small values of r the system is periodic rather than chaotic.

96   CHAPTER 4. THE LORENZ EQUATIONS

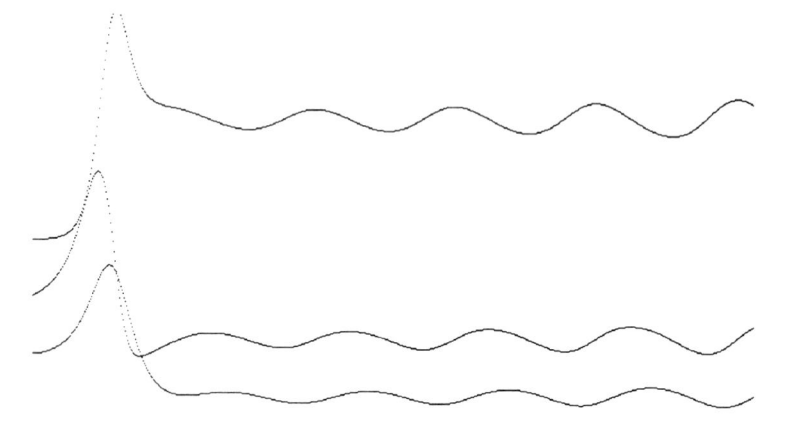

Figure 4-2a. As Figure 4-1a but with different initial conditions.

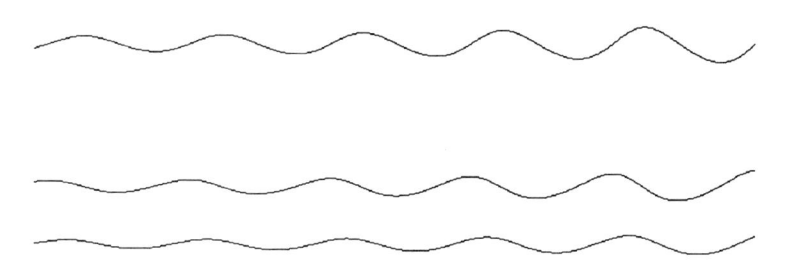

Figure 4-2b. As Figure 4-1b but with different initial conditions.

From plotting the time progression of the three variables, we've learnt the following facts:

1. Some systems are very sensitive to initial conditions, so that tiny changes in initial conditions, too small to show up when we examine the graphs of the equations for the first few hundred iterations, cause vast differences in the behaviour of the system after a few thousand iterations.

2. Within a system of equations that exhibits this chaotic behaviour, there are certain values of constants, etc., which will lead to stable

systems. We saw this by dropping the value of r below the critical value of 24.

3. Detailed examination will show that even in parts of the system showing chaotic behaviour, there is a trace of apparent regularity. You can prove this for yourself by examining areas of the time progression.

Feedback—the common factor

There are a couple of conclusions to be drawn from this. The first is that we've seen this sort of behaviour before in the iterated systems of Chapter 2: regular regions within chaos in bifurcation diagrams, sensitivity to initial conditions, as in Martin's mappings, and a critical value below which chaos does not occur (the k constant in the logistic equation). We've arrived at chaos through two routes, the first via period doubling cascades and the second by the graphical solution of a set of differential equations. However, it is important to note the common factor between these two routes to chaos—*feedback*. In each case, values calculated are used in subsequent calculations, *ad infinitum*.

The Butterfly Effect

The second conclusion to be drawn is a big one, and we need to think about what the equations we're playing with represent. We've looked at population growth in Chapter 2, and here we've examined a system of equations that deal with fluid convection, but were originally intended to partially model a weather system. The differences in the behaviour of the system after several thousand iterations for apparently identical starting conditions indicates that the weather represented by the system of equations would be totally different, despite the apparent similarity of the initial conditions. In practical terms, this has serious implications for us when looking at real-world systems that are modelled by differential equations with a possibility for chaotic behaviour. It means that even a small error in measuring the initial conditions for inputting in to the equations will not be cancelled out but will build up to change the behaviour of the system quite radically.

For example, in the case of a model of the weather system, the information for predicting future weather would come from information such as wind speed, temperature, air pressure, etc., gathered from weather stations. At a specific instant in time, we could gather all these pieces of data, put them into a computer model and watch the future weather evolve. What our model is doing here is extrapolating the future behaviour of the system from a series of snapshot views of the current state of the system—we have no idea what the system is behaving like *between* weather stations. In addition, we don't know whether a snapshot set of measurements from a weather station is truly accurate; what if, for example, there were small local disturbances at some weather stations; wind speed and direction momentarily modified by a large vehicle passing on the road 100 yards away? Or the flapping of a bird's wing? Or the flapping of a butterfly's wing?

All these would vary the initial condition of one parameter in the model, which prior to the advent of chaos theory would have been considered to be a transient effect which would have no long-term effects. Now, of course, we know different, and the term *Butterfly Effect* was termed to graphically describe how even a tiny effect such as the movement of a butterfly's wing might cause vast deviations in the weather in three month's time. The cabbage white that was fluttering around my garden a few minutes ago has probably already had a large impact on the weather, though, of course, we can't tell quite what.

Long-range weather forecasting isn't quite dead yet. The vast array of computer power available to meteorologists has offered a rather neat, if brute-force method of dealing with the problem of the Butterfly Effect. Take some weather data and feed it into a computer model, run your model, and generate as output a weather map of isobars, wind speeds, etc., that the computer expects in, say, one month's time. Now take the original data and make a small difference to the value of one of the initial parameters. Re-run the model for the same month, and get a second weather map. Repeat a few more times, each time making small changes in the input data. If you get weather maps that are totally different, then we know that the initial weather system had the seeds of chaos in it, and that there's no point in trying to predict the weather through the models. Best to go and find similar conditions in the past records and see how they developed, or look at short-term forecasts only. If, however, the maps are similar, then the initial weather system that the data modelled wasn't chaotic, and you stand a reasonable chance of getting some accurate predictions made. It's still not definite that you'll be right, of course, but at least you know whether its worth trying!

The concept of 'butterfly weather' can be extended to other systems; for example, there are few systems in the real world that are not troubled by *noise*. In the scientific sense, noise is any change in the value of a parameter of a system that obscures what you're looking for. The noise in any system might be of a small value, but it is quite possible for that noise, in a suitably chaotic system, to cause the behaviour of the system to change totally from what it would be in the absence of noise.

Further analysis of the equations

Before leaving time progressions, Listing 4-2 shows a potentially useful way of looking at systems like the Lorenz equations. In this method, the x-axis of the graph represents initial values of a parameter, in this case x, and the y-axis represents values arrived at from that start value after a set number of iterations. The technique is as follows:

1. Initialise x, y and z and choose an increment for x. In addition, choose a number of iterations to plot.

2. For the initial values of x, y and z iterate the equations the desired number of times and plot the values obtained against the start x value. This will give a vertical line on the graph, usually broken intermittently.

3. Reinitialise y and z to the start values, then increment x by the step size. Now repeat the iteration to give a second column on the graph.

4. This process can then be repeated, incrementing the start value of x each time.

The resultant graph, an example of which is shown in Figure 4-3a, shows what values of x a particular start value of x will iterate to after so many iterations. This was done with a starting $x = y = z$ of 1 and 200 iterations of each starting x value. The graph is displayed from iteration 1.

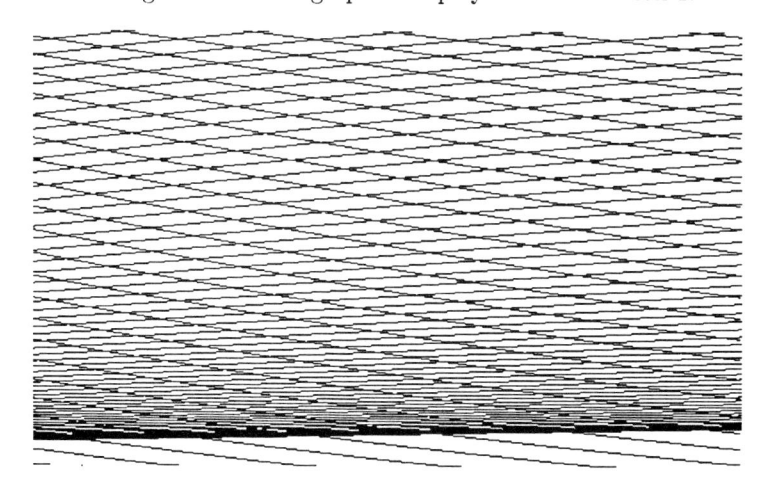

Figure 4-3a. An alternative view of the Lorenz equations.

You can see that for this instance there are certain values of x that have not been visited after 200 iterations. We would expect this, as this

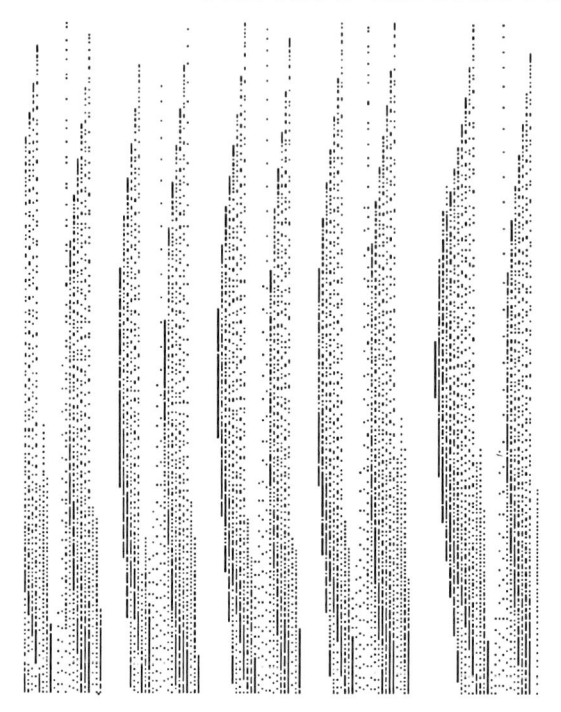

Figure 4-3b. Output from Listing 4-2.

is the area of the equations that shows less sensitivity to initial conditions than later iterations. If adjacent columns on the graph are grossly different, rather than the smooth transitions shown here, then you are in a chaotic area. Such an area is shown in Figure 4-3b, where $x = y = z = 1$, 200 iterations, displayed from iteration 2500. This is slow to draw, so be patient! In each case, the y-axis represents values of x between 0 and the maximum y coordinate for the graphics screen involved divided by 20.

Phase portraits of the Lorenz equations

As we've already seen, phase portraits are a very effective way of examining the behaviour of systems of differential equations, allowing us to see limit cycles, attractors, etc., as well as giving us a good idea about how the system evolves by watching the phase portrait being drawn. All the phase portraits seen so far have been for systems of equations with two variables, x and y. We've got a slight complication here due to the Lorenz equations

PHASE PORTRAITS OF THE LORENZ EQUATIONS

having three variables. How can we plot the three variables on a phase portrait?

Three-dimensional phase space

Well, if we've got a three-dimensional system the obvious way is to describe the system in terms of a three-dimensional phase space. Clearly, as a sheet of paper is only two-dimensional, we've got problems in drawing a three-dimensional object on it, but we can come up with a reasonable representation. Figure 4-4 shows how we might view three-dimensional phase

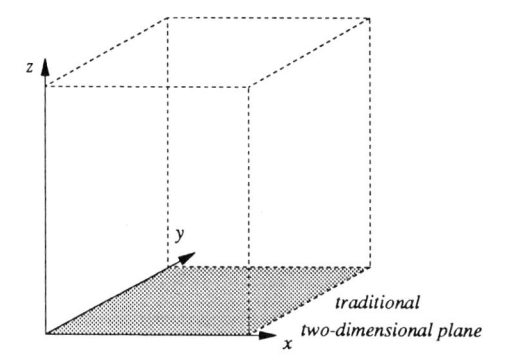

Figure 4-4. Three-dimensional phase space.

space; there are three coordinate axes here, x, y and z. Each point in phase space is defined by specifying values for x, y and z, rather than just for x and y as we have previously done. Again, each point on a three-dimensional phase portrait represents the state of the system at a particular point in time for a set of initial conditions.

Two-dimensional phase portrait

In drawing a phase portrait, there's nothing to stop you using only two dimensions of the three-dimensional system. For example, you could draw a phase portrait of y against x, or z against y, and so on. Such phase portraits are said to be in the yx- or zy-plane, by far the easiest method of representing a three-dimensional attractor on a computer screen.

Like any phase portrait, it's often useful to watch how the phase diagram evolves with time, rather than just viewing the finished item. Listing 4-3 shows a program to plot a two-dimensional phase portrait for the *Lorenz attractor*. As the program stands, the program plots the Lorenz attractor in the zx-plane—a shape that is often called the 'owl face', the 'mask' or 'the butterfly'. If you run the program you'll see how the trajectory

102   *CHAPTER 4. THE LORENZ EQUATIONS*

generated loops around, spirals around one side of the shape and then swings over to the other side, and stays there for a while. Changing the start conditions will cause a similar shape to be generated, but it won't be the same. Additionally, you might choose to display the attractor only after a specified number of iterations by entering a value for the *DisplayAfter* prompt.

Modifications to the program

Figure 4-5 shows three views of the Lorenz attractor. To generate these,

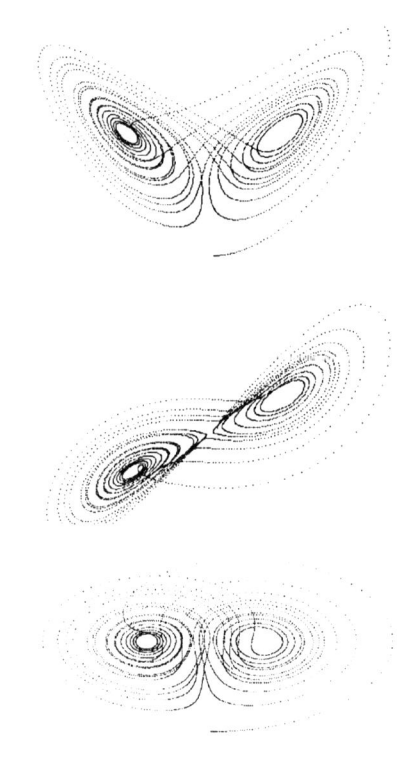

Figure 4-5. Three views of the Lorenz attractor.

simply alter the line that plots the points in the program to change the variables plotted. So, for the yx plot, we'd use the value of the y variable as the y-axis plot, and the value of the x variable for the x-axis plot. You may have to alter the scaling values of x, y and z to get results that fit your display best. The x- and y-axis scaling factors in the PC program are for a VGA screen. You could also modify the program to change the colour of the pixel plotted after a specified number of iterations.

THE STRANGE ATTRACTOR

A further change to the display that can be applied to any program that plots a phase portrait of a system is to modify the code so that after plotting a point on the screen there is a pause and the point is removed from the screen. This simple animation technique gives the illusion of movement and it is quite interesting to watch the particle move around the screen. Listing 4-4 shows the program from Listing 4-3 modified to do this.

Finally, there's nothing to stop you altering the values of the a and b constants in the equations away from the values specified here (10 and 8/3). Although traditionally these values are positive numbers, you can see what happens if you make them negative.

With the default values of a, b and r, no matter what you do with the start values, the three variables assume values that are within a defined area of the phase portrait. The trajectory traces a path around one lobe for a while, then flicks over to the other one, then back again. The trajectory of the system with time therefore constitutes an attractor; but what sort is it? For $r = 28$, the attractor certainly doesn't appear to be a simple limit cycle or point attractor. If it were a point attractor the system would settle down to the point, and if it were a limit cycle some periodic behaviour would be noted.

The strange attractor

The problem is, that as we carry on over a long period of time, our graphical representation of the trajectory gets very crowded, but if we were to examine the trajectory in fine enough detail we would find that the trajectory *never merges with itself*. This is quite important, because if the trajectory were to visit a point in phase space already visited, the next point that the trajectory would visit would be determined from the parameter values at the point that's just been visited, and so on. This would lead to periodic behaviour, and would give us a limit cycle. The trajectories appear to merge, but never do so; the trajectories can be seen to be an infinitesimally small distance apart but *they are still separate*.

No wonder such a beast was called a *strange attractor;* it's a peculiar half-way house between something that's totally unstable, with the trajectory winging its mathematical way off into some far distant corner of phase space, never to be seen again, and a limit cycle, where the trajectory repeats the same path for ever. A few moments thought and examination of the attractor will indicate a few facts about a system that has a strange attractor.

1. Consider two points in phase space, (x, y, z) and (x', y', z'). Now, the difference between these two points, if we view them as initial conditions, might only be a tiny amount but could easily put these two

points on two separate but adjacent trajectories. Despite the proximity of these trajectories, they could end up in totally different parts of phase space after a length of time. This is yet another expression of sensitive dependence upon initial conditions.

2. Because a trajectory around a strange attractor is non-periodic, there are an infinite number of points in phase space that are visited by the trajectory. Of course some of these are so close to one another that for all practical purposes in computer graphics experiments we can't tell the difference between them. It is worth bearing this in mind when generating pictures of strange attractors. In addition, any computer program will generate a finite number of points on the attractor in a certain length of time.

3. Any errors in solving the equations will give a slightly different trajectory than was expected, but will still exhibit the same general form because the trajectory will still be moving around the Lorenz attractor.

The Lorenz attractor is just an example of a whole family of strange attractors that characterise chaotic systems. In the next chapter, we'll look at how these can be generated on a home computer.

BBC BASIC listings

Listing 4-1

```
10 REM Program LorenzEquationTimeProgression
20 REM Listing 4.1
50 REM try a = 10,  b=2.66,  r=28,   x=1,   y=1,    z=1
70 REM  Allow user to enter new values
90 MODE 6
110 d=0.005 : a=10 : b=2.66
120 INPUT "Value of x ",x
130 INPUT "Value of y ",y
140 INPUT "Value of z ",z
150 INPUT "Value of r ",r
160 INPUT "Display after ",DisplayAfter
200 REM  Set graphics mode up
220 MODE 5
230 REM n hold number of time increments passed
250 n=1
270 REPEAT
320   REM Only display the three lines if n is after the value of
330   REM DisplayAfter. If it is, plot three lines top one is z,
340   REM middle is Y and bottom is X.
390   IF n<DisplayAfter THEN GOTO 500
410   GCOL0,1
420   PLOT69,(n-DisplayAfter),(x*4+50)
430   GCOL0,2
440   PLOT69,(n-DisplayAfter),(y*4+100)
450   GCOL0,3
460   PLOT69,(n-DisplayAfter),(z*4+150)
500   REM Now start the Runge-Kutta solution, first for x
520   xinc1=FNEquationX(x,y,z)*d
530   xinc2=FNEquationX(x+xinc1/2,y+d/2,z+d/2)*d
540   xinc3=FNEquationX(x+xinc2/2,y+d,z+d)*d
550   xinc4=FNEquationX(x+xinc3,y+d,z+d)*d
570   REM Now repeat for the Y Equation
590   yinc1=FNEquationY(x,y,z)*d
600   yinc2=FNEquationY(x+yinc1/2,y+d/2,z+d/2)*d
610   yinc3=FNEquationY(x+yinc2/2,y+d,z+d)*d
620   yinc4=FNEquationY(x+yinc3,y+d,z+d)*d
640   REM Finally for z
660   zinc1=FNEquationZ(x,y,z)*d
670   zinc2=FNEquationZ(x+xinc1/2,y+d/2,z+d/2)*d
680   zinc3=FNEquationZ(x+xinc2/2,y+d,z+d)*d
690   zinc4=FNEquationZ(x+xinc3,y+d,z+d)*d
710   REM Only update x,y and z at this stage, as the unaltered
720   REM values of x,y and z are needed for calculations of
730   REM increments to x,y and z above.
760   x=x+(xinc1+2*xinc2+2*xinc3+xinc4)/6
770   y=y+(yinc1+2*yinc2+2*yinc3+yinc4)/6
780   z=z+(zinc1+2*zinc2+2*zinc3+zinc4)/6
800   REM Increment passing of time....
820   n=n+1
860   REM  Repeat until a key is pressed
880 UNTIL (INKEY(1)<>-1)
900 PRINT CHR$(7)
910 G=GET
940 END
970 DEFFNEquationX(xi,yi,zi)=-a*xi+a*yi
990 DEFFNEquationY(xi,yi,zi)= xi*zi+r*xi-yi
1010 DEFFNEquationZ(xi,yi,zi)=-b*zi+xi*yi
```

Listing 4-2

```
10 REM Program LorenzEquationTimeProgression2
50 MODE 6
70 d=0.005
90 REM try a = 10,  b=2.66,  r=28,    x=1,    y=1,    z=1
95 a=10 : b=2.66
110 REM  Allow user to enter new values
130 MODE 6
150 INPUT "Value of x ",x
160 INPUT "Value of y ",y
170 INPUT "Value of z ",z
180 INPUT "Value of r ",r
190 INPUT "Display after ",disp
230 REM  Set graphics mode up
250 MODE 5
290 REM n hold number of time increments passed
300 xpos=1
310 xst=x
320 yst=y
330 zst=x
350 REPEAT
370    REM for each xpos position initialise y and z to the desired
380    REM start values
400    y=yst
410    z=zst
430    REM Make initial value of x equal to current value of x plus
440    REM 0.001 for every step across the x axis
460    x=xst+(xpos*0.001)
480    REM indcremnet x position along graphics screen by 1
500    xpos=xpos+1
520    REM plot 200 iterations for each start value - you could alter
530    REM this to see what happens!
550    FOR n=1 TO (disp+200)
590      REM Now start the Runge-Kutta solution, first for x
610      xinc1=FNEquationX(x,y,z)*d
620      xinc2=FNEquationX(x+xinc1/2,y+d/2,z+d/2)*d
630      xinc3=FNEquationX(x+xinc2/2,y+d,z+d)*d
640      xinc4=FNEquationX(x+xinc3,y+d,z+d)*d
660      REM Now repeat for the Y Equation
680      yinc1=FNEquationY(x,y,z)*d
690      yinc2=FNEquationY(x+yinc1/2,y+d/2,z+d/2)*d
700      yinc3=FNEquationY(x+yinc2/2,y+d,z+d)*d
710      yinc4=FNEquationY(x+yinc3,y+d,z+d)*d
730      REM Finally for z
750      zinc1=FNEquationZ(x,y,z)*d
760      zinc2=FNEquationZ(x+xinc1/2,y+d/2,z+d/2)*d
770      zinc3=FNEquationZ(x+xinc2/2,y+d,z+d)*d
780      zinc4=FNEquationZ(x+xinc3,y+d,z+d)*d
800      REM Only update x,y and z at this stage, as the unaltered
810      REM values of x,y and z are needed for calculations of
820      REM increments to x,y and z above.
860      x=x+(xinc1+2*xinc2+2*xinc3+xinc4)/6
870      y=y+(yinc1+2*yinc2+2*yinc3+yinc4)/6
880      z=z+(zinc1+2*zinc2+2*zinc3+zinc4)/6
910      REM Now plot the iterated value of x against xpos, which
920      REM represents the initial value of x.
940      IF n<disp THEN GOTO 1010
950      PLOT69,xpos*4,x*20
980      REM Increment passing of time by incrementing value of
990      REM 'n' in the for...do loop
1010   NEXT
1030   REM  Repeat until a key is pressed
1050 UNTIL (INKEY(1)<>-1)
1060 PRINT CHR$(7)
```

```
1070 G=GET
1100 END
1130 DEFFNEquationX(xi,yi,zi)=-a*xi+a*yi
1150 DEFFNEquationY(xi,yi,zi)=-xi*zi+r*xi-yi
1170 DEFFNEquationZ(xi,yi,zi)=-b*zi+xi*yi
```

Listing 4-3

```
 10 REM Program LorenzEquation Attractor Plot
 20 REM Listing 4.3
 50 REM try a = 10,  b=2.66,  r=28,   x=1,   y=1,   z=1
 70 REM  Allow user to enter new values
 90 MODE 6
110 d=0.005 : a=10 : b=2.66
120 INPUT "Value of x ",x
130 INPUT "Value of y ",y
140 INPUT "Value of z ",z
150 INPUT "Value of r ",r
160 INPUT "Display after ",DisplayAfter
200 REM  Set graphics mode up
220 MODE 5
230 REM n hold number of time increments passed
250 n=1
270 REPEAT
290   REM Only display a point on the attractor afterthe DisplayAfter
300   REM variable is exceeded.  Change the PutPixel line to change
310   REM the view of the attractor that is displayed.
350   IF n>DisplayAfter THEN PLOT69,(x*20)+500,(z*10)+200
380   REM Now start the Runge-Kutta solution, first for x
400   xinc1=FNEquationX(x,y,z)*d
410   xinc2=FNEquationX(x+xinc1/2,y+d/2,z+d/2)*d
420   xinc3=FNEquationX(x+xinc2/2,y+d,z+d)*d
430   xinc4=FNEquationX(x+xinc3,y+d,z+d)*d
450   REM Now repeat for the Y Equation
470   yinc1=FNEquationY(x,y,z)*d
480   yinc2=FNEquationY(x+yinc1/2,y+d/2,z+d/2)*d
490   yinc3=FNEquationY(x+yinc2/2,y+d,z+d)*d
500   yinc4=FNEquationY(x+yinc3,y+d,z+d)*d
520   REM Finally for z
540   zinc1=FNEquationZ(x,y,z)*d
550   zinc2=FNEquationZ(x+xinc1/2,y+d/2,z+d/2)*d
560   zinc3=FNEquationZ(x+xinc2/2,y+d,z+d)*d
570   zinc4=FNEquationZ(x+xinc3,y+d,z+d)*d
590   REM Only update x,y and z at this stage, as the unaltered
600   REM values of x,y and z are needed for calculations of
610   REM increments to x,y and z above.
640   x=x+(xinc1+2*xinc2+2*xinc3+xinc4)/6
650   y=y+(yinc1+2*yinc2+2*yinc3+yinc4)/6
660   z=z+(zinc1+2*zinc2+2*zinc3+zinc4)/6
680   REM Increment passing of time....
700   n=n+1
740   REM  Repeat until a key is pressed
760 UNTIL (INKEY(1)<>-1)
780 PRINT CHR$(7)
790 G=GET
820 END
850 DEFFNEquationX(xi,yi,zi)=-a*xi+a*yi
870 DEFFNEquationY(xi,yi,zi)=-xi*zi+r*xi-yi
890 DEFFNEquationZ(xi,yi,zi)=-b*zi+xi*yi
```

Listing 4-4

```
 10 REM Program LorenzEquation Attractor Plot (Moving Dot)
 20 REM Listing 4.4
 50 REM try a = 10,  b=2.66,  r=28,   x=1,   y=1,   z=1
 70 REM  Allow user to enter new values
 90 MODE 6
110 d=0.005 : a=10 : b=2.66
120 INPUT "Value of x ",x
130 INPUT "Value of y ",y
140 INPUT "Value of z ",z
150 INPUT "Value of r ",r
160 INPUT "Display after ",DisplayAfter
200 REM  Set graphics mode up
220 MODE 5
230 REM n hold number of time increments passed
250 n=1
270 REPEAT
300   REM Only display a point on the attractor afterthe DisplayAfter
310   REM variable is exceeded.  Change the PutPixel line to change
320   REM the view of the attractor that is displayed.
340   IF n<DisplayAfter THEN GOTO 640
360   GCOL0,1
370   PLOT69,x*10+500,z*5+100
380   PLOT69,x*10+504,z*5+100
390   PLOT69,x*10+504,z*5+104
400   PLOT69,x*10+500,z*5+104
430   REM The following line determines the rate of travel - if too low
440   REM the dot will fly around too fast to see; too high and the dot
450   REM will appear to hop from place to place
470   FOR J%=1 TO 30:NEXT
490   REM Now wipe out the dot with a dot of background colour
510   GCOL0,0
520   PLOT69,x*10+500,z*5+100
530   PLOT69,x*10+504,z*5+100
540   PLOT69,x*10+504,z*5+104
550   PLOT69,x*10+500,z*5+104
620   REM Now start the Runge-Kutta solution, first for x
640   xinc1=FNEquationX(x,y,z)*d
650   xinc2=FNEquationX(x+xinc1/2,y+d/2,z+d/2)*d
660   xinc3=FNEquationX(x+xinc2/2,y+d,z+d)*d
670   xinc4=FNEquationX(x+xinc3,y+d,z+d)*d
690   REM Now repeat for the Y Equation
710   yinc1=FNEquationY(x,y,z)*d
720   yinc2=FNEquationY(x+yinc1/2,y+d/2,z+d/2)*d
730   yinc3=FNEquationY(x+yinc2/2,y+d,z+d)*d
740   yinc4=FNEquationY(x+yinc3,y+d,z+d)*d
760   REM Finally for z
780   zinc1=FNEquationZ(x,y,z)*d
790   zinc2=FNEquationZ(x+xinc1/2,y+d/2,z+d/2)*d
800   zinc3=FNEquationZ(x+xinc2/2,y+d,z+d)*d
810   zinc4=FNEquationZ(x+xinc3,y+d,z+d)*d
830   REM Only update x,y and z at this stage, as the unaltered
840   REM values of x,y and z are needed for calculations of
850   REM increments to x,y and z above.
880   x=x+(xinc1+2*xinc2+2*xinc3+xinc4)/6
890   y=y+(yinc1+2*yinc2+2*yinc3+yinc4)/6
900   z=z+(zinc1+2*zinc2+2*zinc3+zinc4)/6
920   REM Increment passing of time....
940   n=n+1
980   REM  Repeat until a key is pressed
1000 UNTIL (INKEY(1)<>-1)
1020 PRINT CHR$(7)
1030 G=GET
1060 END
```

```
1090 DEFFNEquationX(xi,yi,zi)=-a*xi+a*yi
1110 DEFFNEquationY(xi,yi,zi)=-xi*zi+r*xi-yi
1130 DEFFNEquationZ(xi,yi,zi)=-b*zi+xi*yi
```

Turbo Pascal listings

Listing 4-1

```
Program LorenzEquationTimeProgression;

Uses Crt,Graph,Extend;

Var

x,y,z                       :       Real;
xinc,yinc,zinc              :       Real;
xinc1,xinc2,xinc3,xinc4:            Real;
yinc1,yinc2,yinc3,yinc4:            Real;
zinc1,zinc2,zinc3,zinc4:            Real;
xi,yi                       :       Real;
a,b,r                       :       Real;
xs,ys,zs,rs                 :       String;
temp                        :       Char;
d                           :       Real;
n                           :       Integer;
DisplayAfter                :       Integer;
DisplayAfters               :       String;

Function EquationX(xi,yi,zi : real) : real;
Begin
    EquationX:=-a*xi+a*yi;
end;

Function EquationY(xi,yi,zi : real) : real;
Begin
    EquationY:=-xi*zi+r*xi-yi;
end;

Function EquationZ(xi,yi,zi : real) : real;
Begin
    EquationZ:=-b*zi+xi*yi;
end;

Begin

   d:=0.005;
   a:=10;
   b:=8/3;
   r:=28;
   rs:='28';

   x:=1;
   xs:='1';
   y:=1;
   ys:='1';
   z:=1;
   zs:='1';

{  Allow user to enter new values
}
```

110   CHAPTER 4. THE LORENZ EQUATIONS

```
    TextBackground(Blue);
    ClrScr;
    PrintAt(2,4,'Value of x? ',Green);
    StringEdit(xs,20,50,4,Yellow);
    Val(xs,x,n);
    PrintAt(2,5,'Value of y? ',Green);
    StringEdit(ys,20,50,5,Yellow);
    Val(ys,y,n);
    PrintAt(2,6,'Value of z? ',Green);
    StringEdit(zs,20,50,6,Yellow);
    Val(zs,z,n);

    PrintAt(2,8,'Value of r? ',Green);
    StringEdit(rs,20,50,8,Yellow);
    Val(rs,r,n);
    PrintAt(2,9,'Display after? ',Green);
    StringEdit(DisplayAfters,20,50,9,Yellow);
    Val(DisplayAfters,DisplayAfter,n);

{ Set graphics mode up }

    ChoosePalette;

{ n hold number of time increments passed }

    n:=1;

    Repeat

        Begin

{ Only display the three lines if n is after the value of
  DisplayAfter. If it is, plot three lines; top one is z,
  middle is Y and bottom is X. Again, the Y coordinate
  is altered to fit screen coordinate system. May need to
  change colours used in CGA mode to colours 1,2 and 3. }

        if n>DisplayAfter then begin

            PutPixel(n-DisplayAfter,GetMaxY-trunc(x*4+50),15);
            PutPixel(n-DisplayAfter,GetMaxY-trunc(y*4+100),15);
            PutPixel(n-DisplayAfter,GetMaxY-trunc(z*4+150),15);

        end;

{ Now start the Runge-Kutta solution, first for x }

            xinc1:=EquationX(x,y,z)*d;
            xinc2:=EquationX(x+xinc1/2,y+d/2,z+d/2)*d;
            xinc3:=EquationX(x+xinc2/2,y+d,z+d)*d;
            xinc4:=EquationX(x+xinc3,y+d,z+d)*d;

{ Now repeat for the Y Equation                      }

            yinc1:=Equationy(x,y,z)*d;
            yinc2:=Equationy(x+yinc1/2,y+d/2,z+d/2)*d;
            yinc3:=Equationy(x+yinc2/2,y+d,z+d)*d;
            yinc4:=Equationy(x+yinc3,y+d,z+d)*d;

{ Finally for z  }

            zinc1:=Equationz(x,y,z)*d;
            zinc2:=Equationz(x+xinc1/2,y+d/2,z+d/2)*d;
            zinc3:=Equationz(x+xinc2/2,y+d,z+d)*d;
            zinc4:=Equationz(x+xinc3,y+d,z+d)*d;
```

TURBO PASCAL LISTINGS

```
{ Only update x,y and z at this stage, as the unaltered
  values of x,y and z are needed for calculations of
  increments to x,y and z above.                              }

        x:=x+(xinc1+2*xinc2+2*xinc3+xinc4)/6;
        y:=y+(yinc1+2*yinc2+2*yinc3+yinc4)/6;
        z:=z+(zinc1+2*zinc2+2*zinc3+zinc4)/6;

{ Increment passing of time....}

        n:=n+1;

   end;

{ Repeat until a key is pressed                               }

   until (keyPressed);

   sound(200);
   Delay(30);
   NoSound;
   temp:=Readkey;
   CloseGraph;

end.
```

Listing 4-2

```
Program LorenzEquationTimeProgression2;

Uses Crt,Graph,Extend;

Var

x,y,z                    :        Real;
xinc,yinc,zinc           :        Real;
xinc1,xinc2,xinc3,xinc4: Real;
yinc1,yinc2,yinc3,yinc4: Real;
zinc1,zinc2,zinc3,zinc4: Real;
xi,yi                    :        Real;
a,b,r                    :        Real;
xs,ys,zs,rs,disps        :        String;
temp                     :        Char;
d                        :        Real;
n,disp                   :        Integer;
xpos                     :        Integer;
xst,yst,zst              :        Real;

Function EquationX(xi,yi,zi : real) : real;
Begin
   EquationX:=-a*xi+a*yi;
end;

Function EquationY(xi,yi,zi : real) : real;
Begin
   EquationY:=-xi*zi+r*xi-yi;
end;

Function EquationZ(xi,yi,zi : real) : real;
Begin
   EquationZ:=-b*zi+xi*yi;
end;
```

```
Begin

  d:=0.005;
  a:=10;
  b:=8/3;
  r:=28;
  rs:='28';

  x:=1;
  xs:='1';
  y:=1;
  ys:='1';
  z:=1;
  zs:='1';

{ Allow user to enter new values
}

  TextBackground(Blue);
  ClrScr;
  PrintAt(2,4,'Value of x? ',Green);
  StringEdit(xs,20,50,4,Yellow);
  Val(xs,x,n);
  PrintAt(2,5,'Value of y? ',Green);
  StringEdit(ys,20,50,5,Yellow);
  Val(ys,y,n);
  PrintAt(2,6,'Value of z? ',Green);
  StringEdit(zs,20,50,6,Yellow);
  Val(zs,z,n);

  PrintAt(2,8,'Value of r? ',Green);
  StringEdit(rs,20,50,8,Yellow);
  Val(rs,r,n);
  disps:='1';
  disp:=1;
  PrintAt(2,9,'Display after iteration? ',Green);
  StringEdit(disps,20,50,9,Yellow);
  Val(disps,disp,n);

{ Set graphics mode up }

  ChoosePalette;

{ n hold number of time increments passed }
xpos:=1;
xst:=x;
yst:=y;
zst:=x;

Repeat
  Begin

{ for each xpos position initialise y and z to the desired
  start values }

  y:=yst;
  z:=zst;

{ Make initial value of x equal to current value of x plus
  0.001 for every step across the x axis            }

  x:=xst+(xpos*0.001);

{ indcremnet x position along graphics screen by 1 }
```

```pascal
      xpos:=xpos+1;

{ plot 200 iterations for each start value - you could alter
  this to see what happens!           }

      for n:=1 to (disp+200) do

          Begin

{ Now start the Runge-Kutta solution, first for x }

            xinc1:=EquationX(x,y,z)*d;
            xinc2:=EquationX(x+xinc1/2,y+d/2,z+d/2)*d;
            xinc3:=EquationX(x+xinc2/2,y+d,z+d)*d;
            xinc4:=EquationX(x+xinc3,y+d,z+d)*d;

{ Now repeat for the Y Equation                     }

            yinc1:=Equationy(x,y,z)*d;
            yinc2:=Equationy(x+yinc1/2,y+d/2,z+d/2)*d;
            yinc3:=Equationy(x+yinc2/2,y+d,z+d)*d;
            yinc4:=Equationy(x+yinc3,y+d,z+d)*d;

{ Finally for z }

            zinc1:=Equationz(x,y,z)*d;
            zinc2:=Equationz(x+xinc1/2,y+d/2,z+d/2)*d;
            zinc3:=Equationz(x+xinc2/2,y+d,z+d)*d;
            zinc4:=Equationz(x+xinc3,y+d,z+d)*d;

{ Only update x,y and z at this stage, as the unaltered
  values of x,y and z are needed for calculations of
  increments to x,y and z above.                    }

            x:=x+(xinc1+2*xinc2+2*xinc3+xinc4)/6;
            y:=y+(yinc1+2*yinc2+2*yinc3+yinc4)/6;
            z:=z+(zinc1+2*zinc2+2*zinc3+zinc4)/6;

{ Now plot the iterated value of x against xpos, which
  represents the initial value of x.     }

      if n>=disp then
         PutPixel(xpos,GetMaxY-trunc(x*20),2);

{ Increment passing of time by incrementing value of
  'n' in the for...do loop }

      end;

    end;
{ Repeat until a key is pressed                     }

    until (keyPressed);

    sound(200);
    Delay(30);
    NoSound;
    temp:=Readkey;
    CloseGraph;

end.
```

Listing 4-3

```
Program LorenzAttractor;

{       Draws the Lorenz Attractor }

Uses Crt,Graph,Extend;

Var

x,y,z                       :       Real;
xinc,yinc,zinc              :       Real;
xinc1,xinc2,xinc3,xinc4:            Real;
yinc1,yinc2,yinc3,yinc4:            Real;
zinc1,zinc2,zinc3,zinc4:            Real;
xi,yi                       :       Real;
a,b,r                       :       Real;
xs,ys,zs,rs                 :       String;
temp                        :       Char;
d                           :       Real;
n                           :       Integer;
DisplayAfter                :       Integer;
DisplayAfters               :       String;

Function EquationX(xi,yi,zi : real) : real;
Begin
   EquationX:=-a*xi+a*yi;
end;

Function EquationY(xi,yi,zi : real) : real;
Begin
   EquationY:=-xi*zi+r*xi-yi;
end;

Function EquationZ(xi,yi,zi : real) : real;
Begin
   EquationZ:=-b*zi+xi*yi;
end;

Begin

   d:=0.005;
   a:=10;
   b:=8/3;
   r:=28;
   rs:='28';

   x:=1;
   xs:='1';
   y:=1;
   ys:='1';
   z:=1;
   zs:='1';

   DisplayAfters:='1';
   DisplayAfter:=1;

{ Allow user to enter new values
}

   TextBackground(Blue);
   ClrScr;
   PrintAt(2,4,'Value of x? ',Green);
   StringEdit(xs,20,50,4,Yellow);
   Val(xs,x,n);
```

TURBO PASCAL LISTINGS

```pascal
    PrintAt(2,5,'Value of y? ',Green);
    StringEdit(ys,20,50,5,Yellow);
    Val(ys,y,n);
    PrintAt(2,6,'Value of z? ',Green);
    StringEdit(zs,20,50,6,Yellow);
    Val(zs,z,n);

    PrintAt(2,8,'Value of r? ',Green);
    StringEdit(rs,20,50,8,Yellow);
    Val(rs,r,n);
    PrintAt(2,9,'Display after? ',Green);
    StringEdit(DisplayAfters,20,50,9,Yellow);
    Val(DisplayAfters,DisplayAfter,n);

{ Set graphics mode up }

    ChoosePalette;

{ n hold number of time increments passed }

    n:=1;

    Repeat

        Begin

{       Only display a point on the attractor afterthe DisplayAfter
        variable is exceeded.  Change the PutPixel line to change
        the view of the attractor that is displayed.              }

        if n>DisplayAfter then begin

            PutPixel(trunc(x*10)+trunc(GetMaxX/2),(GetMaxY-trunc(z*5))-100,2);

        end;

{ Now start the Runge-Kutta solution, first for x }

            xinc1:=EquationX(x,y,z)*d;
            xinc2:=EquationX(x+xinc1/2,y+d/2,z+d/2)*d;
            xinc3:=EquationX(x+xinc2/2,y+d,z+d)*d;
            xinc4:=EquationX(x+xinc3,y+d,z+d)*d;

{ Now repeat for the Y Equation                    }

            yinc1:=Equationy(x,y,z)*d;
            yinc2:=Equationy(x+yinc1/2,y+d/2,z+d/2)*d;
            yinc3:=Equationy(x+yinc2/2,y+d,z+d)*d;
            yinc4:=Equationy(x+yinc3,y+d,z+d)*d;

{ Finally for z }

            zinc1:=Equationz(x,y,z)*d;
            zinc2:=Equationz(x+xinc1/2,y+d/2,z+d/2)*d;
            zinc3:=Equationz(x+xinc2/2,y+d,z+d)*d;
            zinc4:=Equationz(x+xinc3,y+d,z+d)*d;

{ Only update x,y and z at this stage, as the unaltered
  values of x,y and z are needed for calculations of
  increments to x,y and z above.                         }

            x:=x+(xinc1+2*xinc2+2*xinc3+xinc4)/6;
            y:=y+(yinc1+2*yinc2+2*yinc3+yinc4)/6;
            z:=z+(zinc1+2*zinc2+2*zinc3+zinc4)/6;
```

116   CHAPTER 4. THE LORENZ EQUATIONS

```
{ Increment passing of time....}

        n:=n+1;

   end;

{ Repeat until a key is pressed                    }

   until (keyPressed);

   sound(200);
   Delay(30);
   NoSound;
   temp:=Readkey;
   CloseGraph;

end.
```

Listing 4-4

```
Program LorenzAttractorParticle;

{       Draws the Lorenz Attractor as a moving dot which follows the
        attractor but does not leave a trail. }

Uses Crt,Graph,Extend;

Var

x,y,z                    :         Real;
xinc,yinc,zinc           :         Real;
xinc1,xinc2,xinc3,xinc4:           Real;
yinc1,yinc2,yinc3,yinc4:           Real;
zinc1,zinc2,zinc3,zinc4:           Real;
xi,yi                    :         Real;
a,b,r                    :         Real;
xs,ys,zs,rs              :         String;
temp                     :         Char;
d                        :         Real;
n                        :         Integer;
DisplayAfter             :         Integer;
DisplayAfters            :         String;

Function EquationX(xi,yi,zi : real) : real;
Begin
   EquationX:=-a*xi+a*yi;
end;

Function EquationY(xi,yi,zi : real) : real;
Begin
   EquationY:=-xi*zi+r*xi-yi;
end;

Function EquationZ(xi,yi,zi : real) : real;
Begin
   EquationZ:=-b*zi+xi*yi;
end;

Begin

   d:=0.005;
   a:=10;
   b:=8/3;
```

TURBO PASCAL LISTINGS   117

```pascal
    r:=28;
    rs:='28';

    x:=1;
    xs:='1';
    y:=1;
    ys:='1';
    z:=1;
    zs:='1';

    DisplayAfters:='1';
    DisplayAfter:=1;

{ Allow user to enter new values
}

    TextBackground(Blue);
    ClrScr;
    PrintAt(2,4,'Value of x? ',Green);
    StringEdit(xs,20,50,4,Yellow);
    Val(xs,x,n);
    PrintAt(2,5,'Value of y? ',Green);
    StringEdit(ys,20,50,5,Yellow);
    Val(ys,y,n);
    PrintAt(2,6,'Value of z? ',Green);
    StringEdit(zs,20,50,6,Yellow);
    Val(zs,z,n);

    PrintAt(2,8,'Value of r? ',Green);
    StringEdit(rs,20,50,8,Yellow);
    Val(rs,r,n);
    PrintAt(2,9,'Display after? ',Green);
    StringEdit(DisplayAfters,20,50,9,Yellow);
    Val(DisplayAfters,DisplayAfter,n);

{ Set graphics mode up }

    ChoosePalette;

{ n hold number of time increments passed }

    n:=1;

    Repeat

        Begin

{       Only display a point on the attractor afterthe DisplayAfter
        variable is exceeded.  Change the PutPixel line to change
        the view of the attractor that is displayed.             }

        if n>DisplayAfter then begin

            PutPixel(trunc(x*10)+trunc(GetMaxX/2),(GetMaxY-trunc(z*5))-100,2);
            PutPixel(trunc(x*10)+trunc(GetMaxX/2)+2,(GetMaxY-trunc(z*5))-100,2);
            PutPixel(trunc(x*10)+trunc(GetMaxX/2)+2,(GetMaxY-trunc(z*5))-100+1,2);
            PutPixel(trunc(x*10)+trunc(GetMaxX/2),(GetMaxY-trunc(z*5))-100+1,2);

{ The following line determines the rate of travel - if too low
  the dot will fly around too fast to see; too high and the dot
  will appear to hop from place to place }

            Delay(45);

{   Now wipe out the dot with a dot of background colour      }
```

```
          PutPixel(trunc(x*10)+trunc(GetMaxX/2),(GetMaxY-trunc(z*5))-100,0);
          PutPixel(trunc(x*10)+trunc(GetMaxX/2)+2,(GetMaxY-trunc(z*5))-100,0);
          PutPixel(trunc(x*10)+trunc(GetMaxX/2)+2,(GetMaxY-trunc(z*5))-100+1,0);
          PutPixel(trunc(x*10)+trunc(GetMaxX/2),(GetMaxY-trunc(z*5))-100+1,0);

      end;

{ Now start the Runge-Kutta solution, first for x }

          xinc1:=EquationX(x,y,z)*d;
          xinc2:=EquationX(x+xinc1/2,y+d/2,z+d/2)*d;
          xinc3:=EquationX(x+xinc2/2,y+d,z+d)*d;
          xinc4:=EquationX(x+xinc3,y+d,z+d)*d;

{ Now repeat for the Y Equation                        }

          yinc1:=Equationy(x,y,z)*d;
          yinc2:=Equationy(x+yinc1/2,y+d/2,z+d/2)*d;
          yinc3:=Equationy(x+yinc2/2,y+d,z+d)*d;
          yinc4:=Equationy(x+yinc3,y+d,z+d)*d;

{ Finally for z }

          zinc1:=Equationz(x,y,z)*d;
          zinc2:=Equationz(x+xinc1/2,y+d/2,z+d/2)*d;
          zinc3:=Equationz(x+xinc2/2,y+d,z+d)*d;
          zinc4:=Equationz(x+xinc3,y+d,z+d)*d;

{ Only update x,y and z at this stage, as the unaltered
  values of x,y and z are needed for calculations of
  increments to x,y and z above.                       }

          x:=x+(xinc1+2*xinc2+2*xinc3+xinc4)/6;
          y:=y+(yinc1+2*yinc2+2*yinc3+yinc4)/6;
          z:=z+(zinc1+2*zinc2+2*zinc3+zinc4)/6;

{ Increment passing of time....}

          n:=n+1;

      end;

{ Repeat until a key is pressed                        }

      until (keyPressed);

      sound(200);
      Delay(30);
      NoSound;
      temp:=Readkey;
      CloseGraph;

end.
```

Chapter 5

Strange attractors

In the previous chapter we saw that the behaviour of the Lorenz equations was caused by the presence of a strange attractor in the phase portrait of the system. In this chapter we'll examine the idea of strange attractors in greater detail. We've already examined a three-dimensional strange attractor, the Lorenz attractor. Let's now look at some of the peculiar properties of strange attractors in greater detail using a simpler system, called the *Hénon attractor*.

The Hénon attractor

Michel Hénon, a French astronomer, published a paper called 'A Two Dimensional Mapping with a Strange Attractor' in 1976. His aim was to create a strange attractor similar to that obtained by Lorenz, but that was simpler to create and handle. However, Hénon used a different approach to find his attractor, in that he used a simple mapping, like the Martin's mappings seen earlier, to generate the attractor rather than a set of differential equations. The mapping used by Hénon was:

$$x_{n+1} = y_n + 1 - a * x_n * x_n$$
$$y_{n+1} = b * x_n$$

with values of $a = 1.4$ and $b = 0.3$. As with all such mappings, start values of x and y are iterated and the resulting x and y values plotted. Listing 5-1 provides a simple tool for investigating the attractor. There are a couple of reasons why this attractor is a useful tool for investigating the behaviour of strange attractors:

1. The calculations involved in iterating these equations are simpler than those performed when solving the differential equations of something

120   CHAPTER 5. STRANGE ATTRACTORS

   like the Lorenz attractor. The practical result of this is that for a
   given number of iterations, the programs will run more quickly.

2. Because we are simply iterating the equations, and not using any
   approximation techniques like Euler or Runge-Kutta, the accuracy of
   the iterations is better. Therefore we will be able to see the finer
   detail of the attractor with greater ease, as we won't have to worry
   about whether what we're viewing is an artifact of the process used
   to solve the differential equations.

The default values in Listing 5-1 will draw the Hénon attractor. The
scaling factors and offsets used are for a VGA screen in the PC version of
the program. You might like to prove for yourself that values much greater
than 1.4 cause some points to escape from the attractor. In our program,
any point with x or y values exceeding 2E6 is considered to have escaped
the attractor, and is bound for infinity, which is the other attractor for
this system. When running the program, a beep indicates that this has
happened. The program will continue plotting points until you press a
key. For $a = 1.4$, you'll find that a range of x and y values will all still
generate the Hénon attractor, thus confirming the fact that the shape drawn
(Figure 5-1) is, in fact, an attractor. Now, if you think back to the Lorenz

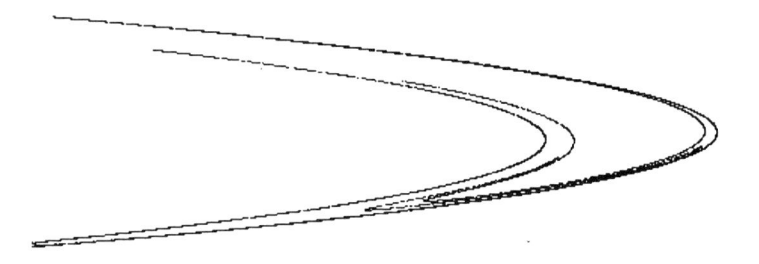

Figure 5-1a. Plot of the Hénon attractor.

attractor, you'll probably remember that a peculiarity of strange attractors
is that points on an attractor that are very close to each other in phase space
(i.e. have similar x and y values) can generate trajectories that are very
different, giving the strange attractor a high sensitivity to initial conditions.
In the Hénon attractor, we can actually see why this is so. Try viewing a
close up of the attractor in Figure 5-1a, with

   X Scaling=2000, Y Scaling=2000, X Offset=200, Y Offset=−200

THE HÉNON ATTRACTOR

Figure 5-1b. Plot of the Hénon attractor.

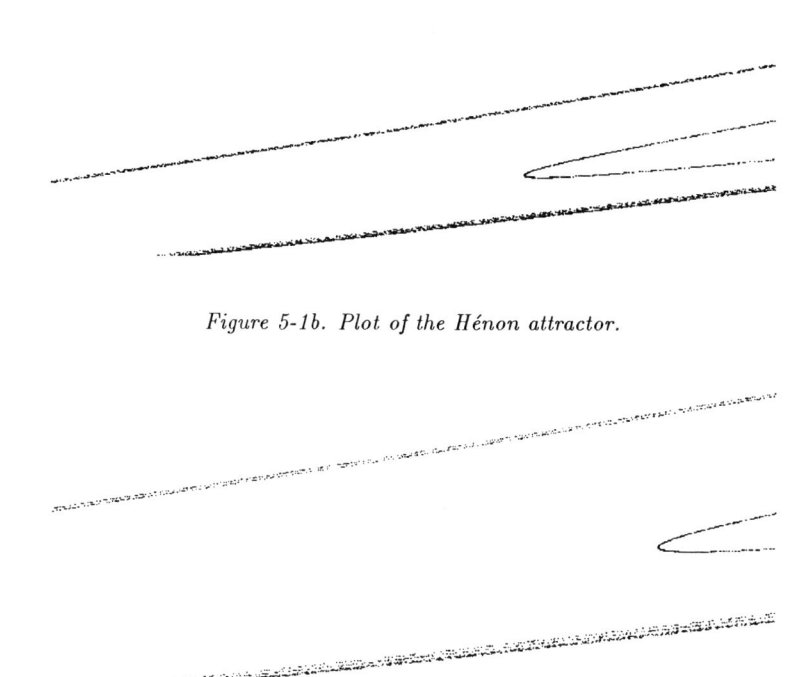

Figure 5-1c. Plot of the Hénon attractor.

and you will see that what looked like single lines in our first view of the attractor now appear to be splitting into two or more lines. This is shown in greater detail in Figure 5-1b:

X Scaling 4000, Y Scale=4000, X Offset=100, Y Offset=−800.

Let the program run for a few minutes until sufficient points have been plotted for you to see the structure. If you were to repeat the process, and gain a greater magnification of the system, then you would find this pattern repeating itself—each line that is seen, representing a trajectory, splits down into finer lines. This is exactly what we supposed happened in the Lorenz attractor. Now, let's gather some facts together here:

1. The values of x and y that fall within the Hénon attractor are all within $-1.5 < x < 1.5$ and $-0.4 < y < 0.4$. If we were to view this area as a plane, then we would clearly have a finite area that is defined by these x and y values.

2. The Hénon attractor can be viewed as a series of points which effectively form trajectories that are followed by sequential (x, y) pairs. Although the final attractor looks like it's a line, the attractor isn't drawn out as a line; points are dotted around the attractor, seemingly at random. However, when the final attractor is drawn we have the appearance of an infinite number of lines as we magnify portions of the attractor. What we have here is *self-similarity;* the system looks the same at whatever magnification you care to apply to it.

These two facts define some further characteristics about a strange attractor—apart from it being a point to which trajectories are attracted, a strange attractor occupies a finite amount of mathematical space and can exhibit self-similarity if you try to magnify different sections of the attractor. Because of the self similarity at all scales within the attractor, it's possible to cram an awful lot of trajectories into a finite area of phase space. This gives us the sensitive dependence on initial conditions in many cases, due to the fact that only a small difference in initial x and y values will select totally different trajectories for those initial values to follow, as we saw with the Lorenz attractor.

Fractal dimensions and strange attractors

This point about self similarity is very important, as it introduces a new concept called the *fractal dimension.* In the real world, we're quite used to saying that something has so many dimensions. Mathematically, we define mathematical objects as having dimensions in the following way:

One dimension: A typical one-dimensional object is a curve in space. It's got just one dimension, which we call *length.* Although when we draw a line we know that it has width as well, in mathematical terms a perfect line has no width, just length. We've dealt with one-dimensional objects quite extensively, when dealing with trajectories in phase space. A trajectory has no width.

Two dimensions: A two-dimensional object is a surface; this has *length* and *width*, but no depth. An example of a surface that we've dealt with is two-dimensional phase space itself, as we saw with the pendulum equations in Chapter 3.

Three dimensions: A three-dimensional object has *length, width* and *depth,* and thus has *volume.*

If we look at a line that curls around itself, yet never crosses itself, and gradually fills up the plane on which we're drawing it—say the computer screen—then that line has at least some of the properties of a two-dimensional surface; it has length, and because of the fact that it is filling a surface it appears to have width as well. However, we know that it's just a line with only one dimension. Such a line is often called a space-filling curve and we've clearly got an example of this with our strange attractor. There we have a line which, at sufficient magnification, is made up of other lines. Thus, at a high enough magnification the line, a one-dimensional object, appears to have two dimensions! This is a little odd, and is described by the term *Hausdorff-Besicovitch dimension*. Whereas the normal dimension is an integer, the HB dimension can be a non-integer; for example, a line having an HB dimension of 1.5 is such that it is better at filling a plane than a normal curve, like a parabola, but is not as good at filling a plane as a surface would be. Any object with an HB dimension different to its normal dimension is said to be *fractal*, a name coined by Benoit Mandelbrot when he started work on these rather odd objects. A further definition of a fractal object is one that shows self-similarity at any level of magnification. So our Hénon attractor is definitely a fractal object, showing self-similarity as we magnify portions of it.

We'll learn more about space-filling curves and fractals in the next chapter, but for now we simply need to note that strange attractors can have a fractal dimension. However, to complicate matters just a bit, not all strange attractors have a fractal dimension, and not all attractors with a fractal dimension are strange.

Other Hénon attractors

Hénon's main area of interest, apart from the exploration of the very guts of strange attractors, was the mathematics of orbits of such things as stars around the centre of the galaxy, asteroids, and other such things. He was particularly interested in computer simulations of what was called the *KAM Theorem*, which is named after the three mathematicians who developed it, Kolmogorov, Arnold and Moser. This theory looked at what happened to stable dynamic systems when they are disturbed in some way. For example, a satellite in orbit around the Earth might have an essentially elliptical orbit, but will often be perturbed by the gravitational effects of the moon, solar wind changes, etc. The KAM theory attempts to see what perturbations to such an orbit will lead to chaos. To simulate this, Hénon took a pair of equations:

$$x_{n+1} = x_n * \cos(a) - (y_n - x_n * x_n) * \sin(a)$$
$$y_{n+1} = x_n * \sin(b) - (y_n - x_n * x_n) * \cos(b)$$

and iterated them. In his work Hénon allowed the values of a and b to be the same.

Listing 5-2 shows a program for iterating this Hénon map. The program plots the mapping for set a and b values for a range of initial x and y values.

Typical examples of the output from the program are shown in Figure 5-2. The essence of these mappings is that outside of the central, ordered area of the mapping chaos reigns with islands of order for certain values of x, y as initial values. You might like to investigate these areas of order, and examine them in greater detail.

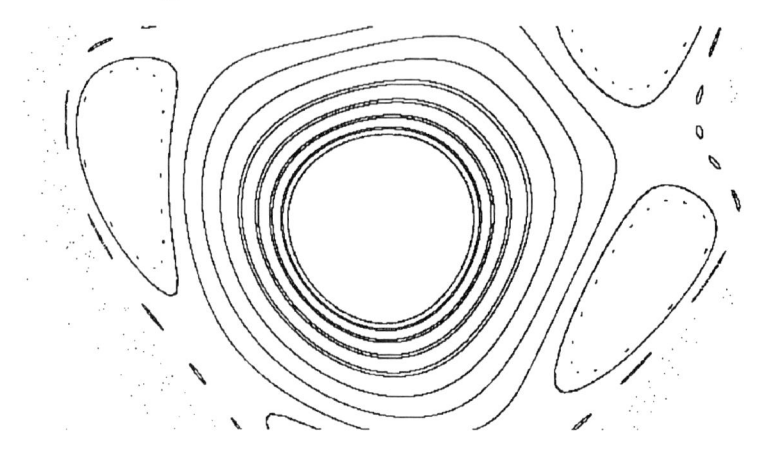

Figure 5-2a. Plot of a Hénon map.

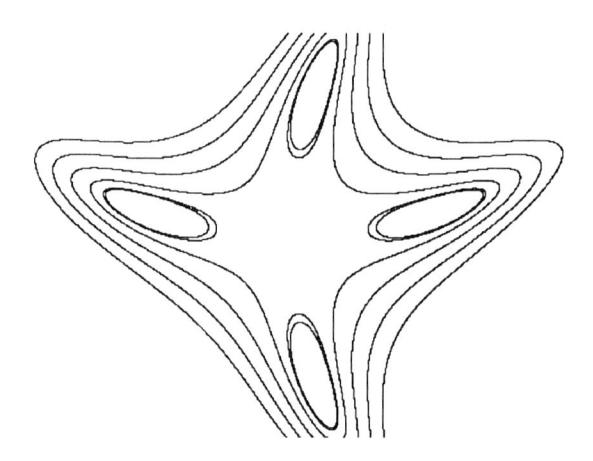

Figure 5-2b. Plot of a Hénon map.

You might like to experiment with derivatives of these mappings. For example, try giving a and b different values. In addition, you might like to try the following pairs of equations in Listing 5-2 rather than those given. They simply replace the two lines containing the equations in Listing 5-2.

STRANGE ATTRACTORS FROM DIFFERENTIAL EQUATIONS

$$x_{n+1} = y_n * \cos(a) - (y_n - x_n * x_n) * \sin(a)$$
$$y_{n+1} = x_n * \sin(b) - (y_n - x_n * x_n) * \cos(b)$$

$$x_{n+1} = y_n * \cos(a) - (y_n - x_n * \cos(x_n))$$
$$y_{n+1} = x_n * \sin(b) + (y_n - x_n * \sin(x_n))$$

$$x_{n+1} = y_n * \cos(a) - (y_n - x_n * \sin(x_n))$$
$$y_{n+1} = x_n * \cos(b) + (y_n - x_n * \cos(x_n))$$

The latter two equations are best displayed with a lower x and y scaling; about 30 or 40 is good. Apart from mappings, there are, as we've already seen, other ways of generating strange attractors, so let's look at these.

Strange attractors from differential equations

The Lorenz equations generate a strange attractor from a set of differential equations. There are several other systems that can do this as well, so I would like to give you the basic details of putting these systems into a simple program to investigate the behaviour of strange attractors. In programming terms, the main points to remember about writing programs to generate strange attractors from differential equations are:

1. Use a method of solving the equations that is as accurate as possible. The Runge-Kutta method is to be preferred over Euler, for example.

2. Use a small value for d, but be wary of making it too small and producing a slow program. Large values of d can also cause problems with accuracy; I've found 0.001 to 0.01 give good starting values of d.

3. Start off by exploring these systems with the values given. Some systems of equations are prone to 'blowing up' and causing numerical overflow with parameters outside a given range of values.

4. In three-dimensional systems, like the Lorenz attractor, you will have to decide which pair of parameters to plot as a phase portrait, unless you decide to try a projection of the three-dimensional phase portrait.

5. In many systems, you may need to allow the transients in the system to die out before plotting the attractor; this is easily done by only plotting points after a specified number of iterations. Failure to do this will obscure the picture of the attractor.

6. Use point-plotting rather than line drawing commands; line-drawing commands can often give a false impression of what's happening. In some cases, line-drawing commands will give an accurate picture of

what the attractor looks like, but only if point $n+1$ is very close to point n.

7. Colour can be useful. I find it interesting to change colour every few hundred iterations.

8. The scaling of the attractors depends very much on the attractor itself, the values used in the equations as constants and initial values and also the type of display used. You may find it useful to experiment with different scaling factors to get an image that fills the screen as much as possible.

The Rossler equations

The Rossler attractor is generated by three differential equations like the Lorenz attractor, but exhibits simpler but more revealing behaviour. The equations are:

$$\frac{dx}{dt} = -(y+z)$$
$$\frac{dy}{dt} = a*y+x$$
$$\frac{dz}{dt} = b+r*x-r*c$$

Again, a, b and r are constants of the system, and I've dèliberately labelled them to be the same as the corresponding constants in the Lorenz equations and r can be seen as the parameter that controls whether or not the system goes in to chaos.

Listing 5-3 shows a Runge-Kutta solution for the Rossler equations, and a good start point is for you to set $x = y = z = 1$, $a = b = 0.2$ and $r = 8$. This will exhibit chaotic behaviour (Figure 5-3) after the transients are allowed to die away (a couple of thousand iterations will suffice). What is interesting about the Rossler system is that if you set the value of r to 2 or so, a single limit cycle can be seen. Increasing the value of r leads to a series of period doublings until chaos finally sets in. This is very familiar; the period doubling mechanism was the way by which the logistic equation became chaotic, so the Rossler system indicates to us graphically the link between the chaotic graphs obtained by iterating simple equations and strange attractors.

You may like to try setting a and b to other values and increasing the value of r still further; this will give a shape known as the *Rossler funnel*, which is shown in Figure 5-4. This program can be modified to accept any set of differential equations that do not have a forcing term (see below). If you're interested in this set of equations, a rather nice paper on the Rossler

STRANGE ATTRACTORS FROM DIFFERENTIAL EQUATIONS 127

Figure 5-3. Plot of a solution to the Rossler equations.

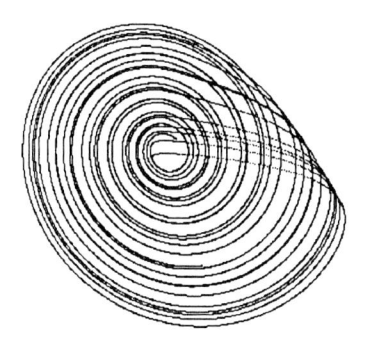

Figure 5-4. The Rossler funnel.

system is 'Power Spectral Analysis of a Dynamical System', by Crutchfield, Farmer, Packard, Shaw, Jones and Donnelly, *Physics Letters* 1980.

The forced Duffing oscillator

This is an example of a forced, periodic system and was first put forward by Ueda. This is a system where a naturally oscillating system is kicked occasionally. If these kicks arrive at the right time and are of the right magnitude then the system can go from being a well-behaved periodic system to a chaotic one. The Duffing equations are classical equations in physics that describe the activity of an oscillating system in the real world where friction or other losses takes place. It thus has relevance to physicists as a real problem, rather than a mathematical curiosity like the Rossler equations. The forced Duffing system is where a forcing signal is applied. The equations for this system are:

$$\frac{dx}{dt} = y$$

$$\frac{dy}{dt} = -a*y - x*x*x + b*\cos(c*t)$$

and the program in Listing 5-4 will generate the attractor for this system.

The program can be modified to accommodate any system of equations that contains a forcing term. Here, it's $b*\cos(t)$, where t is a measure of time and the cos function is used to generate a periodic forcing waveform from the t variable, which is incremented by d for each iteration. The constant b controls the magnitude of the forcing waveform, and c controls the frequency of the forcing waveform. With systems like this, it is again important to allow all transients to die out by iterating the equations several times before plotting the points.

Other strange attractors

Other systems of equations have been examined and found to contain strange attractors. These include things such as models of the mechanism that produces the Earth's magnetic field, the equations describing the kinetics of chemical reactions and even aspects of the way in which your body chemistry works! In the sections that follow, I'll list a few examples of these systems.

The Earth's magnetic field

The set of equations put forward by Chillingworth and Holmes to model the mechanism that creates the Earth's magnetic field can be reduced to a set of three equations that give rise to a strange attractor very similar to the Lorenz attractor. These equations are:

$$\frac{dx}{dt} = a*y - a*x$$

$$\frac{dy}{dt} = z*x - y$$

$$\frac{dz}{dt} = b - x*y - z*r$$

If you put these equations in to the Lorenz equation programs from the last chapter, then a zy plot of the attractor will give a similar picture to

OTHER STRANGE ATTRACTORS 129

the Lorenz attractor. Some typical values quoted in *Chaos*, edited by A. V. Holden, for generating these areas are $a = 14.6$, $b = 1$ and $r = 5$. This generates x, y and z values in the range -12 to $+12$. As can be seen from the two-lobed attractor, a trajectory will flip from one lobe to the other as time passes. Interestingly enough, this is the sort of behaviour that the Earth's magnetic field exhibits over thousands of years, with the north and south magnetic poles flipping over every now and again, which is analogous to the trajectory flipping between lobes of the attractor.

The Van der Pohl oscillator

The Van der Pohl oscillator is a model of a forced electronic oscillator, described by the following equations:

$$\frac{dx}{dt} = -a*y + x*(1 - r*r)$$

$$\frac{dy}{dt} = a*x + y*(1 - r*r) + f$$

where a and r represent parameters of the electronic circuit under examination and f is a term representing the forcing signal—a bit like the $b*\cos(t)$ term we met earlier when discussing the forced Duffing system.

It's quite important to realise that there are real-world systems around that contain strange attractors in them, as well as mathematical models that describe these systems. Let's look briefly at a couple of systems that have provided experimenters with useful information about chaos.

Cardiac rhythms

We tend to look at the heart as a regular, rhythmic pump, and for most of the time that's a fair enough assessment. However, there are occasions when the regularity of the heartbeat becomes disrupted by chaotic behaviour. The heart is a naturally oscillating system, beating between 50 and 100 times per minute under most conditions. However, on occasion, *cardiac arrhythmias* can occur where the regular action of the heart is disrupted. We've already seen how we might do this for other oscillating systems by providing a forcing signal. The heart muscle contraction is coordinated by an area known as the *pacemaker*, which basically acts as timekeeper for the heart. It's thought that chaotic cardiac activity might originate from other parts of the heart muscle acting as pacemakers and sending out signals which are out of synchronisation with the proper pacemaker. This chaotic activity of heart tissue, known as fibrillation, might also be caused by problems in the conduction of electrical signals across the surface of the heart resulting in parts of the heart starting off a new beat before it's really ready.

Disease epidemics

As we might expect from our previous experiments on the logistic equation, disease spread can exhibit chaotic dynamics. Some of the classic work in this field was done by studying the numbers of cases of measles amongst citizens of New York over a period of time. The system of equations used is based on what is called the SIR model:

$$\frac{dS}{dt} = -r*S*I$$

$$\frac{dI}{dt} = r*S*I - a*R$$

$$\frac{dR}{dt} = a*I$$

where S is the number of susceptible people in the population, I the people carrying the disease who can spread it to others—infectives—and R is the number of people who have been in some way removed from the scenario (either dead, recovered and immune, in isolation hospitals, naturally immune, etc.), r is a constant, the infection rate, that determines the rate at which people are infected, and a is the removal rate, which represents the rate at which individuals are removed from the scenario. The actual model used in the New York measles study was more complex than this, in that it used an additional parameter, the number of people exposed to the illness but not yet infectious. This models the incubation period of the disease, which in the first model is not taken in to account. A further feature added to the disease model was a modified infection rate $b*\cos(t)$, where t is the time into the epidemic being modelled. This will act to make the infection rate periodic.

Arms races

A final model, discussed in some detail in *Modelling with differential equations*, by Burghes and Borrie, is designed to model the arms race between two alliances, countries or superpowers. You might like to model this and see what you can find. The system of equations is:

$$\frac{dx}{dt} = k_1*y - c_1*x + g_1$$

$$\frac{dy}{dt} = k_2*x - c_2*y + g_2$$

where x and y are a measure of the ability of the nations to fight wars. The rate at which this changes can be estimated as being dependant upon the current perceived threat from the other nations, as given by the k_1*y and k_2*y parts of the equation, the limiting factor placed by arms expenditure, etc., given by c_1*x and c_2*y and a constant term described as 'underlying grievances' between the two countries. Here, g_1 is a measure of the grievances felt by country 1 against country 2.

STRANGE ATTRACTORS FROM REAL DATA

You might like to add other terms here; how about adding a term to model the deterrence effect of modern weapons held by one side but not the other? Or another term to indicate the attitude of the people (added if the people are willing to fight, subtracted if they aren't). The depressing aspect of this model is that it's very easy to find instabilities which lead to x and y shooting off to infinity, which would indicate an ever escalating arms race and eventual war.

Strange attractors from real data

We've so far looked at searching for strange attractors in systems that are actually mathematical models of the real world. However, there's nothing to stop us looking at real data and treating it in a similar way to that already discussed to see if we can spot any strange attractors. For example, if we had the means to do so we could measure the angular velocity and displacement from a centre point of a *real* pendulum, and plot the values on a graph. Of course we wouldn't expect to see phase portraits as clean cut and clear as those from the mathematical models, but we would expect to be able to see the pictures of any strange attractors present.

So far, when discussing phase portraits we've looked at systems of equations or real-world situations with two or more variables. In the case of two variables, we simply plot one against the other. In the case of systems of a higher order, we can either create a two-dimensional representation of a three-dimensional system, or plot just two of the variables of the system on the screen. What if we have a system that only has one variable and we wish to examine it to see if a strange attractor is present? Clearly, just plotting the value of x against time wouldn't do the trick; that would simply produce the time progression for the system. For example, imagine the dripping tap system described briefly in Chapter 11. Here we don't have a mathematical model to give us any data; what we do have is a series of time intervals between water drops leaving the tap. What we can do is to create some more *observable parameters* from this simple set of data points. For example, let's suppose we've got a set of time intervals of:

$$1.2 \quad 2.1 \quad 1.3 \quad 4.5 \quad 2.3$$

Now, suppose you think that there might be a two-dimensional strange attractor lurking in the data somewhere. What we do is take pairs of these intervals, and plot them. The pairs plotted would be as follows:

$$(1.2, 2.1) \quad (2.1, 1.3) \quad (1.3, 4.5) \quad (4.5, 2.3)$$

and so on for the whole set of data. These pairs of values correspond to the x, y pairs we've seen in other systems. If we wanted to look for a three-dimensional attractor in the data, then we'd assemble sets of three points, which here would be:

$$(1.2, 2.1, 1.3) \quad (2.1, 1.3, 4.5) \quad (1.3, 4.5, 2.3)$$

and these would correspond to the x, y and z values found in systems like the Lorenz attractor. This method, often known as creating false observables, was developed initially by Robert Shaw in the United States for use with his dripping tap experiment. It is a very powerful tool for pulling attractors out of the noise of experimental data. In an apparently random set of values from a real system, you might find an attractor, which would appear on a graph generated from the data as a concentration of dots in specific areas of the graph; truly random data would give a scatter of dots across the graph. This has proved to be quite a valuable experimental tool. For example, workers have examined such things as Stock Market prices and other economic indicators using this technique to see if there is some sort of strange attractor in the economic system.

How can strange attractors be found?

To finish off this chapter, I'll summarise briefly the factors to look at when you're investigating a set of equations on a computer for chaos. The first thing to point out is that there are analytical methods of examining sets of equations and determining whether the system they represent will be chaotic or not. However, that's rather beyond the scope of this book, so I'll simply provide some pointers for examining systems of equations on the computer.

1. The more interacting variables there are in a system of differential equations, the higher the possibility of chaotic behaviour in that system.

2. Don't forget that whether a system is chaotic or not often depends upon the value of one or more constants in the equations. We saw this in the case of the Lorenz equations, where the value of r made a considerable difference to whether the system showed a strange attractor. You may need to vary the values of constants, therefore, to see if a strange attractor is generated.

3. When you come across a set of equations, look at them carefully to see if they are anything like the equations we've seen in this and previous chapters that do show chaotic behaviour. For example, a brief examination of the equations describing the Earth's magnetic field showed them to be of similar form to the Lorenz equations, so we might expect to find some chaotic behaviour.

HOW CAN STRANGE ATTRACTORS BE FOUND?

4. For some equation sets, you may see that the graphs drawn all look like a single-point attractor. Of course, it might be that this is the case for all initial variable values over a wide range of constants in the equations. Alternatively, it may be that the x and y scaling chosen to display the graph are too small; try increasing the scaling factors.

5. A system that is driven by another system is quite likely to exhibit chaos in certain circumstances, as in the case of the forced oscillators that we've already met.

6. If you are processing data from real systems, then don't forget that you may need to use the fake observable technique to realise any attractor present. A practical example of this is shown in Chapter 11.

BBC BASIC listings

Listing 5-1

```
 10 REM Program HenonAttractor
 50 REPEAT
 70    REM Try a=1.4, b=0.3, x=y=1, xscale=yscale=xoffset=yoffset=500
 90    MODE 6
100    PRINT "Henon Attractor"
110    INPUT "Start X ",x
120    INPUT "Start Y ",y
130    INPUT "Value of A ",a
140    INPUT "Value of B ",b
150    INPUT "X Scaling ",xs
160    INPUT "Y Scaling ",ys
170    INPUT "X Offset  ",xc
180    INPUT "Y Offset  ",yc
200    MODE 5
220    REPEAT
240      REM Next two lines evaluate the (r+1) state of x and y.
260      x1=y+1-(a*x*x)
270      y1=b*x
300      REM Check values of x1,y1 to avoid overflow
320      IF ABS(x1)<2E6 THEN x=x1
330      IF ABS(y1)<2E6 THEN y=y1
350      REM Now plot the point if within limits set - beep if outside set
360      REM limits
380      IF (ABS(x1)<2E6) AND (ABS(y1)<2E6) THEN PLOT69,(xc+x1*xs),(yc+y1*ys)
         ELSE PRINT CHR$(7)
420    UNTIL (INKEY(1)<>-1)
430    *FX15,1
440    CH$=GET$
450 UNTIL (ch$="X") OR (ch$="x")
460 END
```

Listing 5-2

```
 10 REM Program HenonMaps
 70 MODE 6
 80 REM Try x scale=y scale=400, a=b=1.3
100 INPUT "Value of A:  ",a
110 INPUT "Value of B:  ",b
120 INPUT "X Scale     ",xs
130 INPUT "Y Scale     ",ys
140 MODE 5
150 xc=500
160 yc=500
180 REM This program will loop around until all the x,y pairs of initial
190 REM values are dealt with, or until you press a key. There will be
200 REM a delay after the key press in some cases until the inner loop is
210 REM executed
230 starty=-0.2
240 REPEAT
260   startx=-0.2
270   x=startx
280   y=starty
290   REPEAT
320     REM Now evaluate evaluate the (n+1) state of x and y 1200 times.
340     FOR g=1 TO 1200
360       x1=x*COS(a)-(y-x*x)*SIN(a)
370       y1=x*SIN(b)+(y-x*x)*COS(b)
400       REM Check values of x1,y1 to avoid overflow
```

```
420        IF ABS(x1)<2E6 THEN x=x1
430        IF ABS(y1)<2E6 THEN y=y1
450        REM Now plot the point if within limits set - beep if outside set
460        REM limits
480        IF (ABS(x1)<2E6) AND (ABS(y1)<2E6) THEN
490          PLOT69,(xc+x1*xs),(yc+y1*ys)
520        NEXT
530        REM of number of iterations
540        startx=startx+0.05
550        T=INKEY(1)
560      UNTIL (startx>0.8) or (T<>-1)
570      starty=starty+0.05
580    UNTIL (starty>0.8) or (T<>-1)
600    G=GET
610    END
```

Listing 5-3

```
 10 REM Program RosslerAttractor
 30 REM    Draws the Rossler Attractor }
 50 REM Try a=0.2, b=0.2, r=2, x=y=z=1
 70 d=0.005
 80 a=0.2
 90 b=0.2
110 REM Allow user to enter new values
130 MODE 6
140 INPUT "X Value ",x
150 INPUT "Y Value ",y
160 INPUT "Z Value ",z
170 INPUT "R Value ",r
180 INPUT "Display After ",DisplayAfter
200 MODE 5
230 REM n hold number of time increments passed
250 n=1
270 REPEAT
300    REM Only display a point on the attractor afterthe DisplayAfter
310    REM variable is exceeded. Change the PutPixel line to change
320    REM the view of the attractor that is displayed.
340    IF n>DisplayAfter THEN PLOT69,(x*50)+500,(y*50)+500
360    REM Now start the Runge-Kutta solution, first for x
380    xinc1=FNEquationX(x,y,z)*d
390    xinc2=FNEquationX(x+xinc1/2,y+d/2,z+d/2)*d
400    xinc3=FNEquationX(x+xinc2/2,y+d,z+d)*d
410    xinc4=FNEquationX(x+xinc3,y+d,z+d)*d
430    REM Now repeat for the Y Equation
450    yinc1=FNEquationY(x,y,z)*d
460    yinc2=FNEquationY(x+yinc1/2,y+d/2,z+d/2)*d
470    yinc3=FNEquationY(x+yinc2/2,y+d,z+d)*d
480    yinc4=FNEquationY(x+yinc3,y+d,z+d)*d
500    REM Finally for z
520    zinc1=FNEquationZ(x,y,z)*d
530    zinc2=FNEquationZ(x+xinc1/2,y+d/2,z+d/2)*d
540    zinc3=FNEquationZ(x+xinc2/2,y+d,z+d)*d
550    zinc4=FNEquationZ(x+xinc3,y+d,z+d)*d
570    REM Only update x,y and z at this stage, as the unaltered
580    REM values of x,y and z are needed for calculations of
590    REM increments to x,y and z above.
620    x=x+(xinc1+2*xinc2+2*xinc3+xinc4)/6
630    y=y+(yinc1+2*yinc2+2*yinc3+yinc4)/6
640    z=z+(zinc1+2*zinc2+2*zinc3+zinc4)/6
660    REM Increment passing of time....
680    n=n+1
710    REM Repeat until a key is pressed
```

136   *CHAPTER 5. STRANGE ATTRACTORS*

```
730 UNTIL (INKEY(1)<>-1)
750 PRINT CHR$(7)
760 G=GET
780 END
820 DEFFNEquationX(xi,yi,zi)=-(yi+zi)
830 DEFFNEquationY(xi,yi,zi)=xi+a*yi
840 DEFFNEquationZ(xi,yi,zi)=b+zi*(xi-r)
```

Listing 5-4

```
 10 REM Program RungeKuttaForcedDuffingOscillator
 20 REM Program to draw Forced Duffing Oscillator equations
 30 REM by the Runge-Kutta method for use in the Chaos book.
150 d=0.005
170 REM Try Transient=100, a=0.2, b=20, c=1
190 t=0
220 REM  Allow user to enter new values
260 MODE 6
270 INPUT "Value of a? ",a
280 INPUT "Value of b? ",b
290 INPUT "Value of c? ",c
300 INPUT "Transient delay time? ",transient
320 REM  Set graphics mode up
340 MODE 5
350 REM Initialise x and y values
370 x=1
380 y=1
420 REPEAT
460    REM Only plot after the transients have decayed
480    IF (t>transient) THEN PLOT69,(x*20)+500,(y*20)+500
500    REM Now start the Runge-Kutta solution, first for x
520    xinc1=FNEquationX(x,y)*d
530    xinc2=FNEquationX(x+xinc1/2,y+d/2)*d
540    xinc3=FNEquationX(x+xinc2/2,y+d)*d
550    xinc4=FNEquationX(x+xinc3,y+d)*d
570    REM Now repeat for the Y Equation
590    yinc1=FNEquationY(x,y)*d
600    yinc2=FNEquationY(x+yinc1/2,y+d/2)*d
610    yinc3=FNEquationY(x+yinc2/2,y+d)*d
620    yinc4=FNEquationY(x+yinc3,y+d)*d
660    x=x+(xinc1+2*xinc2+2*xinc3+xinc4)/6
670    y=y+(yinc1+2*yinc2+2*yinc3+yinc4)/6
690    REM  Increase value of t
710    t=t+d
740    REM  Repeat until a key is pressed
760 UNTIL (INKEY(1)<>-1)
780 PRINT CHR$(7)
790 G=GET
820 END
860 DEFFNEquationX(xi,yi)=yi
880 DEFFNEquationY(xi,yi)=-a*yi-xi*xi*xi+b*COS(c*t)
```

Turbo Pascal listings

Listing 5-1

```
Program HenonAttractor;

Uses Crt,Graph,Extend;

Var
    d,x,x1,y1                :    Real;
    y,xc,yc,xs,ys,a,b        :    Real;
    g                        :    Integer;
    ch                       :    Char;
    xt,yt,is,sx,sy           :    String;
    yst,xst,ycs,xcs,as,bs    :    String;
Begin
  sx:='1';
  sy:='1';
  xst:='200';
  yst:='200';
  xcs:='200';
  ycs:='100';
  as:='1.4';
  bs:='0.3';

  Repeat
   TextBackground(Blue);
   ClrScr;
   PrintAt(2,2,'Henon Attractor, Implementation Joe Pritchard, 1989',Yellow);
   PrintAt(2,6,'Start X:',Green);
   StringEdit(sx,20,20,6,Yellow);
   Val(sx,x,g);
   PrintAt(2,7,'Start Y:',Green);
   StringEdit(sy,20,20,7,Yellow);
   Val(sy,y,g);
   PrintAt(2,8,'A:',Green);
   StringEdit(as,20,20,8,Yellow);
   Val(as,a,g);
   PrintAt(2,9,'B:',Green);
   StringEdit(bs,20,20,9,Yellow);
   Val(bs,b,g);

   PrintAt(2,11,'X Scaling:',Green);
   StringEdit(xst,20,20,11,Yellow);
   Val(xst,xs,g);
   PrintAt(2,12,'Y Scaling:',Green);
   StringEdit(yst,20,20,12,Yellow);
   Val(yst,ys,g);
   PrintAt(2,13,'X Offset:',Green);
   StringEdit(xcs,20,20,13,Yellow);
   Val(xcs,xc,g);
   PrintAt(2,14,'Y Offset:',Green);
   StringEdit(ycs,20,20,14,Yellow);
   Val(ycs,yc,g);

   ChoosePalette;

   Repeat
      Begin
{ Next two lines evaluate the (r+1) state of x and y.           }

         x1:=y+1-(a*x*x);
```

```
            y1:=b*x;

{ Check values of x1,y1 to avoid overflow }

            if abs(x1)<2E6 then
              x:=x1;
            if abs(y1)<2E6 then
              y:=y1;

{           Now plot the point if within limits set - beep if outside set
            limits     }

            if (abs(x1)<2E6) and (abs(y1)<2E6) then
              PutPixel(trunc(xc+x1*xs),trunc(yc+y1*ys),2)
            else
              Begin
                Sound(100);
                Delay(45);
                NoSound;
              end;

        end;
    until (KeyPressed);
    ch:=readkey;
    CloseGraph;
 until (ch='X') or (ch='x');
end.
```

Listing 5-2

```
Program HenonMaps;

Uses Crt,Graph,Extend;

Var
   x1,y1,x,y              :      Real;
   xs,ys,xc,yc,a,b        :      Real;
   g                      :      Integer;
   ch                     :      Char;
   xt,yt                  :      String;
   yst,xst,as,bs          :      String;
   startx,starty          :      Real;

Begin
  xst:='400';
  yst:='400';
  as:='1.3';
  bs:='1.3';

  TextBackground(Blue);
  ClrScr;
  PrintAt(2,2,'Henon Mapping, Implementation Joe Pritchard, 1990',Yellow);
  PrintAt(2,8,'A:',Green);
  StringEdit(as,20,20,8,Yellow);
  Val(as,a,g);
  PrintAt(2,9,'B:',Green);
  StringEdit(bs,20,20,9,Yellow);
  Val(bs,b,g);

  PrintAt(2,11,'X Scaling:',Green);
  StringEdit(xst,20,20,11,Yellow);
  Val(xst,xs,g);
  PrintAt(2,12,'Y Scaling:',Green);
```

TURBO PASCAL LISTINGS

```
    StringEdit(yst,20,20,12,Yellow);
    Val(yst,ys,g);

    ChoosePalette;
    xc:=trunc(GetMaxX/2);
    yc:=trunc(GetMaxY/2);

{ This program will loop around until all the x,y pairs of initial
  values are dealt with, or until you press a key. There will be
  a delay after the key press in some cases until the inner loop is
  executed }

    starty:=-0.2;
    Repeat
      Begin
        startx:=-0.2;
        x:=startx;
        y:=starty;
        Repeat
          Begin

{ Now evaluate evaluate the (n+1) state of x and y 1200 times. }

            for g:=1 to 1200 do begin

              x1:=x*cos(a)-(y-x*x)*sin(a);
              y1:=x*sin(b)+(y-x*x)*cos(b);

{ Check values of x1,y1 to avoid overflow }

              if abs(x1)<2E6 then
                x:=x1;
              if abs(y1)<2E6 then
                y:=y1;

{         Now plot the point if within limits set - beep if outside set
          limits           }

              if (abs(x1)<2E6) and (abs(y1)<2E6) then
                PutPixel(trunc(xc+x1*xs),GetMaxY-trunc(yc+y1*ys),2);

            end;         { of number of iterations }
          end;
          startx:=startx+0.05;
          until (startx>0.8) or (KeyPressed);
      end;
      starty:=starty+0.05;
      until (starty>0.8) or (KeyPressed);

    ch:=readkey;
    CloseGraph;
end.
```

Listing 5-3

```
Program RosslerAttractor;

{       Draws the Rossler Attractor }

Uses Crt,Graph,Extend;

Var
```

```
  x,y,z                     :       Real;
  xinc,yinc,zinc            :       Real;
  xinc1,xinc2,xinc3,xinc4:          Real;
  yinc1,yinc2,yinc3,yinc4:          Real;
  zinc1,zinc2,zinc3,zinc4:          Real;
  xi,yi                     :       Real;
  a,b,r                     :       Real;
  xs,ys,zs,rs               :       String;
  temp                      :       Char;
  d                         :       Real;
  n                         :       Integer;
  DisplayAfter              :       Integer;
  DisplayAfters             :       String;

Function EquationX(xi,yi,zi : real) : real;
Begin
   EquationX:=-(yi+zi);
end;

Function EquationY(xi,yi,zi : real) : real;
Begin
   EquationY:=xi+a*yi;
end;

Function EquationZ(xi,yi,zi : real) : real;
Begin
   EquationZ:=b+zi*(xi-r);
end;

Begin

   d:=0.005;
   a:=0.2;
   b:=0.2;
   r:=2;
   rs:='2';

   x:=1;
   xs:='1';
   y:=1;
   ys:='1';
   z:=1;
   zs:='1';

   DisplayAfters:='1';
   DisplayAfter:=1;

{  Allow user to enter new values
}

   TextBackground(Blue);
   ClrScr;
   PrintAt(2,4,'Value of x? ',Green);
   StringEdit(xs,20,50,4,Yellow);
   Val(xs,x,n);
   PrintAt(2,5,'Value of y? ',Green);
   StringEdit(ys,20,50,5,Yellow);
   Val(ys,y,n);
   PrintAt(2,6,'Value of z? ',Green);
   StringEdit(zs,20,50,6,Yellow);
   Val(zs,z,n);

   PrintAt(2,8,'Value of r? ',Green);
   StringEdit(rs,20,50,8,Yellow);
   Val(rs,r,n);
```

TURBO PASCAL LISTINGS   141

```
    PrintAt(2,9,'Display after? ',Green);
    StringEdit(DisplayAfters,20,50,9,Yellow);
    Val(DisplayAfters,DisplayAfter,n);

{ Set graphics mode up }

    ChoosePalette;

{ n hold number of time increments passed }

    n:=1;

    Repeat

        Begin

{       Only display a point on the attractor afterthe DisplayAfter
        variable is exceeded.  Change the PutPixel line to change
        the view of the attractor that is displayed.           }

        if n>DisplayAfter then begin

{       Modify if needed for your display screen      }

            PutPixel(trunc(x*10)+trunc(GetMaxX/2),(GetMaxY-trunc((y*10)+GetMaxY/2)),2);

        end;

{ Now start the Runge-Kutta solution, first for x }

            xinc1:=EquationX(x,y,z)*d;
            xinc2:=EquationX(x+xinc1/2,y+d/2,z+d/2)*d;
            xinc3:=EquationX(x+xinc2/2,y+d,z+d)*d;
            xinc4:=EquationX(x+xinc3,y+d,z+d)*d;

{ Now repeat for the Y Equation                     }

            yinc1:=Equationy(x,y,z)*d;
            yinc2:=Equationy(x+yinc1/2,y+d/2,z+d/2)*d;
            yinc3:=Equationy(x+yinc2/2,y+d,z+d)*d;
            yinc4:=Equationy(x+yinc3,y+d,z+d)*d;

{ Finally for z  }

            zinc1:=Equationz(x,y,z)*d;
            zinc2:=Equationz(x+xinc1/2,y+d/2,z+d/2)*d;
            zinc3:=Equationz(x+xinc2/2,y+d,z+d)*d;
            zinc4:=Equationz(x+xinc3,y+d,z+d)*d;

{ Only update x,y and z at this stage, as the unaltered
  values of x,y and z are needed for calculations of
  increments to x,y and z above.                    }

            x:=x+(xinc1+2*xinc2+2*xinc3+xinc4)/6;
            y:=y+(yinc1+2*yinc2+2*yinc3+yinc4)/6;
            z:=z+(zinc1+2*zinc2+2*zinc3+zinc4)/6;

{ Increment passing of time....}

            n:=n+1;

        end;

{ Repeat until a key is pressed                    }
```

```
until (keyPressed);

sound(200);
Delay(30);
NoSound;
temp:=Readkey;
CloseGraph;

end.
```

Listing 5-4

```
10 REM Program RungeKuttaForcedDuffingOscillator
20 REM Program to draw Forced Duffing Oscillator equations
30 REM by the Runge-Kutta method for use in the Chaos book.
150 d=0.005
170 REM Try Transient=100, a=0.2, b=20, c=1
190 t=0
220 REM   Allow user to enter new values
260 MODE 6
270 INPUT "Value of a? ",a
280 INPUT "Value of b? ",b
290 INPUT "Value of c? ",c
300 INPUT "Transient delay time? ",transient
320 REM  Set graphics mode up
340 MODE 5
350 REM Initialise x and y values
370 x=1
380 y=1
420 REPEAT
460    REM Only plot after the transients have decayed
480    IF (t>transient) THEN PLOT69,(x*20)+500,(y*20)+500
500    REM Now start the Runge-Kutta solution, first for x
520    xinc1=FNEquationX(x,y)*d
530    xinc2=FNEquationX(x+xinc1/2,y+d/2)*d
540    xinc3=FNEquationX(x+xinc2/2,y+d)*d
550    xinc4=FNEquationX(x+xinc3,y+d)*d
570    REM Now repeat for the Y Equation
590    yinc1=FNEquationY(x,y)*d
600    yinc2=FNEquationY(x+yinc1/2,y+d/2)*d
610    yinc3=FNEquationY(x+yinc2/2,y+d)*d
620    yinc4=FNEquationY(x+yinc3,y+d)*d
660    x=x+(xinc1+2*xinc2+2*xinc3+xinc4)/6
670    y=y+(yinc1+2*yinc2+2*yinc3+yinc4)/6
690    REM  Increase value of t
710    t=t+d
740    REM  Repeat until a key is pressed
760 UNTIL (INKEY(1)<>-1)
780 PRINT CHR$(7)
790 G=GET
820 END
860 DEFFNEquationX(xi,yi)=yi
880 DEFFNEquationY(xi,yi)=-a*yi-xi*xi*xi+b*COS(c*t)
```

Chapter 6

The fractal link

When chaotic dynamic systems were first being explored, *fractal geometry* wasn't thought to have much in common with the world of bifurcation diagrams or attractors. However, as we saw in the last chapter, there is a connection between the two, deep inside the structure of strange attractors. In this chapter, we'll take a brief detour to look at fractal geometry in a little more detail and explore some simple fractal shapes.

If you look around at natural and man-made objects, you will soon notice a few things. The first is that man-made objects are designed using essentially regular shapes, whereas natural objects are very irregular. As Benoit Mandelbrot pointed out in *The Fractal geometry of nature*, mountains are not shaped like cones, and clouds are not spheres. Cones, spheres, cubes and the like are all traditional, Euclidean geometry shapes, created by man for use in our own structures; there's nothing special about them. As well as the irregularity, there is what is best called a 'roughness' about natural objects. And, if you do look at natural objects from different distances, you'll see that they also exhibit self-similarity. That is, a mountain observed from a mile away will exhibit a similar jaggedness to part of the mountain peak observed at close quarters. This isn't exhibited, for example, by a sphere; if I look at a sphere from a distance, then look at it close up, there is a difference.

This concept of self-similarity is often described in terms of what's been called the *coastline of Britain* problem. If you start off with a typical atlas map of the world, and use various means to measure the perimeter of the coast of mainland Britain, then you'll get a particular value. Now, turn to the pages showing a larger scale map, and repeat the exercise. You'll find that the larger scale map shows more of what Slartibartfast, in the *Hitchhiker's guide to the galaxy*, by Douglas Adams, called the 'crinkly bits'. If you measure the new perimeter by including all the crinkly bits, the bays and coves and inlets, then the distance obtained will be larger. Going to a set of Ordnance Survey maps for the coast will give an even longer distance, because the detail shown on the map will be greater. Now, if we forsake our maps and start walking around the coast, we'd have an even longer distance to cover, as we could follow the absolute line of the coast to a couple of feet. And, if we had the services of an ant, the distance covered would be even greater, because the ant could explore the gaps between pebbles! Taking this to an extreme situation, it's possible to argue that the coastline of a country, if viewed in extreme close up, would constitute an infinitely long line enclosing a finite area. In addition, a highly magnified view of a coast line would show similar crinkliness to the coast line as a whole. This self-similarity is a common feature of many natural shapes. This type of self-similarity has been described as *statistical self-similarity*, as the similarity exhibited is *not* precise and the shape does not repeat itself exactly at different scales. However, it's close enough to be recognisably the same.

For example, clouds exhibit a statistical self-similarity, as do trees and plants such as ferns. If you don't believe me, go out and look for yourself. I have to admit, that since becoming a bit of a fractophile I also spend a lot of time looking at my rather moth-eaten potted fern! Indeed, it seems that the natural world is built out of shapes exhibiting these properties. These shapes are all *fractal*, as defined in the last chapter, having an HB dimension that's different to their Euclidean dimension. To start examining the weird and wonderful world of fractals, it's a good idea to build one and look at it in greater detail. The simplest shape to start with is called the *Koch snowflake*. This shape hasn't got anything to do with chaotic systems as explored in previous parts of the book, but will prepare us to handle the fractal nature of some other iterative systems that we will explore later in the book. Mathematically generated fractal shapes are exactly self-similar; that is, they are made up of the same shapes at different scales.

It's interesting to note that these shapes were discovered long before computers and their graphics abilities, and that all the ground work was done before anyone had even dreamt of chaotic systems. In addition, fractal shapes were deemed to be so far out of the normal mathematical establishment that they were called pathological curves, and it's even reported that mathematicians cast doubt on the mental stability of some of the original workers in the field!

The Koch curve

This curve was described by Helge von Koch in the early 1900s. It is created by the following sequence of steps:

1. Take an equilateral triangle, and build another equilateral triangle in the middle of each side of the shape, the new triangle having a base

length of 1/3 of the length of the side.

2. Repeat *ad infinitum*.

This simple recipe has in it two steps that are at the heart of fractal geometry; the self-similarity at all scales (if we repeated the operation an infinite number of times, a magnified view of any section of the Koch curve would look the same as any other part at any magnification), and the same step is repeated an infinite number of times to generate the fractal. The process that draws the curve actually calls itself. This is called *recursion*. It should be noted that these mathematically constructed shapes don't actually become fractal until the number of iterations involved is infinite. Obviously, we're not going to be able to wait that long, so we effectively get an approximation to a fractal shape. Anyway, before going into detail on the theory side, let's build a Koch snowflake. This is simply three Koch curves joined together. Listing 6-1 shows how this is done, and you will see that there are two new programming techniques used in this listing.

Recursion

Recursion is the name given to the process in which a procedure or function in a computer program calls itself (direct recursion) or calls another routine which then calls itself (indirect recursion). In the Koch curve procedure, the routine works by decrementing *depth* and only draws a line when *depth* reaches 0. If you go through the procedure, you'll see how the routine works; it's not transparently obvious at first, so be patient! A call to the curve generator requires two parameters; the *depth*, which gives you a measure of the 'crinkliness' of the resulting curve, and the *length*, which is effectively the length of the starting line. You will find that, with any recursive routine used to draw fractal curves, the following is true:

1. The greater the depth parameter is, the longer the routine will take to run.

2. The greater the depth parameter is, the more 'crinkly' the curve will be.

3. For a given depth, some values of length do not appear to work. In general terms, the higher the value of depth, the higher the value of length will need to be.

4. In this particular routine, a large length parameter will give a large snowflake on the screen, unless the drawing routines are modified to scale the X and Y distances drawn.

The process of recursion will clearly give rise to self-similarity; after all, if we repeat a task using parameters that have been modified by that task, we're bound to get some self-similarity along the way. It's a little like the repeated iterations we used to generate the various mappings from the logistic equation, but there the equation didn't call itself; it was used repeatedly to process data that it had already processed.

Recursive programming

I can't resist the pun; 'Here there be dragons!' (see the dragon curves, later in this chapter). Recursive programming can be a bit of a minefield in some languages due to the construction of the language interpreter or compiler. The routines listed here that use recursion have been tested on PCs and BBC Microcomputers, using Turbo Pascal and BBC BASIC, but tread warily on other dialects and languages. The reason for this is that many languages are designed and implemented with recursion very far down the list of desirable qualities. When a recursive routine is used, a set of calls to that routine is stored in the computer's stack, an area of memory designed to keep track of where the computer is in your BASIC or Pascal program. Some dialects don't like this at all, and either generate an error message or, worse still, crash the computer. So, here are a few pointers for pain-free recursion.

1. *Always* save your work before test running a program. Murphy's Law will always make sure that if you test a recursive program before saving it you will crash the computer.

2. Consult your language compiler/interpreter manual. Some languages allow you to have some control over stack size, etc., so if the manual offers guidance, follow it.

3. PC languages frequently have a stack size limited to 64k, irrespective of the available RAM on the computer. This offers a finite limit on the depth of recursion you can go to before the stack fills up. Stack overflows are frequently fatal error; you will be unable to finish the program run, although the language will frequently return you to a safe place. Some compilers offer the chance of turning off stack error checks; don't, until you've got some experience.

4. When testing a program, start with a small depth of recursion just to see what happens.

5. Within procedures that are to be called recursively, define no more local variables than are actually needed for the correct execution of the program. Don't forget that each time the procedure is called

recursively, a new set of local variables are defined which could easily eat up memory.

6. Many languages have dirty tricks associated with them for increased speed. I try to avoid them all the time when programming; I *always* avoid any dirty techniques when writing recursive code.

Turtle Graphics

The process of drawing fractal curves and other fractal shapes is not easily accomplished using what might be termed 'conventional' drawing methods. Usually, for drawing a curve, we use some formula or other, call it a few hundred times to get a series of plottable points, and then plot the points. Despite the apparent simplicity of the Koch curve shown in Figure 6-1, there is no algebraic formula that will define the position of points on the curve! This is another common feature about fractal curves; they may look

a. the Koch snowflake

b. initiator

generator

Figure 6-1. Generation of the Koch snowflake.

simple, but to draw them requires a new way of looking at geometry and computer graphics. A useful technique known as *Turtle Graphics* has been shown to be a simple and effective way of drawing these shapes. The original 'turtle' was actually a computer-controlled buggy with a pen, driven by a computer running a language called *Logo*. This was designed as an educational tool, allowing children to explore mathematics in concrete terms of shape drawing, whilst also giving an insight into programming and the mental processes involved in that field.

In Turtle Graphics, there are a collection of commands which tell the turtle, or a screen representation of it, what to do. So, *Forward*100 will send the turtle moving forwards 100 turtle units, which might be inches on the floor, or pixels on the screen. A command such as *Turn*90 would turn the turtle 90° clockwise, and so on. By combining these commands together into procedures, shapes of great complexity could be drawn.

Here I've defined a couple of Turtle Graphics commands in the Pascal toolkit, and use them to draw curves. *Front* is the command for *Forward*.

(The reason for this is simple; there are some languages that seem to object to the use of the name *Forward* as a procedure or function name, and I got rather fed up with the errors! The *Turn* command takes an argument in degrees and turns the turtle through that number of degrees clockwise (or, if the argument is negative, anticlockwise). The 0°point is at 3 on a clock face, +90° at 6, 180°at 9, 270°at 12 and 360°back at 3.

The curve itself

If you run the program listing given, the snowflake will be drawn on the screen. It consists of three Koch curves drawn one after another, after turning the turtle. The Koch snowflake that we've drawn has clearly got a finite length; I could measure each side, and add the lengths together, and that would give me a total length for the perimeter of the snowflake. This is because I've only developed the snowflake to a specified depth of recursion. If I took the recursion depth to infinity, then the shape resulting would be truly fractal. In fact, the curve that would result from this would be a little odd, because the direction of the curve would change at each point in the curve! This isn't too easy to visualise, but you can see how we might end up there if you modify the depth variable in Listing 6-1.

The Koch curve, which would be drawn by calling the Koch routine once with a suitable depth and length parameter, is a useful experimental beast for us: It is so simple, but it reflects the behaviour of other fractal curves. The first thing we can do is to examine a method of quantifying the fractal dimension of these shapes. So far, we've said that something like the Koch curve would have a fractal dimension of between 1 and 2; but what would it be?

We'll start off by looking at two important ideas about fractal shapes. They all have what is called an *initiator* and a *generator*. The initiator for a Koch curve is shown in Figure 6-1b—it's simply a straight line. The generator for a Koch snowflake would be an equilateral triangle. The initiator of a fractal shape is the Euclidean shape that gives rise to the fractal curve. The generator, on the other hand, is what the initiator is deformed into as the fractal curve is developed. The generator consists of a series of line segments. Here, for example, the generator for both curve and snowflake is shown in Figure 6-1b. It consists of four line segments, and in the fractal generation process each straight line segment of the curve would be replaced by a generator like this. Then, if the process were allowed to continue, the line segments of the generator would in turn be replaced by new generators! An examination of the generator for the Koch curve will show that it has four segments—call this number n—and that each segment is 1/3 the length of the initiator—call this number l, where, for this case,

$l = initiator length/3$. The fractal dimension of a curve can be given by:
$$d = \log(n)/\log(1/l)$$
This is a general relationship for many simple fractal curves. For the Koch curve, we can stick some figures in and arrive at:
$$d = \log(4)/\log(1/(1/3))$$
$$d = \log(4)/\log(3)$$
$$d = 1.26218$$

This calculation of the fractal (more formally, Hausdorff-Bessicovitch) dimension won't work for all types of fractal, but will for the sorts of fractal curves that we'll be discussing in this chapter. The greater the value of d, the more irregular is the fractal being described. For example, a fractal curve with a d value of 1.5 would be more irregular than the Koch curve. It's interesting to note that estimates of the d value for a typical coastline are about the same as that for a Koch curve.

As already mentioned, the Koch curve, when the recursion is taken to infinity, would have a length of infinity. If we think about this a little more, then you can see how, as the HB dimension increases from 1 to 2 then any line drawn will come to look more and more like a surface; i.e. it will start filling up more and more of the two-dimensional plane. Such curves, with relatively high HB dimensions, are called *space-filling curves*. However, these curves still remain lines rather than surfaces; breaking the curve in one place will give two separate curves; with a true surface this wouldn't happen. There are many variants of the Koch curve that have been published in books and magazines; the simplest is to replace the equilateral triangle generator of the Koch curve with a square, thus giving a generator with five segments rather than the normal four. This will give a slightly different curve with a higher HB dimension.

The path the turtle takes when creating the generator for this curve is as follows:
$$Front(l/3)$$
$$Turn(-90)$$
$$Front(l/3)$$
$$Turn(90)$$
$$Front(l/3)$$
$$Turn(90)$$
$$Front(l/3)$$
$$Turn(-90)$$
$$Front(l/3)$$

where l is the length of the initiator line. You might like to modify the program given to generate this curve, as shown in Listing 6-2. When experimenting with this curve, the same rules apply as did with the triangular version. You might also like to try generating a square version of the snowflake, or altering the distance in one of the *Front* commands to a different value.

Fractal Brownian motion

The Koch curve, whilst providing a useful model for understanding fractals, isn't used to model anything in the real world. Let's now look at a fractal curve generating method that can model some natural phenomena. One thing that can be modelled by a fractal curve is Brownian motion. This is the random motion taken by a very small particle in a fluid of some sort as it is bombarded by the motion of the molecules of the fluid itself. The traditional demonstration was to take something like Indian ink and make a solution in water. Then, under a microscope, the motions of the ink particles could be followed as they were knocked around by the invisible water molecules. Brownian motion is interesting as a fractal system for the following reasons:

1. It's real; you can see it under a microscope, and the model created as a fractal curve can accurately reflect what is seen in real life.

2. It's a fractal generated by random means; in the case of the Koch curve we saw a specific course of action laid down to generate the curve. Here, we can use random numbers to help generate the fractal curve. Such curves are called *random fractal curves*.

3. It's an example of a statistically self-similar system.

An in-depth analysis of this subject can be found in *The Science of fractal images* edited by Peitgen and Saupe. In that work you'll find many algorithms to generate this curve, and Listing 6-3 is based on the *Random Cuts BM* algorithm they describe. On running the program with the default values you should see a jagged line generated. Lower values of *wiggle* give a less bumpy ride. For higher values you may need to alter the value of the variable *scale*. Note that Gaussian random numbers are used. All this means is that the random numbers returned by the random number function are modified slightly to give a range of numbers that follows a Gaussian distribution. Other methods of generating this type of curve include what is called *mid-point displacement*, where a line is drawn and the mid-point of the line is pushed up or down (selected randomly) by a randomly selected amount. Each of the two halves is then subjected to the same treatment, and so on *ad infinitum*. This recursive process again gives rise to a fractal representation of Brownian motion.

One thing to note about the generation of random fractals is that the calculation of the HB dimension is clearly not as straightforward as it was

CANTOR SET

for the Koch curve. However, a value can be calculated mathematically, but that's beyond the scope of this book.

Cantor set

So far, the fractal curves that we've looked at have HB dimensions between 1 and 2, so they fall somewhere between a line and a surface. We can go in the other direction, however, and generate patterns that have HB dimensions between 0 (representing a collection of points) and 1. Such shapes are often called dusts, because they have the appearance of a sprinkling of dust particles on a surface. The simplest example of this is called the *Cantor set*, which was discovered by Georg Cantor in the nineteenth century. The recipe for producing a Cantor dust, or Cantor discontinuum, as it is occasionally called, is quite simple.

1. Take the numbers between 0 and 1, and represent them as a line.

2. Remove the central third, leaving, if you like, 0 to 0.333 ... and 0.6666 ... to 1.

3. Repeat this process *ad infinitum*.

Again, taking the process to infinity is needed to generate the true Cantor dust; doing the operation a finite number of times will generate approximates to the dust. The implications of this method of generation are that you will end up with an infinite number of points, arranged in clusters, that have a total length of zero! Now, what is the HB dimension of this set? Well, we can calculate it

$$d = \log(n)/\log(1/l)$$

where n is the number of segments in the generator of the fractal curve and l is the length of each of the segments as a fraction of the full length of the initiator. When n is 2, there are two segments generated, and $l = 1/3$ makes each segment 1/3 of the full length of the initiator, which is a straight line. Thus, the HB dimension is:

$$\begin{aligned} d &= \log(2)/\log(1/1/3) \\ &= 0.301/0.477 \\ &= 0.631 \end{aligned}$$

This makes the Cantor set a fractal dust with properties intermediate between those of scattered points and a full line. It is *not* just a regular scattering of points! Not surprisingly, changing the size of the fraction removed will generate dusts with different fractal dimensions. In addition, it's possible to have a random Cantor dust, where the fraction removed each time is randomly generated, sometimes a third, sometimes a quarter, and so on. You will often read of the Cantor set being *disconnected;* this is a mathematical term referring to the fact that it is a dust. Within the clusters there is self-similarity—each cluster is, after all, arrived at by taking out the middle third in the case of the standard Cantor set. In the case of the random Cantor set, there would be statistical self-similarity rather than exact self-similarity. Listing 6-4 shows how a simple dust can be generated. However, we're limited by the screen so you'll soon see that the pattern no longer changes. Each line of dots is a step of removal of thirds.

As far as we're concerned, the Cantor set has another interesting property; it provides us with a link to chaotic systems. To see how this is, we need to consider the work of Benoit Mandelbrot as he dealt with the problem of telephone line noise for IBM. When computers communicate over telephone lines, there is always going to be some noise. Whenever a particularly loud burst of electronic noise occurs, data might be lost and the transmitting end of the link will have to send that chunk of data again. The more errors, the slower the link will be at transmitting a given amount of data. The particular problem that Mandelbrot was asked to look at was that of intermittent noise bursts which defied the efforts of the engineers to cure them. To start with, the problems came in bursts—you might have the odd hour with no problems, then a burst of noise, then no problems again. Funnily enough, on examining the period of time with noise problems, you'd find periods of time with no noise, and others with noise. Going down further into the noisy parts of the hour, you find periods of time with no noise and periods with noise.

What Mandelbrot did was to treat the intermittent problem of noise in the system as a Cantor set, with the noise instead of the line that we drew. Noise-free periods are represented by the segments taken out of the line. Ultimately, we get a dust of noise clusters. The result in practical terms was that the engineers were going to have to learn to live with these noise clusters and develop means of catching the errors and correcting them quickly. From our point of view, though, we have what was on the surface an essentially chaotic event, the random noise, being modelled mathematically by simply treating the events as a Cantor dust in *time* instead of space. The Cantor set provides us with a method of looking at *intermittency*, that is, events happening on an occasional basis. Many chaotic systems exhibit intermittency. Remember the logistic equation, and how periodic areas popped up from chaotic regions of the bifurcation diagrams? Well, that's an example of intermittency. Here we see that we can model intermittency in some systems using a fractal object. The nature of noise on a phone line is also an example of a *discontinuous* phenomenon, where a smooth transition from one state to another does not exist; you've either got noise there or you haven't. In some ways, this is analogous to the discontinuity observed as the logistic equation slips into chaotic behaviour after a sequence of period doubling.

More importantly, though, the Cantor set allows us to tie up an annoying feature about strange attractors. In a strange attractor, the trajectory does not intersect itself, as that would constitute a closed loop and indicate a periodic system. However, we know that these strange attractor trajectories contain points that are very close together without touching each other, as shown in the Hénon attractor. This can all be explained if we treat the cross section of a strange attractor as a Cantor set. After all, the Cantor dust is as close to nothing as you can get whilst still being a single mathematical object, so that would allow trajectories of a strange attractor to apparently merge together, as in the Lorenz attractor, without actually intersecting. Indeed, one of the definitions of a strange attractor is that it is *a non-intersecting line of infinite length with an infinite number of loops having a Cantor set as its cross section*. A further link between chaotic systems and fractals is found in the phrase *infinite length*—this line is, after all, enclosing a finite volume or area of phase space, and so will, like the Koch curve, have a fractal dimension that is different to its Euclidean dimension.

Other fractal curves

There are lots of other fractal curves that we can write programs to generate, and there's no way that I can cover all of them in the space available. Instead, I'll look at a few samples of the many types of curve available to us, and provide pointers for generating other curves.

Peano

The Peano is a rather different type of fractal curve to the Koch curve, in that it has an HB dimension of 2. For this reason, it is often called a *plane-filling* curve, after all, that's exactly what it does. It will ultimately fill the plane in which it is drawn. One implication of this is that the traditional Peano curve will intersect itself, unlike the Koch curve. There are variations of the Peano curve that do not intersect in this way; these, however, will have a slightly lower HB dimension. Listing 6-5 shows a routine for generating a Peano curve of the traditional sort, and Figure 6-2 shows the initiator and generator.

Again, I've used Turtle Graphics routines to draw the shape. You should be able to see how the generator is traversed by looking at the *Peano* procedure in the program. You will see that the development for the Peano curve is not easy to follow; this was one reason why non-intersecting curves

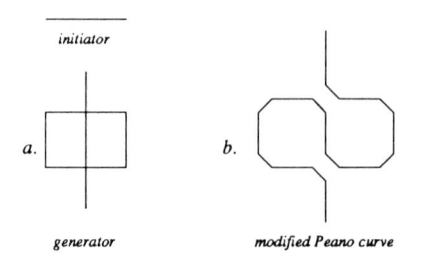

Figure 6-2. Generation of a Peano curve.

(Figure 6-2b) have been studied, as they allow the way in which the curve develops to be easily seen.

One interesting experiment that you might care to try is to use the Peano curve to save the contents of a graphics display screen. The curve fills a plane, and if the plane is a computer display then the curve would eventually address each location on the display. Now, if we look at each pixel visited by the curve and note its colour in an array of integers, we will have a list of numbers that defines the colour of each screen point visited by the curve. Thus, to re-create the screen, all we need to do is to provide the same start point as was used when the array was generated, the same length and recursion depth, and start the program running so that at each point visited the program sets the visited pixel on the screen to the colour held in the corresponding position in the array of numbers. Of course, this is a rather inefficient way of storing screens, but it offers some interesting features:

1. The array of numbers is meaningless without knowledge of the fractal curve parameters used to generate the numbers from the screen image. This is a simple form of encryption.

2. The resolution of the saved image is varied by altering the number of pixels visited on the original screen. This can be done by changing the recursion and length parameters.

Listing 6-6 shows a practical demonstration of this. As you can see, I've used a bold, simple image; these experiments work best with this simple approach. This simple system here models a more complex approach taken by a group of British Government scientists who have explored the use of Peano curves for the storage of data in a similar way to this, but they also include run-length coding algorithms to allow compression of stored data. Due to the amount of storage required for the image, I've not included a BBC BASIC version of the program.

Dragon and C curves

Some fractal curves generated in this way create very convoluted shapes that are called *dragons*. This is a little confusing, as there is a set of fractal patterns created by iteration (see Chapters 7 and 8) that are also called dragons, despite the different way of generating them. In this section I'll examine a couple of dragon curves that are generated recursively by applying Turtle Graphics techniques to an initiator and a generator. Why are they called dragons? Well, some of the curves produced do have a passing resemblance to heraldic dragons. The C curves are so called because they look like highly elaborate letter Cs. Starting off with the C curves, the recipe is quite simple:

$$Front(l)$$
$$Turn(-90)$$
$$Front(l)$$
$$Turn(90)$$

Listing 6-7 draws a C curve. You may like to try varying the angles turned, as well as the depth of recursion and the length of the sides drawn. If you want a spiky C curve, simply make both turn commands turn through either −90° or 90°.

As to the dragon curves, these are slightly more complex, in that to draw them in the same way as we've drawn the previous curves we'd need to alternate between two separate generators as the curve is drawn. Listing 6-8 draws a dragon curve based upon an algorithm published in *Dynamical systems and fractals*, by Becker and Dorfler. The curve itself can be modified by varying the angle through which the turtle turns; certain angles give dragons that look surprisingly like rivers seen from the air.

Sierpinski carpet

The *Sierpinski carpet* is quite interesting. Rather than draw a Sierpinski curve, which consists of a line, I chose to include a Sierpinski carpet. Although I've drawn the carpet using recursive procedural techniques, you can also generate a Sierpinski carpet by a technique using what are called *iterated function systems* (IFS). This technique is stunningly simple but quite spectacular, and I'll be considering it in greater detail in Chapter 9. However, for now I'll simply consider the generation of a carpet using recursive techniques.

There are two types of Sierpinski curve that you'll come across in text books; these are the *triadic curve*, which ultimately gives triangular shapes, and the *quadric curve*, which ultimately gives rise to square shapes. Listing 6-9 shows a program to generate a quadric Sierpinski carpet. The triadic carpet is the one that usually gets the coverage, so I hope that this redresses the balance somewhat!

Drawing with the L-language

We've seen in this chapter that a simple way of drawing fractal curves is to use Turtle Graphics, and effectively write instructions for the turtle to follow. Of course, in each of the programs listed there was a different set of turtle instructions needed to generate the desired fractal shape. In the book *The Science of fractal images*, Dietmar Saupe introduces a very useful idea in what he calls *string rewriting systems* or *L-systems* for drawing fractal curves. This is simply a means of storing instructions for the generation of fractal images as a series of characters. Those of you who've had experience of Turtle Graphics in the wider context of the Logo language will recognise this idea as being central to the concept of that language, that more complex operations can be built up from simple operations.

The basic idea of the L-Language is that a fractal curve is generated from a character string in the following steps:

1. An initial character string, called the axiom, is put together to represent the initial state of the system. This is analogous to the initiator that we used when developing the Koch curve above. This axiom consists of one or more characters.

2. A series of production rules (at least one) is then developed that operates on the axiom and replaces each character with a sequence of L-language commands that control a screen turtle. These commands are simple commands like *Forward*, *Turn*, etc. The set of commands that the program listed here understands includes

    F   Move one step forward (the length of a turtle step is pre-defined in the program) and draw a line.

    M   Move one step forward but *do not* draw a line.

    +   Turn to the right by a pre-defined number of degrees.

    −   Turn to the left by a pre-defined number of degrees, usually the same as defined for +.

    R   Turn through 180° and point back the way the turtle came.

    [Save the current state of the turtle (x position, y position and direction in which the turtle is pointing) on a stack (see below).

    ]   Remove the turtle state from the top of the stack and make the turtle assume that state.

DRAWING WITH THE L-LANGUAGE

Other characters can be used in the axiom string as *procedure names*, which will be expanded by the L-language into a sequence of the above listed commands. The *stack* is a commonly used storage device in computer languages; think of it as a stack of plates on a table top. If you think of each plate as referring to a turtle state, then you can see that the only turtle state we have easy access to is the one at the top of the stack. Storing a turtle state simply puts it on the top of the stack (called *pushing* a state on to the stack), thus hiding the turtle state which was last at the top of the stack. Retrieving the pushed turtle state from the top of the stack (*popping* the turtle state) will remove the pushed state and return the stack to its original state. The stack thus allows any number of turtle states to be stored and retrieved in the same order as that in which they were stored.

3. The production rule(s) are applied a specified number of times, thus expanding the axiom string into a string of turtle commands. For example, if we had an axiom of F, and a production rule of F is replaced by $F + F$, and specified that we wanted to apply the rule twice, we'd get the following:

axiom	F
rule applied once	$F + F$
rule applied twice	$F + F + F + F$

4. The resulting expanded string is then passed to a simple interpreter which converts the turtle commands in the string into activity on the screen.

You might ask, 'Why Bother? I can generate fractal curves by simply writing short programs'. However, the strong point of the L-language system is that only one program needs to be written to do the drawing of many fractals; all that is changed is the axiom, the production rule and the number of times that the rule is applied.

Programming L-language systems

In theory, programming of L-systems is quite straightforward, as it is simply an exercise in string processing. However, there are a couple of points to watch out for, based mainly around the language used to write the system.

1. String sizes in many programming languages are limited to 255 characters maximum, and although this will still allow you to write simple axioms and production rules, complex rules or a large number of applications of the production rule could easily generate a final string for passing to the Turtle Graphics interpreter that exceeds the maximum allowable string size for the language in use. There are ways

around this, the simplest being to use a block of memory and use the *POKE* and *PEEK* operators of BASIC to gain direct access to RAM. Your BASIC language manual will show you how to set aside an area of memory for this type of use. In Pascal, you will need to set up a data structure to suit these requirements.

2. It's often useful, if you're writing a system to be used by other people, to make upper and lower case letters identical in function; that is, if a production string contained both f and F, the same operation should be done each time. This makes life easier for the users, as they don't have to bother about case sensitivity. The listings here do this for the commands, but not for procedure names.

3. As in many programs listed in this book, get the angles in either degrees or radians; most people who aren't mathematicians tend to think in terms of degrees.

4. When running, the expansion of the axiom into the command string may take some time—at least a couple of seconds even on fast machines.

Listing 6-10 shows a simple L-language interpreter that incorporates these features but is, however, limited to a maximum 255-character command string. This needs to be expanded somewhat for complex fractal curves to be drawn, but I'll leave that as an exercise for you! Of course, for such a system to be useful, you need to provide it with axioms and production rules. I've hard-coded these for a Koch curve into the program, but they can be altered by changing the lines of code. Due to the 255-character command line limit in this program, the curve will not be very long. In fact, it's a good idea to check roughly how far the axiom will be expanded after the number of applications of the production rule that you've specified before trying to plot the curve; it's surprising how fast this can get out of hand! The value *theta* gives the value by which the turtle will be turned on each left or right command, and the variable *forwardstep* gives the value by which the turtle will be moved by F or M commands. The variable *application* gives the number of times that the production rule will be applied to the axiom before the curve is drawn. The details for the Dragon curve are:

$$\begin{aligned} axiom \quad &: \quad d \\ angle \quad &: \quad 90° \\ production\ rule \quad &: \quad d \text{ is replaced by } e + gf+ \\ & g \text{ is replaced by } -fd - y \end{aligned}$$

The production details for other curves can be found by experimentation or from Suape's book mentioned at the start of this section. One point to

FRACTALS IN THE REAL WORLD   159

remember is that the more complex curves will not become totally apparent with only 255 drawing steps; many more are needed. For example, Saupe describes the production rules for plant-like structures as well as a variety of space-filling curves.

The L-language gives us a third way of drawing fractal curves; we've already seen the procedural method, used for the Sierpinski curve and the Turtle Graphics method used for the Koch curve. However, the L-language still generates fractal curves using the techniques of recursion explored earlier in this chapter, and also begins to demonstrate to us a very important point. Here we have a method of drawing different fractal objects, such as Koch curves, dragon curves and even *bushes* that uses one simple program and a set of rules for each different shape. Now, the interesting thing is that, despite the apparent complexity of the resultant fractal images the curve can be described in its entirety by a string of maybe 30 characters giving the axiom, production rule and the number of times the production rule is to be applied. This is an incredible feat of image compression, coding the information required to generate a Koch snowflake into less than an average sentence. This is possiblu due to the self-similarity of fractal curves, and this method of describing images will only work for curves or shapes that are fractal in nature. We'll encounter a similar phenomenon in Chapter 9 when we look at iterated function systems.

Fractals in the real world

To complete this chapter, I thought that I would briefly examine some of the places that the fractal curves explored in this chapter turn up in the real world, as well as some applications of fractals like those we've seen. In a later chapter, we'll explore other, more complex, fractal systems and examine their practical uses. It is often a surprise for programmers to learn that their interesting shapes also have some practical significance and do occasionally show up in the real world! The self-similarity exhibited by the real-world fractals is statistical; there are no naturally occurring exactly self-similar fractals.

Rain and snow

One feature of clouds has already been discussed in this chapter; they are, like many natural objects, fractal. Indeed, film-makers have used fractal routines to produce graphical images of clouds. Shaun Lovejoy discovered by studying both satellite and radar images of clouds that the fractal dimension of clouds is the same (about 1.33) at a variety of scales. The practical upshot of this is that it's hard to get any idea of the size of a cloud if you're shown a photograph of one with nothing to act as a scale.

It could be a small cloud seen close up, or a large one seen from a distance. In addition, he found that the fall of rain tends to occur in bursts of rain, at irregular intervals. This is a similar mechanism to the Cantor set model of transmission-line noise put forward by Mandelbrot. He also determined that the perimeters of areas on to which rain is falling are also fractal, rather than smooth.

A microscopic examination of snowflakes also exhibits fractal properties, and we can model snowflakes very crudely with a fractal curve, as we saw in Listing 6-1. The reason why our Koch snowflake doesn't look much like a normal snowflake is that a snowflake is statistically self-similar, rather than exactly self-similar. Also, the method by which a snowflake is built is an example of crystal growth, rather than a deterministic process. An examination of a frost-covered window in cold weather will show another form of fractal object.

A final example of a fractal object in meteorology is the structure of fork lightning. Close examination of photographs of this type of lightning indicate the existence of many fingers of lightning off the main branches.

Geography

We've already seen how a coastline might be viewed as a fractal shape. The most obvious example is probably the fjord coast of Norway, but even apparently smooth coasts consisting of sandy beaches will have a fractal dimension. It shouldn't surprise us, therefore, to find that other geographical objects have a fractal dimension as well. The coastlines of lakes will clearly have this feature, as do mountain ranges, cliff faces and other natural boundary points. Mandelbrot observed that the path of rivers can be modelled by fractal curves, with an HB dimension of between 1.2 and 1.3. The windier the rivers are, then the higher their fractal dimension. A useful rule of thumb when looking out for geographical or geological fractals is to look at pictures that do not include any means of scaling. A common characteristic of a shape that can be modelled by a fractal curve is that, without any other information, you cannot tell its size in the real world; an isolated picture of a rock might be in someone's garden, or at the top of an escarpment in the Alps!

If you want to experiment with generating models of coastlines using fractal curves, you could start with the fractal Brownian motion routine listed above. Alternatively, try Listing 6-11; this is a modified Koch curve generator which uses random numbers to give the turn angle and lengths traversed by the turtle. Plots from this program are prone to suffer from doubling back on themselves, but you can get some quite realistic looking coasts with some perseverance.

Biology

Plants are very fractal; in Chapter 9 we'll examine how models of trees and ferns can be synthesised using fractal techniques, and so it's not surprising that plants can be said to have a fractal dimension to at least parts of their structure. The easiest example is in the leaves; a fern leaf is often used as an example of a perfect fractal object, but leaves of other plants, such as oak trees or holly bushes, also have a fractal dimension. The root and branch systems of trees and plants also exhibit fractal dimensions. A further interesting point is in the fractal nature of leaf surfaces when considered as an environment in which creatures can live; the pitted surface of leaves gives a larger than expected surface area that can be exploited by creatures small enough to fit.

In the animal kingdom things like the bronchial structure in the lungs and the capillary system are fractal. A similar system is the spiracle system in insects, responsible for getting oxygen into the insect body. There is a very good biological reason for these structures to exhibit a fractal structure, as they all need to create a large surface area (for gas and chemical exchanges) within a frequently small volume. For example, it's been estimated that the bronchial system of a human lung would, if opened out, have an area equivalent to that of a soccer pitch. The exchange of materials is aided at the cellular level by cells frequently having rough walls to maximise their surface area. A similar argument applies to the digestive tract of mammals, where the cells of the gut are covered in villi, finger-like structures that offer a larger area for absorption than the cells themselves would otherwise.

The cell walls of living things are also fractal, as are the protein molecules essential to life. For example, a molecule designed for carrying materials around the body has a fairly high fractal dimension as it needs to adsorb other molecules onto its surface and hold them. Protein molecules, like cell walls, have HB dimensions between 2 and 3 (that is, exhibiting properties between those of a surface and a three-dimensional object.

Fluid flow

Fluid travelling through other fluids, or the same fluid at a different speed, also exhibits a fractal structure. You can see this by simply looking at the turbulence in rivers, or even the smoke from a chimney; the boundary between the smoke and the air is a fractal boundary. A classic example of the flow of fluid within fluid is demonstrated in a piece of equipment called a *Hele-Shaw* cell, which was developed around the turn of the century to allow the exploration of fluid-fluid boundaries. It consists, in essence, of two plates of glass separated by a thin gap into which one fluid can be placed; another fluid is then forced into the cell. The resultant patterns, which consist of 'fingers' of the forced fluid penetrating the other fluid, have a fractal dimension. There might be economic implications in this, as oil is often extracted from oil bearing deposits by forcing water in to the oil bearing rocks—a giant Hele-Shaw.

Catalysts and enzymes

Catalysts and enzymes are both chemical middlemen; they facilitate a chemical reaction to take place without changing their own chemical structure in the process. The general differences between the two are that enzymes, which mediate the chemical processes in living things, are made of protein and are deactivated or destroyed by temperatures outside a very precise range. The function of both is the same, however, to provide a jig into which the molecules that are to react together can fit and react without extreme conditions.

A typical catalyst might be a finely divided metal powder, or a sponge through which the chemical reactants can be pumped. An efficient catalyst is one that promotes the largest number of the desired chemical reactions for the smallest amount of catalyst. This usually requires a large surface area, and the best way to get that is to have a surface with a fractal dimension. In action, the catalyst adsorbs onto its surface one of the chemicals in the reaction mixture, and effectively holds it still until a molecule of the other reactant comes along to react. Clearly, the more adsorption that can take place, the better.

With regard to enzymes, specific active sites exist that hold reacting molecules in a particular orientation so that the desired reaction can take place. The enzyme molecules have a fractal surface which, incidentally, is lower than that of other protein molecules. This reflects the fact that once a reaction has taken place on an enzyme, the newly formed molecule must be able to get away fairly easily. A highly fractal surface might not allow this to happen.

Aggregation

There are certain physical processes where aggregation is an important part of the mechanism. Aggregation is the process by which a shape is built up by particles hitting an existing cluster of particles and sticking. This process is what occurs when particles build up on air filters, or other such processes where small particles can accumulate together. This process, called *diffusion limited aggregation* (DLA) can be modelled with a computer program using simulated particles that move according to an algorithm like that used to simulate Brownian motion earlier in this chapter. As soon as one of these particles comes orthogonally adjacent to a seed pixel on the screen, it sticks to the seed and thus starts the aggregation process, thus eventually forming an irregular, spidery aggregation. This shape is fractal; if you take two points joined through the aggregation, the straight-line distance is shorter than the path traversed by following the particles between the two particles of interest; this is analogous to the coastline problems mentioned above.

Astronomy

A fine example of a system that can be modelled by a Cantor set is to be found in the rings of the planet Saturn. Although the rings of the planet were discovered very soon after the development of the telescope, their nature was only unravelled over the years as better instruments were built and, eventually, satellites were used to take photographs of the ring structure. This eventually resolved the rings into a collection of smaller rings within rings, through which you could actually see the stars. A cross-section of the rings would give us a pattern that looks very like a Cantor dust.

A further example is to be found in the distribution of matter through space; it's not regular, but clumpy. Matter in the universe tends to be gathered together in the collections of stars that we call galaxies. There is stuff in between, but not that much, and various astronomers have created models of the universe in which the galaxies follow a fractal arrangement in terms of distribution. There are problems with proving this, however, in that the dust, etc., between galaxies, as well as a variety of other problems, makes the accurate measurement of the position and speed of galaxies difficult.

BBC BASIC listings

Listing 6-1

```
 10 REM Program VonKochSnowFlake
 30 REM Generates the Von Koch Snowflake.
 80 MODE 4
 90 REM Set up initial TurtleX and TurtleY values.
110 TurtleX=90
120 TurtleY=90
130 MOVE TurtleX,TurtleY
140 TurtleTheta=0
160 REM The snow flake is drawn by drawing three Von Koch curves, turning
170 REM through 120' between each curve.
200 PROCVonKoch(3,300)
210 PROCTurn(120)
220 PROCVonKoch(3,300)
230 PROCTurn(120)
240 PROCVonKoch(3,300)
250 PROCTurn(120)
270 END
320 DEFPROCVonKoch(depth,length)
340 REM This procedure generates a Von Koch curve via recursion. A turtle
350 REM graphics approach (see text) is used to generate this curve.
370 IF depth=0 THEN PROCFront(length) : GOTO 460
380 PROCVonKoch(depth-1,(length/3))
390 PROCTurn(-60)
400 PROCVonKoch(depth-1,(length/3))
410 PROCTurn(120)
420 PROCVonKoch(depth-1,(length/3))
430 PROCTurn(-60)
440 PROCVonKoch(depth-1,(length/3))
460 ENDPROC
490 DEFPROCFront(distance)
500 TurtleX=( distance*COS(TurtleTheta*3.14/180)) + TurtleX
510 TurtleY=( distance*SIN(TurtleTheta*3.14/180)) + TurtleY
520 DRAW TurtleX,TurtleY
530 ENDPROC
550 DEFPROCTurn(theta)
560 TurtleTheta=(TurtleTheta + theta) MOD 360
570 ENDPROC
```

Listing 6-2

```
 10 REM Program VonKochSquareCurve
 30 REM Generates a Von Koch Curve.
 60 MODE 4
 90 TurtleX=90
100 TurtleY=500
110 MOVE TurtleX,TurtleY
120 TurtleTheta=0
150 PROCVonKoch(5,800)
180 END
220 DEFPROCVonKoch(depth,length)
240 REM This procedure generates a Von Koch curve via recursion. A turtle
250 REM graphics approach (see text) is used to generate this curve.
270 IF depth=0 THEN PROCFront(length) : GOTO 380
280 PROCVonKoch(depth-1,(length/3))
290 PROCTurn(-90)
300 PROCVonKoch(depth-1,(length/3))
310 PROCTurn(90)
320 PROCVonKoch(depth-1,(length/3))
```

BBC BASIC LISTINGS

```
330 PROCTurn(90)
340 PROCVonKoch(depth-1,(length/3))
350 PROCTurn(-90)
360 PROCVonKoch(depth-1,(length/3))
380 ENDPROC
430 DEFPROCFront(distance)
440 TurtleX=( distance*COS(TurtleTheta*3.14/180)) + TurtleX
450 TurtleY=( distance*SIN(TurtleTheta*3.14/180)) + TurtleY
460 DRAW TurtleX,TurtleY
470 ENDPROC
490 DEFPROCTurn(theta)
500 TurtleTheta=(TurtleTheta + theta) MOD 360
510 ENDPROC
```

Listing 6-3

```
 10 REM Program FractalBrownianMotion
 30 REM Generate Fractal Curve Simulating Brownian Motion
 40 REM Algorithm from Peitgen and Saupe.
 60 DIM Brown(200)
110 Wiggle=20
120 Scale=4
130 PROCCalculateBrownian
140 MODE 4
150 PROCPlotArray
160 END
230 DEFPROCGauss(dummy)
250 REM This function generates random numbers between 0 and 'wiggle' that
260 REM have a Gaussian Distribution.  Most PC random number generators don't.
280 LOCAL GaussAdd,GaussFac
300 sum=0
310 GaussAdd=SQR(12)
320 GaussFac=(2*GaussAdd)/(4*Wiggle)
330 FOR j=1 TO 4
340    sum=sum+RND(Wiggle)-1
350 NEXT
360 Gauss=GaussFac*sum-GaussAdd
380 ENDPROC
410 DEFPROCPlotArray
420 REM Now plot the contents of the array of numbers.
440 MOVE 0,500
450 FOR k=1 TO 200
460    DRAW k*3,Brown(k)*Scale+500
470 NEXT
480 ENDPROC
510 DEFPROCCalculateBrownian
530 REM Calculate array of numbers, place in array called Brown
550 FOR k=0 TO 200
560    Brown(k)=0
570 NEXT
580 T=RND(-TIME)
590 N=200
600 MaxSteps=200
610 FOR i=1 TO MaxSteps
620    k0=RND(N)-1
630    k1=k0+(N/2)-1 : IF k1>200 THEN k1=200
640    PROCGauss(1)
650    A=Gauss
660    FOR k=k0 TO k1
670       IF (k<N) THEN Brown(k)=Brown(k)+A ELSE Brown(k)=Brown(k)-A
680    NEXT
690 NEXT
700 ENDPROC
```

Listing 6-4

```
10 REM Generates Cantor Set.
20 MODE 5
30 depth=8
40 length=600
50 x1=100
60 x2=700
70 y1=10
80 y2=100
90 GCOL0,1
100 PROCbar(x1,y1,x2,y2)
110 PROCCantor(x1,y1,depth,length)
120 END
130 DEFPROCbar(xa1,ya1,xa2,ya2)
140 FOR I%=ya1 TO ya2
150   MOVE xa1,I% : DRAW xa2,I%
160 NEXT I%
170 ENDPROC
180 :
190 DEFPROCCantor(x1,y1,depth,length)
200 LOCAL seg_length,x3,x4
210 seg_length=INT(length/3)
215 x3=x1+seg_length
220 x4=x1+seg_length*2
225 GCOL0,0
230 PROCbar(x3,INT((depth-1)*10),x4,INT((depth-1)*10-2))
240 IF depth=0 THEN ENDPROC
250 PROCCantor(x1,INT((depth-1)*10),depth-1,seg_length)
260 PROCCantor(x3,INT((depth-1)*10),depth-1,seg_length)
270 PROCCantor(x4,INT((depth-1)*10),depth-1,seg_length)
300 ENDPROC
```

Listing 6-5

```
10 REM Program PeanoCurve
30 REM Generates a Peano Curve.
70 MODE 4
100 TurtleX=90
110 TurtleY=500
120 MOVE TurtleX,TurtleY
130 TurtleTheta=0
150 PROCPeano(3,400)
180 END
200 DEFPROCPeano(depth,length)
210 REM This procedure generates a Peano curve via recursion. A turtle
220 REM graphics approach (see text) is used to generate this curve.
230 IF depth=0 THEN PROCFront(length) : GOTO 430
240 PROCPeano(depth-1,(length/3))
250 PROCTurn(-90)
260 PROCPeano(depth-1,(length/3))
270 PROCTurn(90)
280 PROCPeano(depth-1,(length/3))
290 PROCTurn(90)
300 PROCPeano(depth-1,(length/3))
310 PROCTurn(90)
320 PROCPeano(depth-1,(length/3))
330 PROCTurn(-90)
340 PROCPeano(depth-1,(length/3))
350 PROCTurn(-90)
360 PROCPeano(depth-1,(length/3))
370 PROCTurn(-90)
380 PROCPeano(depth-1,(length/3))
```

```
390 PROCTurn(90)
400 PROCPeano(depth-1,(length/3))
430 ENDPROC
480 DEFPROCFront(distance)
490 TurtleX=( distance*COS(TurtleTheta*3.14/180)) + TurtleX
500 TurtleY=( distance*SIN(TurtleTheta*3.14/180)) + TurtleY
510 DRAW TurtleX,TurtleY
520 ENDPROC
540 DEFPROCTurn(theta)
550 TurtleTheta=(TurtleTheta + theta) MOD 360
560 ENDPROC
```

Listing 6-6

Because of limited memory in the BBC Micro, Listing 6-6 is only available as a Turbo Pascal program.

Listing 6-7

```
10 REM Program CCurveGenerator
30 REM Generates a C Curve.
60 MODE 4
70 TurtleX=500
80 TurtleY=500
90 MOVE TurtleX,TurtleY
100 TurtleTheta=0
120 PROCCCurve(9,6)
150 END
200 DEFPROCCCurve(depth,length)
210 IF depth=0 THEN PROCFront(length) : GOTO 280
220 PROCCCurve(depth-1,length)
230 PROCTurn(-90)
240 PROCCCurve(depth-1,length)
250 PROCTurn(90)
280 ENDPROC
320 DEFPROCFront(distance)
330 TurtleX=( distance*COS(TurtleTheta*3.14/180)) + TurtleX
340 TurtleY=( distance*SIN(TurtleTheta*3.14/180)) + TurtleY
350 DRAW TurtleX,TurtleY
360 ENDPROC
380 DEFPROCTurn(theta)
390 TurtleTheta=(TurtleTheta + theta) MOD 360
400 ENDPROC
```

Listing 6-8

```
10 REM Program MakeDragonGenerator
50 MODE 4
60 TurtleX=500
70 TurtleY=500
80 MOVE TurtleX,TurtleY
90 TurtleTheta=0
110 PROCMakeDragon(10,5)
140 END
190 DEFPROCMakeDragon(depth,length)
200 IF depth=0 THEN PROCFront(length) : GOTO 260
210 IF depth>0 THEN PROCMakeDragon(depth-1,length) : PROCTurn(-90) :
    PROCMakeDragon(-(depth-1),length)
```

```
230 IF depth<0 THEN PROCMakeDragon(-(depth+1),length) : PROCTurn(90) :
    PROCMakeDragon(depth+1,length)
260 ENDPROC
290 DEFPROCFront(distance)
300 TurtleX=INT( distance*COS(TurtleTheta*3.14/180)) + TurtleX
310 TurtleY=INT( distance*SIN(TurtleTheta*3.14/180)) + TurtleY
320 DRAW TurtleX,TurtleY
330 ENDPROC
350 DEFPROCTurn(theta)
360 TurtleTheta=(TurtleTheta + theta) MOD 360
370 ENDPROC
```

Listing 6-9

```
10 REM Generates Cantor Set.
20 MODE 5
30 depth=3
40 length=600
50 x1=100
60 x2=700
70 y1=0
80 y2=600
90 GCOL0,1
100 PROCbar(x1,y1,x2,y2)
110 PROCSierpinski(x1,y1,x2,y2,depth,length)
120 END
130 DEFPROCbar(xa1,ya1,xa2,ya2)
140 FOR I%=ya1 TO ya2
150   MOVE xa1,I% : DRAW xa2,I%
160 NEXT I%
170 ENDPROC
180 :
190 DEFPROCSierpinski(x1,y1,x2,y2,depth,length)
200 LOCAL seg_length,x3,y3,x4,y4
210 seg_length=INT(length/3)
220 x3=x1+seg_length
230 y3=y1+seg_length
240 x4=x2-seg_length
250 y4=y2-seg_length
260 GCOL0,0
270 PROCbar(x3,y3,x4,y4)
280 IF depth=0 THEN ENDPROC
290 PROCSierpinski(x1,y1,x3,y3,depth-1,seg_length)
300 PROCSierpinski(x3,y1,x4,y3,depth-1,seg_length)
310 PROCSierpinski(x4,y1,x2,y3,depth-1,seg_length)
320 PROCSierpinski(x1,y3,x3,y4,depth-1,seg_length)
330 PROCSierpinski(x4,y3,x2,y4,depth-1,seg_length)
340 PROCSierpinski(x1,y4,x3,y2,depth-1,seg_length)
350 PROCSierpinski(x3,y4,x4,y2,depth-1,seg_length)
360 PROCSierpinski(x4,y4,x2,y2,depth-1,seg_length)
370 ENDPROC
```

Listing 6-10

```
10 REM Program L_LanguageSystem
30 REM Program to generate Fractal Curves based on the L-Language System
40 REM described in 'The Science of Fractal Images.
70 DIM stackX(30), stackY(30), stackA(30),prodrule$(10),rule_code$(10)
100 MODE 4
110 REM Next block of statements initialises the turtle
```

BBC BASIC LISTINGS
169

```
 130 theta=60               : REM angle size for this axiom and production
     rule
 140 forwardstep=20         : REM forward step for this axiom and production
     rule
 150 TurtleTheta=0
 160 TurtleX=200
 170 TurtleY=150
 190 REM The following axiom and production rule is for a Koch Curve.
 210 axiom$="F"
 220 prodrule$(1)="F"
 230 rule_code$(1)="F-F++F-F"
 240 number_rules=1
 250 application=2
 270 MOVE TurtleX,TurtleY
 280 PROCExpandAxiomString
 290 PROCTurtleInterpret
 320 END
 360 DEFPROCFrontTurtle
 370 TurtleX=( forwardstep*COS(TurtleTheta*3.14/180)) + TurtleX
 380 TurtleY=( forwardstep*SIN(TurtleTheta*3.14/180)) + TurtleY
 390 DRAW TurtleX,TurtleY
 400 ENDPROC
 420 DEFPROCRightTurnTurtle
 430 TurtleTheta=(TurtleTheta + theta) MOD 360
 440 ENDPROC
 470 DEFPROCMoveTurtle
 490 TurtleX=( forwardstep*COS(TurtleTheta*3.14/180)) + TurtleX
 500 TurtleY=( forwardstep*SIN(TurtleTheta*3.14/180)) + TurtleY
 510 MOVE TurtleX,TurtleY
 520 ENDPROC
 540 DEFPROCLeftTurnTurtle
 550 TurtleTheta=(TurtleTheta - theta) MOD 360
 560 ENDPROC
 580 DEFPROCPushTurtle
 590 stack_pos=stack_pos+1
 600 stackX(stack_pos)=TurtleX
 610 stackY(stack_pos)=TurtleY
 620 stackA(stack_pos)=TurtleTheta
 630 ENDPROC
 650 DEFPROCPopTurtle
 660 TurtleX=StackX(stack_pos)
 670 TurtleY=StackY(stack_pos)
 680 TurtleTheta=StackA(stack_pos)
 690 stack_pos=stack_pos-1
 700 ENDPROC
 720 DEFPROCReverseTurtle
 730 TurtleTheta=TurtleTheta+180
 740 ENDPROC
 760 DEFPROCTurtleInterpret
 770 REM This procedure takes the string in command$ and extracts from it the
     turtle
 780 REM graphics commands and acts upon them to draw the image required.
 790 FOR i=1 TO LEN(command$)
 810   current$=MID$(command$,i,1)
 820   IF (current$="F") OR (current$="f") THEN PROCFrontTurtle
 830   IF (current$="M") OR (current$="m") THEN PROCMoveTurtle
 840   IF (current$="R") OR (current$="r") THEN PROCReverseTurtle
 850   IF (current$="+") THEN PROCRightTurnTurtle
 860   IF (current$="-") THEN PROCLeftTurnTurtle
 870   IF (current$="[") THEN PROCPushTurtle
 880   IF (current$="]") THEN PROCPopTurtle
 890 NEXT
 900 ENDPROC
 940 DEFPROCExpandAxiomString
 960 REM This currently expands a string to no more than 255 characters. This
```

170   CHAPTER 6. THE FRACTAL LINK

```
         could
 970 REM be extended by direct RAM access or a different data structure.
1010 REM Loop around for 'application' times to expand the axiom$ the correct
         number
1020 REM of times
1040 FOR i=1 TO application
1060   command$=""
1070   FOR j=1 TO LEN(axiom$)
1090     current$=MID$(axiom$,j,1)
1100     AxiomExtended=0
1120     FOR k=1 TO number_rules
1140       IF (current$=prodrule$(k)) THEN command$=command$+rule_code$(k) :
             AxiomExtended=1
1160     NEXT
1170     IF (AxiomExtended=0) THEN command$=command$+current$
1190   NEXT
1210   REM Old axiom$ string is now replaced by the expanded command$ string,
         and the
1220   REM expanded command$ string is now processed.    Y
1240   axiom$=command$
1290 NEXT : ENDPROC
```

Listing 6-11

```
  10 REM Program RandomVonKoch
  50 REM Generates a Von Koch Coastline.
 150 MODE 4
 190 REM Set up initial TurtleX and TurtleY values. These variables
 200 REM defined in the
 210 REM Extend Library.  Then move to start position, and set initial
 220 REM Turtle
 230 REM Direction to 0'.
 290 TurtleX=500
 310 TurtleY=500
 330 MOVE TurtleX,TurtleY
 350 TurtleTheta=0
 390 REM Randomize function used to seed random number generator.
 430 T=RND(-TIME)
 450 PROCVonKoch(4,6000)
 510 END
 570 DEFPROCVonKoch(depth,length)
 610 REM This procedure generates a random Von Koch curve via recursion.
 630 REM graphics approach (see text) is used to generate this curve.
 670 IF depth=0 THEN PROCFront(length) : ENDPROC
 710 REM Random number function used to generate the length and angle.
 750 PROCVonKoch(depth-1,(RND((length/3))))
 770 PROCTurn(-(RND(60)-1))
 790 PROCVonKoch(depth-1,(RND((length/3))))
 810 PROCTurn((RND(120)))
 830 PROCVonKoch(depth-1,(RND((length/3))))
 850 PROCTurn((-RND(60)))
 870 PROCVonKoch(depth-1,(RND((length/3))))
 910 ENDPROC
 970 DEFPROCFront(distance)
 990 TurtleX=( distance*COS(TurtleTheta*3.14/180)) + TurtleX
1010 TurtleY=( distance*SIN(TurtleTheta*3.14/180)) + TurtleY
1030 DRAW TurtleX,TurtleY
1050 ENDPROC
1070 DEFPROCTurn(theta)
1090 TurtleTheta=(TurtleTheta + theta) MOD 360
1110 ENDPROC
```

Turbo Pascal listings

Listing 6-1

```
Program VonKochSnowFlake;

{       Generates the Von Koch Snowflake.        }

Uses Crt,Graph,Extend;

Procedure VonKoch(depth,length : integer);

{         This procedure generates a Von Koch curve via recursion. A turtle
          graphics approach (see text) is used to generate this curve.    }
Begin
   if depth=0 then Begin
      Front(length);
     end
   else
     begin
       VonKoch(depth-1,trunc(length/3));
       Turn(-60);
       VonKoch(depth-1,trunc(length/3));
       Turn(120);
       VonKoch(depth-1,trunc(length/3));
       Turn(-60);
       VonKoch(depth-1,trunc(length/3));

   end;
end;

Begin

   ChoosePalette;

{ Set up initial TurtleX and TurtleY values.  These variables defined in the
  Extend Library.  Then move to start position, and set initial Turtle
  Direction to 0'.
  }

   TurtleX:=90;
   TurtleY:=90;
   MoveTo(TurtleX,TurtleY);
   TurtleTheta:=0;

{ The snow flake is drawn by drawing three Von Koch curves, turning
  through 120' between each curve.                                   }

   VonKoch(3,300);
   Turn(120);
   VonKoch(3,300);
   Turn(120);
   VonKoch(3,300);
   Turn(120);

end.
```

Listing 6-2

```
Program VonKochSquareCurve;

{       Generates a Von Koch Curve.      }

Uses Crt,Graph,Extend;

Procedure VonKoch(depth,length : integer);

{        This procedure generates a Von Koch curve via recursion.  A turtle
         graphics approach (see text) is used to generate this curve.     }

Begin
   if depth=0 then Begin
      Front(length);
    end
   else
     begin
       VonKoch(depth-1,trunc(length/3));
       Turn(-90);
       VonKoch(depth-1,trunc(length/3));
       Turn(90);
       VonKoch(depth-1,trunc(length/3));
       Turn(90);
       VonKoch(depth-1,trunc(length/3));
       Turn(-90);
       VonKoch(depth-1,trunc(length/3));
   end;
end;

Begin
   ChoosePalette;

{ Set up initial TurtleX and TurtleY values.  These variables defined in the
  Extend Library.  Then move to start position, and set initial Turtle
  Direction to 0'.
  }

   TurtleX:=90;
   TurtleY:=GetMaxY;
   MoveTo(TurtleX,TurtleY);
   TurtleTheta:=0;

   VonKoch(5,800);
end.
```

Listing 6-3

```
Program FractalBrownianMotion;

{ Generate Fractal Curve Simulating Brownian Motion
  Algorithm from Peitgen and Saupe.                            }

Uses Crt,Graph,Extend;

Var    Brown           :    Array[1..200] of real;
       N               :    Integer;
       MaxSteps        :    Integer;
       k,k0,k1,step:        Integer;
       A               :    Real;
       i,j             :    Integer;
       Scale           :    Real;
```

TURBO PASCAL LISTINGS

```
            Wiggle       :   Integer;
            sum          :   Real;

Function Gauss(dummy : real) : Real;

{ This function generates random numbers between 0 and 'wiggle' that
  have a Gaussian Distribution.  Most PC random number generators don't. }

Var GaussAdd        :   Real;
    GaussFac        :   Real;

Begin
    sum:=0;
    GaussAdd:=sqrt(12);
    GaussFac:=(2*GaussAdd)/(4*Wiggle);
    for j:=1 to 4 do
        sum:=sum+random(Wiggle);
    Gauss:=GaussFac*sum-GaussAdd;
end;

Procedure PlotArray;

{ Now plot the contents of the array of numbers.           }

Begin
   ChoosePalette;
   MoveTo(0,trunc(GetMaxY/2));
   for k:=1 to 200 do begin
       LineTo(k*3,GetMaxY-(trunc(Brown[k]*Scale)+(trunc(GetMaxY/2))));
    end;
end;

Procedure CalculateBrownian;

{ Calculate array of numbers, place in array called Brown }

Begin
    for k:=0 to 200 do
        Brown[k]:=0;
    Randomize;
    N:=200;
    MaxSteps:=200;
    for i:=1 to MaxSteps do
        Begin
            k0:=random(N);
            k1:=k0+trunc(N/2)-1;
            A:=Gauss(1);
            for k:=k0 to k1 do begin
                if (k<N) then
                    Brown[k]:=Brown[k]+A
                else
                    Brown[k]:=Brown[k]-A;
            end;
      end;
end;

Begin
    Wiggle:=20;
    Scale:=4;
    CalculateBrownian;
    PlotArray;
end.
```

Listing 6-4

```
Program CantorSet;

{       Generates a Cantor Set       }

Uses Crt,Graph,Extend;

var         x1,x2,x3,x4         : integer;
            y1,y2               : integer;
            length              : integer;
            seg_length          : integer;
            depth               : integer;

Procedure Cantor(x1,y1,depth,length : integer);

Var     seg_length,x3,x4        : integer;

Begin
   seg_length:=trunc(length/3);
   x3:=x1+seg_length;
   x4:=x1+trunc(2*seg_length);
   SetColor(1);
   SetFillStyle(1,0);
   Bar(x3,trunc((depth-1)*10),x4,trunc((depth-1)*10-2));
   if depth<>0 then begin
      Cantor(x1,trunc((depth-1)*10),depth-1,seg_length);
      Cantor(x3,trunc((depth-1)*10),depth-1,seg_length);
      Cantor(x4,trunc((depth-1)*10),depth-1,seg_length);

   end;

end;

Begin

   ChoosePalette;

{  Note that a too high value of depth will give a rather odd effect due
   to lack of screen resolution.     }

   depth:=8;
   length:=200;
   x1:=100;
   x2:=300;
   y1:=10;
   y2:=100;

   SetColor(1);
   Bar(x1,y1,x2,y2);

   Cantor(x1,y1,depth,length);

end.
```

Listing 6-5

```
Program PeanoCurve;

{       Generates a Peano Curve.     }

Uses Crt,Graph,Extend;
```

TURBO PASCAL LISTINGS   175

```pascal
Procedure Peano(depth,length : integer);

{         This procedure generates a Peano curve via recursion. A turtle
          graphics approach (see text) is used to generate this curve.   }

Begin
   if depth=0 then Begin
      Front(length);
    end
   else
     begin
       Peano(depth-1,trunc(length/3));
       Turn(-90);
       Peano(depth-1,trunc(length/3));
       Turn(90);
       Peano(depth-1,trunc(length/3));
       Turn(90);
       Peano(depth-1,trunc(length/3));
       Turn(90);
       Peano(depth-1,trunc(length/3));
       Turn(-90);
       Peano(depth-1,trunc(length/3));
       Turn(-90);
       Peano(depth-1,trunc(length/3));
       Turn(-90);
       Peano(depth-1,trunc(length/3));
       Turn(90);
       Peano(depth-1,trunc(length/3));

  end;
end;

Begin

   ChoosePalette;

{ Set up initial TurtleX and TurtleY values. These variables defined in the
  Extend Library. Then move to start position, and set initial Turtle
  Direction to 0'.
  }

   TurtleX:=90;
   TurtleY:=trunc(GetMaxY/2);
   MoveTo(TurtleX,TurtleY);
   TurtleTheta:=0;

   Peano(3,400);

end.
```

Listing 6-6

```pascal
Program PeanoCurveScreenSaver;

{       Generates a Peano Curve.        }

Uses Crt,Graph,Extend;

Var               saved : array[1..8000] of byte;
                  current_point : integer;
                  i,j           : integer;

Procedure SaveData(distance : Integer);
```

```
Begin
   TurtleX:=trunc( distance*cos(TurtleTheta*3.14/180)) + TurtleX;
   TurtleY:=trunc( distance*sin(TurtleTheta*3.14/180)) + TurtleY;
   saved[current_point]:=GetPixel(TurtleX,TurtleY);
end;

Procedure WriteData(distance : Integer);

Begin
   TurtleX:=trunc( distance*cos(TurtleTheta*3.14/180)) + TurtleX;
   TurtleY:=trunc( distance*sin(TurtleTheta*3.14/180)) + TurtleY;
   PutPixel(TurtleX,TurtleY,saved[current_point]);
end;

Procedure Turn(theta : integer);
Begin
   TurtleTheta:=(TurtleTheta + theta) mod 360;
end;

Procedure SavePeano(depth,length : integer);

{         This procedure generates a Peano curve via recursion.
          and uses it to save a screen portion. A turtle
          graphics approach (see text) is used to generate this curve.     }

Begin
   if depth=0 then Begin
      SaveData(length);
      current_point:=current_point+1;
    end
    else
      begin
      SavePeano(depth-1,trunc(length/3));
      Turn(-90);
      SavePeano(depth-1,trunc(length/3));
      Turn(90);
      SavePeano(depth-1,trunc(length/3));
      Turn(90);
      SavePeano(depth-1,trunc(length/3));
      Turn(90);
      SavePeano(depth-1,trunc(length/3));
      Turn(-90);
      SavePeano(depth-1,trunc(length/3));
      Turn(-90);
      SavePeano(depth-1,trunc(length/3));
      Turn(-90);
      SavePeano(depth-1,trunc(length/3));
      Turn(90);
      SAvePeano(depth-1,trunc(length/3));

  end;
end;

Procedure WritePeano(depth,length : integer);

{         This procedure generates a Peano curve via recursion.
          and uses it to write a screen portion. A turtle
          graphics approach (see text) is used to generate this curve.     }

Begin
   if depth=0 then Begin
      WriteData(length);
      current_point:=current_point+1;
    end
```

TURBO PASCAL LISTINGS

```
    else
      begin
        WritePeano(depth-1,trunc(length/3));
        Turn(-90);
        WritePeano(depth-1,trunc(length/3));
        Turn(90);
        WritePeano(depth-1,trunc(length/3));
        Turn(90);
        WritePeano(depth-1,trunc(length/3));
        Turn(90);
        WritePeano(depth-1,trunc(length/3));
        Turn(-90);
        WritePeano(depth-1,trunc(length/3));
        Turn(-90);
        WritePeano(depth-1,trunc(length/3));
        Turn(-90);
        WritePeano(depth-1,trunc(length/3));
        Turn(90);
        WritePeano(depth-1,trunc(length/3));
      end;
end;

Begin

   ChoosePalette;

{ Set up initial TurtleX and TurtleY values.  These variables defined in the
  Extend Library.  Then move to start position, and set initial Turtle
  Direction to 0'.
  }

   for i:=1 to 4000 do
      saved[i]:=0;

   Circle(200,100,40);
   FloodFill(200,100,15);

   TurtleX:=90;
   TurtleY:=trunc(GetMaxY/2);
   MoveTo(TurtleX,TurtleY);
   TurtleTheta:=0;
   current_point:=1;
   SavePeano(4,400);

   ClearDevice;

   TurtleX:=90;
   TurtleY:=trunc(GetMaxY/2);
   MoveTo(TurtleX,TurtleY);
   TurtleTheta:=0;
   current_point:=1;
   WritePeano(4,400);

end.
```

Listing 6-7

```
Program CCurveGenerator;

{       Generates a C Curve.       }

Uses Crt,Graph,Extend;
```

```
Procedure CCurve(depth,length : integer);

Begin
  if depth=0 then Begin
     Front(length);

  end
  else
    begin

      CCurve(depth-1,length);
      Turn(-90);
      CCurve(depth-1,length);
      Turn(90);

  end;
end;

Begin

  ChoosePalette;

{ Set up initial TurtleX and TurtleY values.  These variables defined in the
  Extend Library.  Then move to start position, and set initial Turtle
  Direction to 0'.
}

  TurtleX:=trunc(GetMaxX/2);
  TurtleY:=trunc(GetMaxY/2);
  MoveTo(TurtleX,TurtleY);
  TurtleTheta:=0;

  CCurve(9,6);

end.
```

Listing 6-8

```
Program MakeDragonGenerator;

{        Generates a Dragon Curve.
         Based on algorithm in Becker and Dorfler  }

Uses Crt,Graph,Extend;

Var   theta           : integer;
      direction       : integer;

Procedure MakeDragon(depth,length : integer);

Begin
  if depth=0 then
     Front(length)
  else
    begin

      if depth>0 then begin
         depth:=depth-1;
         MakeDragon(depth,length);
         turn(-theta);
         MakeDragon(-(depth),length);
      end;
```

TURBO PASCAL LISTINGS   179

```pascal
        if depth<0 then begin
           depth:=depth+1;
           MakeDragon(-depth,length);
           turn(theta);
           MakeDragon(depth,length);
        end;

   end;
end;

Begin

   ChoosePalette;

{ Set up initial TurtleX and TurtleY values.  These variables defined in the
  Extend Library.   Then move to start position, and set initial Turtle
  Direction to 0'.
  }

   TurtleX:=trunc(GetMaxX/2);
   TurtleY:=trunc(GetMaxY/2);
   MoveTo(TurtleX,TurtleY);
   TurtleTheta:=0;

{  The value of theta determines the way in which the dragon grows; try a
   few different values.       }

   theta:=90;
   direction:=1;

   MakeDragon(9,5);

end.
```

Listing 6-9

```pascal
Program SierpinskiCarpet;

{       Generates a Sierpinski Carpet       }

Uses Crt,Graph,Extend;

var         x1,x2,x3,x4         : integer;
            y1,y2,y3,y4         : integer;
            length              : integer;
            seg_length          : integer;
            depth               : integer;

Procedure Sierpinski1(x1,y1,x2,y2,depth,length : integer);

Var    seg_length,x3,y3,x4,y4           : integer;

Begin
   seg_length:=trunc(length/3);
   x3:=x1+seg_length;
   y3:=y1+seg_length;
   x4:=x2-seg_length;
   y4:=y2-seg_length;
   SetColor(0);
   SetFillStyle(1,0);
   Bar(x3,y3,x4,y4);
   if depth<>0 then begin
```

```
           Sierpinski1(x1,y1,x3,y3,depth-1,seg_length);
           Sierpinski1(x3,y1,x4,y3,depth-1,seg_length);
           Sierpinski1(x4,y1,x2,y3,depth-1,seg_length);
           Sierpinski1(x1,y3,x3,y4,depth-1,seg_length);
           Sierpinski1(x4,y3,x2,y4,depth-1,seg_length);
           Sierpinski1(x1,y4,x3,y2,depth-1,seg_length);
           Sierpinski1(x3,y4,x4,y2,depth-1,seg_length);
           Sierpinski1(x4,y4,x2,y2,depth-1,seg_length);

      end;

end;

Begin

    ChoosePalette;

  { Note that a too high value of depth will actually 'eat away' all
    of the carpet!     }

    depth:=3;
    length:=200;
    x1:=100;
    x2:=300;
    y1:=0;
    y2:=200;

    SetColor(1);
    Bar(x1,y1,x2,y2);

    Sierpinski1(x1,y1,x2,y2,depth,length);
end.
```

Listing 6-10

```
Program L_LanguageSystem;

  { Program to generate Fractal Curves based on the L-Language System
    described in 'The Science of Fractal Images.                    }

  Uses         Crt,Graph,Extend;

  Var          stackX,stackY,stackA    :   array[1..30] of integer;
               axiom                   :   string;
               prodrule                :   array[1..30] of string;
               rule_code               :   array[1..30] of string;
               number_rules            :   integer;
               command                 :   string;
               theta,forwardstep       :   Integer;
               application             :   integer;
               current                 :   string;
               stack_pos               :   integer;
               i,j,k                   :   integer;
               AxiomExtended           :   boolean;

Procedure FrontTurtle;
Begin
   TurtleX:=trunc( forwardstep*cos(TurtleTheta*3.14/180)) + TurtleX;
   TurtleY:=trunc( forwardstep*sin(TurtleTheta*3.14/180)) + TurtleY;
   LineTo(TurtleX,TurtleY);
end;
```

TURBO PASCAL LISTINGS

```pascal
Procedure RightTurnTurtle;
Begin
   TurtleTheta:=(TurtleTheta + theta) mod 360;
end;

Procedure MoveTurtle;
Begin
   TurtleX:=trunc( forwardstep*cos(TurtleTheta*3.14/180)) + TurtleX;
   TurtleY:=trunc( forwardstep*sin(TurtleTheta*3.14/180)) + TurtleY;
   MoveTo(TurtleX,TurtleY);
end;

Procedure LeftTurnTurtle;
Begin
   TurtleTheta:=(TurtleTheta - theta) mod 360;
end;

Procedure PushTurtle;
Begin
   stack_pos:=stack_pos+1;
   stackX[stack_pos]:=TurtleX;
   stackY[stack_pos]:=TurtleY;
   stackA[stack_pos]:=TurtleTheta;
end;

Procedure PopTurtle;
Begin
   TurtleX:=StackX[stack_pos];
   TurtleY:=StackY[stack_pos];
   TurtleTheta:=StackA[stack_pos];
   stack_pos:=stack_pos-1;
end;

Procedure ReverseTurtle;
Begin
   TurtleTheta:=TurtleTheta+180;
end;

Procedure TurtleInterpret;
Begin

{  This procedure takes the string in command and extracts from it the turtle
   graphics commands and acts upon them to draw the image required.        }

for i:=1 to length(command) do begin

   current:=copy(command,i,1);
   if (current='F') or (current='f') then
      FrontTurtle;
   if (current='M') or (current='m') then
      MoveTurtle;
   if (current='R') or (current='r') then
      ReverseTurtle;
   if (current='+') then
      RightTurnTurtle;
   if (current='-') then
      LeftTurnTurtle;
   if (current='[') then
      PushTurtle;
   if (current=']') then
      PopTurtle;

   end;

end;
```

```
Procedure ExpandAxiomString;

{       This currently expands a string to no more than 255 characters.  This
        could
        be extended by direct RAM access or a different data structure.  }

Begin

{  Loop around for 'application' times to expand the axiom the correct number
   of times      }

  for i:=1 to application do begin

      command:='';
      for j:=1 to length(axiom) do begin

          current:=copy(axiom,j,1);
          AxiomExtended:=FALSE;

          for k:=1 to number_rules do begin

              if (current=prodrule[k]) then begin
                 command:=command+rule_code[k];
                 AxiomExtended:=TRUE;
              end;
          end;
          if (AxiomExtended=FALSE) then
             command:=command+current;

      end;

{  Old axiom string is now replaced by the expanded command string, and the
   expanded command string is now processed.    }

      axiom:=command;
   end;

end;

begin

ChoosePalette;

{  Next block of statements initialises the turtle   }

theta:=60;                 { angle size for this axiom and production rule }
forwardstep:=20;           { forward step for this axiom and production rule }
TurtleTheta:=0;
TurtleX:=200;
TurtleY:=150;

{ The following axiom and production rule is for a Koch Curve.      }

axiom:='F';
prodrule[1]:='F';
rule_code[1]:='F-F++F-F';
number_rules:=1;
application:=8;

MoveTo(TurtleX,TurtleY);
ExpandAxiomString;
TurtleInterpret;

end.
```

Listing 6-11

```pascal
Program RandomVonKoch;

{       Generates a Von Koch Coastline.       }

Uses Crt,Graph,Extend;

Procedure VonKoch(depth,length : integer);

{         This procedure generates a random Von Koch curve via recursion. A
          turtle
          graphics approach (see text) is used to generate this curve.     }

Begin
   if depth=0 then Begin
      Front(length);
     end
   else
     begin

{ Random number function used to generate the length and angle.   }

       VonKoch(depth-1,trunc(random(trunc(length/3))));
       Turn(-trunc(random(60)));
       VonKoch(depth-1,trunc(random(trunc(length/3))));
       Turn(trunc(random(120)));
       VonKoch(depth-1,trunc(random(trunc(length/3))));
       Turn(trunc(-random(60)));
       VonKoch(depth-1,trunc(random(trunc(length/3))));

   end;
end;

Begin

   ChoosePalette;

{ Set up initial TurtleX and TurtleY values. These variables defined in the
  Extend Library. Then move to start position, and set initial Turtle
  Direction to 0'.
  }

   TurtleX:=90;
   TurtleY:=trunc(GetMaxY/2);
   MoveTo(TurtleX,TurtleY);
   TurtleTheta:=0;

{ Randomize function used to seed random number generator.
 }
   Randomize;
   VonKoch(4,6000);

end.
```

Chapter 7

The Mandelbrot set

It has been called the most complex object known to man, and hailed as the most beautiful mathematical object there is. Exhibitions of computer-generated art from this object have toured the world, and the shape shown in the colour plates of this book adorns many a book cover, record sleeve, tee-shirt and computer display. Along with the twisted figure of eight of the Lorenz attractor, the Mandelbrot set has almost become a trademark of chaos. For us, the Mandelbrot set represents another meeting point of chaotic systems and fractal geometry. In this chapter, I'll explore exactly what the set is, how we can generate it, and other avenues of exploration based on the techniques used to generate the set.

The Mandelbrot set, illustrated in the colour plates of this book, and beautifully documented in *The Beauty of fractals* by Peitgen and Saupe, is a fractal object exhibiting the following characteristics:

1. It is extremely complicated, giving great complexity at all levels of magnification. A close-up view of the set will give us a complicated close up, and if you magnify sections of the set millions of times you'll still find infinite complexity.

2. Although it exhibits great complexity at many scales of magnification, it does not exhibit the exact self-similarity of the fractal curves described in the previous chapter. In fact, so rich is the variety of images to be found within the Mandelbrot set that there's not any statistical self-similarity either. The set combines infinite complexity whilst maintaining some similarity.

3. Deep within the Mandelbrot set you will find shapes that are almost the same as the parent set—almost, but not quite. Each of these babies exhibits similar characteristics to the parent.

4. The Mandelbrot set is generated by an iterative process that can be coded in a very short program. The function that is iterated is a function involving *complex numbers*, as we'll see in detail below.

5. The Mandelbrot set itself, in strict mathematical terms, is the collection of black points within the brightly coloured fringe of the generated image. The brightly coloured area surrounding the set is called the *boundary*. The area outside the boundary consists of points that are not within the Mandelbrot set. Even within the boundary region, magnification of areas of the boundary will reveal areas that do belong to the Mandelbrot set.

Complex numbers

The Mandelbrot set is generated by iterating a function that contains complex numbers. So far, all the numbers we've dealt with have been what are called *real numbers*. These are the numbers that we deal with in day-to-day arithmetic, such as 1.2, 432, −65, and so on. We can stick these numbers in equations, for example, and work out the values of unknown parameters:

$$x + y = 5$$
$$x * y = 6$$

is fairly easily solved if we say $x = 2$, $y = 3$. Similarly, for the equation $x * x = 1$, the solution is given by $x = 1$. However, we have a problem if we consider an equation like:

$$x * x = -1$$

If you try getting the square root of a negative number on your PC or calculator, there is likely to be the computing equivalent of 'Oh no you don't!'. It just can't be done. In fact, you can't describe the square root of a negative number in terms of the real numbers that we've dealt with so far. However, problems of finding a solution for equations involving the square roots of negative numbers have been around for some considerable time in mathematics, so it was essential that a method of representing a solution for such equations was obtained. This came in the form of *imaginary numbers*, and the simplest imaginary number is called i, and is defined as the square root of −1. Thus:

$$i * i = -1$$

You can combine i with other numbers, to give expressions like $3 * i$, $i + i$, $i - i$ and so on. By the way, just to make life difficult you may find i called j in engineering textbooks! The use of i isn't immediately obvious, but if you look at standard textbooks for electronics and other branches of physics, you'll find that it's absolutely essential. You very rarely find

COMPLEX NUMBERS

imaginary numbers on their own. More often than not an imaginary number is combined with a real number to create what is called a *complex number* such as:

$$1 + 3i$$
$$4 - 4i$$

A complex number consists of two parts; a real part and an imaginary part. In mathematics, complex numbers are often represented by the letter z, so when you see this letter used to represent a complex number it's important to remember that z actually represents something like:

$$z = a + b * i$$

where a is the real part of the number and $b * i$ is the imaginary part of the number. You can do arithmetic with complex numbers, and in doing so you treat the numbers in the following way, assuming $z_1 = a + i * b$ and $z_2 = c + i * d$.

$$\begin{aligned}
z_1 + z_2 &= (a+c) + i*(b+d) \\
z_1 - z_2 &= (a-c) + i*(b-d) \\
z_1 * z_2 &= (a*c - b*d) + i*(a*d + b*c) \\
z_1 / z_2 &= ((a*c + b*d)/(c*c + d*d)) + \\
&\quad i*((b*c - a*d)/(c*c + d*d))
\end{aligned}$$

One thing that you'll have noticed is that the dreaded i turns up in the right-hand side of these equations. A further way of representing complex numbers is to use what is called the *Gaussian* or *Argand* plane representation. Here, a graph is drawn with the real part of the number represented

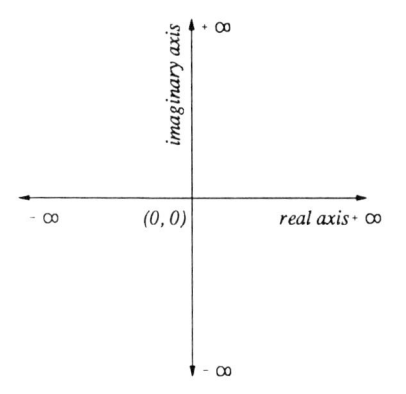

Figure 7-1. The complex plane.

by the x-axis, and the imaginary part of the number represented by the

y-axis of the graph (see Figure 7-1). In this representation, we can add two numbers together by simply plotting z_1 on the plane using its real and imaginary parts; then, using the position of z_1 as the new start point, we can plot the real and imaginary parts of z_2 as offsets from that point (see Figure 7-2). The *absolute* value of a complex number represented in this way is given by the distance between the origin point of the plane $(0,0)$ and the number. This can also be obtained by use of Pythagoras' Theorem, giving:

$$|z| = \sqrt{a*a + b*b}$$

where $|z|$ is the absolute value of the complex number, a is the real part and b is the imaginary part of the number. To give an example of this, the absolute value of $2 + 4i$ is given by:

$$\begin{aligned} |z| &= \sqrt{2*2 + 4*4} \\ &= \sqrt{4 + 16} \\ &= \sqrt{20} \\ &= 4.47 \end{aligned}$$

As we've just seen, we can do arithmetic with complex numbers just as with real numbers, so it is quite easy to see how we could have an iterative system that consists of repeated iterations of a function involving complex

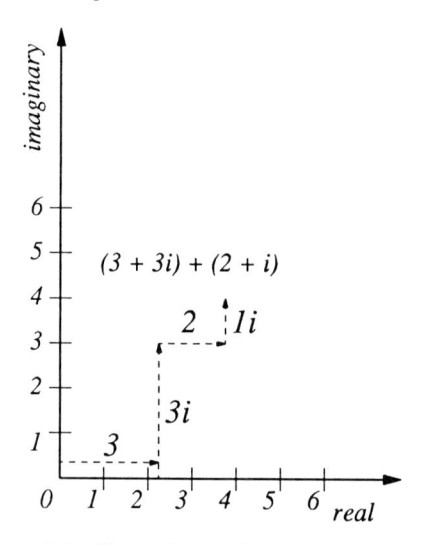

Figure 7-2. The addition of two complex numbers.

numbers. Let's see what happens when we do this. You may remember in Chapter 2 that we iterated the function:

$$x_{n+1} = x_n^2$$

COMPLEX NUMBERS

That is, we square a real number, x, and then use that number as the seed for the next iteration. As you may recall, the actual trajectory of the iterated function depended upon the initial x value; for example, $x > 1$ resulted in the attractor being at infinity, $x < 1$ resulted in the attractor being at 0 and $x = 1$ resulted in the attractor at 1. The iteration of complex functions was something mathematicians had wondered about for some time, but it was not feasible to work on this subject in any great detail until computers came along due to the complexity of the equations used when iterating the functions, as indicated by the rules of addition, subtraction, multiplication and division shown above. Having said that, some work was done in the early years of this century (see Chapter 8), but the recent explosion of interest in the iteration of complex functions started taking place in the 1970s, when Benoit Mandelbrot started iterating a complex function on IBM mainframe computers. The function that is iterated to give the Mandelbrot set is very simple:

$$z_{n+1} = z_n^2 + c$$

where z is a complex number which is initialised to a starting value then allowed to alter as the function is iterated, and c is a complex number that stays constant for a particular sequence of iterations.

Listing 7-1 iterates this function for a user-entered value of c and initial $z = 0$. The function is iterated ten times and the absolute value (calculated above) of z after each iteration is displayed. The method used to solve the squaring of z is a special case of the multiplication of complex numbers outlined above. We can use this program to see what happens when we iterate the function in a similar way to the simple program we used to start investigating the logistic equation in Chapter 2. Here, however, we're not plotting graphs.

You might like to try putting different real and imaginary values of c into the program to see what happens. A logical approach is to feed the program a series of real c values whilst keeping the imaginary part of c equal to 0, then try a series of imaginary c values whilst keeping the real part of c equal to 0. Finally, try mixing real and imaginary values. The sort of picture that you'll get is as follows:

Real	Imag	trend in absolute z value
0	0	0
1		infinity
−1		alternate between 1 and 0
0.3		sequence obtained tending towards 1 and beyond
0		Trend in absolute m value
	1	alternate between 1 and $\sqrt{2}$
	−1	alternate between 1 and $\sqrt{2}$

With values between 0 and 1, the picture obtained is more interesting, and with both real and imaginary parts of c between 0 and 1 some very interesting and apparently chaotic sequences are obtained. We shouldn't really be surprised about this, because the iteration of a complex function is not exactly a trivial process. Thus, as with iteration of functions using real numbers, the iteration of complex functions can also give rise to chaotic sequences. This simple program also gives us some useful insights into programming for iterating complex functions, as follows:

1. As with functions involving the iteration of real numbers, the results returned by the function can become very large and easily outstrip the numerical resolution of the computer.

2. It is important to remember that we are calculating the $n+1$ state from the nth state in all these iterative processes. This means that when we are making these calculations, we have to make the calculations of the new values using the old values; we have to take care that we don't calculate a new imaginary part of a number and use the *new* imaginary value to calculate the corresponding new real number. We should, of course, be calculating the new real number using the *old* imaginary part of the number. This is most important, as the results obtained from experiments with these functions will be quite different if you don't take this into account.

These results indicate that these complex functions, when iterated, give rise to attractors in the same way as functions gave rise to Hénon or Martin's mappings, as seen earlier in this book. So, what we can do is to look at these attractors by plotting the real and imaginary parts of z calculated as the iterations proceed. Listing 7-2 is a modification of Listing 7-1 that plots the trajectories that result from this iterative process, and Figure 7-3 shows a typical result from this program, by plotting the real part of z as the x coordinate and the imaginary part of z as the y coordinate. The results obtained from different real and imaginary values of c are different, and small changes in values will give different images. As with many dynamic systems, the final results are interesting but so is the process by which the final image is constructed. In addition, you may find that, as with all chaotic systems, different computers or languages may give similar but slightly different results due to rounding errors, etc.

The Mandelbrot set

OK, so I took some time in getting here, but we're now ready to start examining the Mandelbrot set in some detail. We've already seen how,

THE MANDELBROT SET

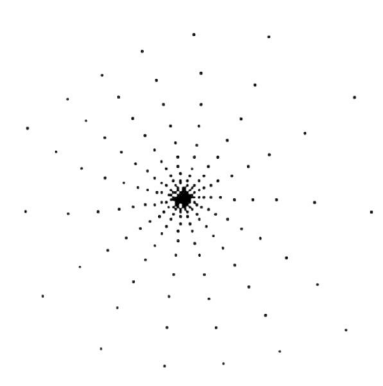

Figure 7-3. An example iteration of a complex function.

for certain c values and $z = 0$, the function $z * z + c$ can be iterated to generate sequences of numbers that behave chaotically or tend towards a specific attractor—infinity. In Listing 7-2 we drew the chaotic trajectories of some of these sequences. There is another way of looking at these iterative sequences, and that is to examine them in terms of what is called the escape time for the c value. You'll often read of sequences escaping towards infinity, and the escape time is a measure of the number of iterations needed for this to happen to a sequence for a particular c value. This can be done as follows:

1. Split the computer screen into a grid, centred on $(0,0)$, representing real numbers on the x-axis (say from -2 to $+2$), and imaginary numbers on the y-axis, from say -2 to $+2$. This is reminiscent of Figure 7-1.

2. Select a point on this grid, representing a particular complex number, the real part being represented by the x position and the imaginary part being represented by the y position. This complex number we will call c.

3. Set a complex number, z, to 0 and iterate the function $z^2 + c$ until the absolute value of z goes off to infinity, or, after a pre-specified number of iterations, is still fairly small.

4. If the function has iterated to infinity, then plot the point on the screen corresponding to the real and imaginary values of c in a colour

that is related in some way to the number of iterations needed to iterate to infinity. If the number is still small after iterating it for the pre-determined number of steps, then we assume that this value of c will not iterate to infinity, so colour the corresponding point black.

5. Repeat this step for all points on the grid.

Although we've talked about infinity in this algorithm, in practical terms we can say that if the absolute value of the complex number generated exceeds 2, then it will iterate off to infinity. This process generates the *Mandelbrot set*, as shown in the colour plates, which is the dark area of the screen representation of the complex plane representing those c values that did not go shooting off to infinity. The coloured areas around the set indicate areas where c values go off to infinity after a number of iterations corresponding to a colour on the grid. The Mandelbrot set is thus defined as the set of complex numbers, c, for which $z^2 + c$ does not go to infinity after an infinite number of iterations. The set itself is rather boring when delineated in this way; after all, it's just black! However, the border of the set, where values of c take varying amounts of time to wander off to infinity, is the fascinating part of the image. You will often find the Mandelbrot set referred to as the *M* set in mathematical textbooks. By the way, some writers generate the Mandelbrot set by iterating the function:

$$z = z^2 - c$$

rather than the one we've been using. This generates a set like the one we've seen, but flipped around the imaginary axis going through the origin of the complex plane (0,0). This can occasionally be confusing at first glance, so if you see a Mandelbrot set that looks the wrong way around, then the chances are that it has been generated from this function.

Listing 7-3 provides a program to generate the Mandelbrot set, and before describing its use I'll detail some of the programming tricks used. One point to note is that this routine is *not* optimised for speed in any way. In fact, it's written so that you can see what's happening. Some useful points to note are:

1. The value *deltaRnum* is the amount by which the real part of c is incremented from the initial real value. It is calculated by taking the difference between the maximum real value entered and the minimum real value entered and dividing this by the maximum x coordinate value of the screen.

2. The value *deltaImaginary* is used to increment the imaginary part of c in a similar fashion.

3. Two repeat loops are used (in the Pascal program) to scan the area of the complex plane of interest a column at a time. In each set of

iterations, x and y, representing the real and imaginary parts of z, are both set to 0 before the iteration is started.

4. The colours assigned to each point, dependent upon the escape time for the value of c under examination, are specified by increasing the value of the variable *colour* by 1 at each iteration. Should the program take more iterations than there are colours in the palette of the graphics mode in use, then we use the *colour* variable *modulo* the number of colours to give a value that is within the range of colours available. This is very simple, and later in this chapter you'll see how we can improve matters by taking a little care about how colours are chosen.

5. The variable *max_iterations* determines how many iterations are gone through before the program decides that the point under consideration is not going to escape to infinity. The default value of 256 suggested in this program is often adequate, although for higher magnifications of smaller areas of the set a higher value will give better results. Low values of this parameter will give crude, low-resolution images.

6. The two loops within this program can be exited by pressing a key while the set generation is in progress; at this point, a screen image is saved on the disc which can be recovered and displayed using the *LoadScreen* function from the Turbo Pascal software library.

On running the program, use the default values prompted by the program. These allow the whole Mandelbrot set to be drawn. Figure 7-4 shows how the set maps on to the Argand plane. This diagram can be useful when you wish to draw just small portions of the Mandelbrot set. Be warned; the set drawing process as documented here takes a considerable length of time. Methods of speeding up this process are described below.

Nomenclature

The Mandelbrot set has effectively been mapped by different workers, and the different parts of the graphical representation as seen in the plates have been given different names. In addition, you'll come across various names given to the set itself. Two common names are the *Gingerbread Man*, particularly popular with European workers and the fairly descriptive *Beetle*.

194   CHAPTER 7. THE MANDELBROT SET

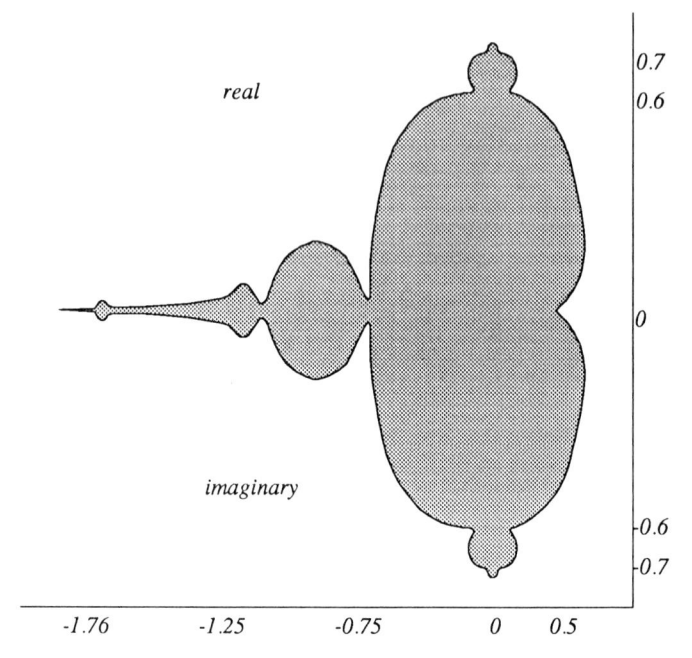

Figure 7-4. A mapping of the Mandelbrot set on to the complex plane.

Finite attractor basin

This is the black area in the centre of the full set. It has a fractal boundary, as we will see when we investigate magnified views of the Mandelbrot set. It is called a basin because it is an area within which a point tends towards a certain attractor—in this case the value of the points stay finite, and follow trajectories as explored by Listings 7-1 and 7-2. It is also called the *cardioid*, after its heart-shaped appearance.

Equipotential lines

The coloured bands outside the immediate boundary of the Mandelbrot set represent areas in which the points all take the same amount of time to escape to infinity. These are called *equipotential* lines, or *contour* lines. There is a method of illustrating the Mandelbrot set in which the escape times are shown as vertical columns in a pseudo-three-dimensional image, and in this sort of representation you can see why these areas are called contour lines, as lines appear like the slope of a hill, getting steeper as the boundary is approached.

Buds

These are the baby Mandelbrot sets that are stuck on to the side of the main, central part of the set. The boundary regions of these buds often give very attractive images when magnified.

Satellites

The Mandelbrot set sits in the middle of the complex plane; surrounding it at some distance from the main set are smaller areas that look like the Mandelbrot set but are apparently unconnected to it. There is considerable mathematical evidence to suggest that these satellites are linked to the main set by filaments that are only visible under extreme magnification. The most obvious satellite is that on the real axis (imaginary part=0) at about +1.8 along the real axis. This is clearly visible in Plate 1 and this region merits further investigation.

Filaments

A *filament* is a stringy extension of the set, like the one that extends along the Real axis. It's been suggested that infinitesimally small filaments may join satellites to the main set itself, but these are most unlikely to be seen in any computer experiments due to the great magnification that is theoretically required.

Magnifying the Mandelbrot set

The program in Listing 7-3 allows you to home in on specific areas of the set by defining the maximum real and imaginary areas of the complex plane to scan in order to create a magnified image of part of the set. The smaller the area of the complex plane represented on the computer graphics screen (i.e., the smaller the distance between the maximum and minimum real and imaginary values) the larger the magnification. Magnification allows us to see deep into the detailed structure of the Mandelbrot set, and by doing this we can see the fractal nature of it and its infinite complexity. When magnifying areas of the set you may also find that you need to change the maximum number of iterations to get a high-resolution picture of a magnified portion of the set.

There are limitations to the magnification that can be obtained from a particular computer or language. These are set by the mathematical precision available, but typically will allow magnifications of a couple of hundred thousand times before problems arise in most high-level languages. The use of maths coprocessors can help here, as the coprocessor will have a higher level of precision in most cases than will the language by itself.

Problems with mathematical resolution will manifest themselves as they do in all chaotic systems; images will result that are artifacts, rather than real images of the mathematical object under examination.

The technique of magnification of areas of the complex plane in this way is quite important in investigations of the Mandelbrot set and of Julia sets (Chapter 8). You might like to consider making the task easier in your own programs by offering the user a means of, for example, defining a square area on the screen and having the program sort out the real and imaginary values corresponding to the edges of the defined box.

Interesting areas

It's not easy to point out interesting areas of the Mandelbrot set, as there are so many—an infinite number, in fact. Each point within the boundary of the set can be magnified to generate a brand new image, the detail of which will depend upon the area chosen and the magnification applied. Within the set, magnifications of different areas will produce images such as those shown in Plates 2 to 6. Also indicated here are the maximum and minimum real and imaginary parts of c used to generate the image, along with the maximum number of iterations used. Don't forget it's the boundary of the set where the stunning images can be found, so investigate there first. Here are a few pointers to use when investigating areas of the set yourself.

1. Use a series of increasing magnifications to home in on potentially interesting areas. A useful tool is a sheet of transparent acetate onto which has been photocopied the grid from a sheet of 1 or 2 mm graph paper. This can be laid on the computer screen and will allow you to read off real and imaginary values that can be entered to define an area to magnify.

2. Areas within the Mandelbrot set (i.e. the black area) will take the longest time to calculate, as in these instances you will iterate the maximum number of times before the program determines that the point is within the set. This is particularly the case with higher magnifications which, as I've already mentioned, benefit from a larger value of $max\_iterations$.

3. When examining small areas of the set (i.e. using large magnification) you may sometimes see less detail than you expect. There are two possible reasons for this; the first is that you've entered the area to be displayed incorrectly; the second is that the $max\_iterations$ figure was too low.

Colour selection and aesthetics

Once you've got a program that generates Mandelbrot set images, you can start playing with the image itself. There are two main things that determine how visually appealing an image is. These are the aspect ratio of the displayed image—the ratio of height to width—and the colours chosen. The aspect ratio can be specified when entering the ranges of real and imaginary areas to plot; in general, I like to keep the difference between the maximum and minimum real and imaginary values the same but this does give a slightly elongated image on the screen. You can, of course, create a square image by using the same number of pixel points in the x direction that you have in the y direction.

With regard to colour, the approach taken in Listing 7-3 is very simple, but better effects can be obtained by adopting a slightly more complicated arrangement for defining what colour to paint a particular pixel. Clearly, images are best served by video systems that will give as wide a range of colours as possible, but even two-colour images can be quite effective. Whilst on the subject of screen images, it's useful to note that the resolution of the graphics screen of your computer will limit the detail you can see in the set under magnification, and there will come a point where an increasing value of *max_iterations* will not necessarily lead to an increased resolution.

The most obvious way to make better use of the available colours is to use an array containing colours to plot, and then select a colour from the array dependent upon the number of iterations needed for a particular point to escape. The colours in the array can be ordered in a pattern that gives a more pleasing combination than that produced by simply using the colours in the order in which the PC palette supplies them. Listing 7-4 shows this in action (the PC version being for a 16-colour EGA/VGA screen), and when run gives a rather different view of the set (see colour plates) which I find quite attractive. A further modification, giving an overall image much like the one in Plate 1, is given by modifying the colour-selection part of the program to use the mod function to select a colour. In the listings, the variable *colour* counts the number of iterations the point has undergone before escape. On the PC version, the term:

    plotcolour[trunc(colour/10)]

is replaced by:

    plotcolour[trunc(colour MOD 10)]

An alternative way of selecting colours is to apply a function to the number of iterations needed for the point to escape. The simplest function to try here is probably a logarithm. For example, using $LN(colour + 1)$ to select the colour will choose a colour based on the natural logarithm of the iteration number. The +1 term is needed as otherwise, when a point fails to escape, the way in which the program is written would lead to a $LN(0)$ operation being attempted, which will generate an error on most computers. Try this approach; it certainly is well worth it to maximise the effect that an image can have.

Playing around with colours can also be done by using a screen grabber to save the finished image to a file format capable of being loaded into a computer graphics package; you can then modify the colours of the finished image in the package. This method is slightly different to the methods used above, as they can show us something about the dynamics of the construction of the Mandelbrot set—this method is strictly aesthetic!

The latter point is well worth bearing in mind. The iteration number in its raw state tells us about the length of time that that point took to escape towards infinity; if we use functions like MOD, then we are likely to get areas of the screen image with the same colour that took *different* numbers of iterations to escape to infinity. The more colours you have available, the less the risk of this is, but with 16 colours on VGA/EGA we're going to encounter it. The 255 colours offered by the enhanced VGA modes (if supported by your programming language/computer graphics card) offer us better images.

Time-saving steps

Once you've played around with this program, you'll find that it's rather slow. There are quicker algorithms for investigating the Mandelbrot set, but I don't intend to cover them in this book. They are detailed in some of the other works listed in the Bibliography. In terms of speeding up the programs shown, staying with the rather simple brute-force algorithm employed therein, you might like to consider the following.

Maths coprocessor

A coprocessor will make a considerable difference here, due to the floating point calculations involved. In addition, the use of a maths coprocessor will allow the use of higher resolution in the calculations. Of course, this assumes two things:

the first is that you are using a PC (maths coprocessors aren't available for most other home computers) and the second is that the language used actually recognises the presence of and supports the coprocessor. Not all language interpreters or compilers do this.

Repeating instructions

A good rule is that any instructions in a computer program should only be executed the minimum number of times needed to get the result desired. In other words, make sure there is no superfluous code within loops, especially the inner loops of programs. In the Mandelbrot set program listed, there are three loops, one for the column, one for the row and the third loop to iterate the specified point. Thus, an instruction in the inner loop will be executed up to:

    maxrow * maxcol * max_iterations

times in the process of running a Mandelbrot set generation. So, for a 640 × 320 screen with 256 iterations per point, an instruction in the inner loop will be executed up to 52,500,000 times during the plotting of the set. Of course, this isn't usually the case as many points in the image are outside the Mandelbrot set and so will iterate to infinity quite quickly. However, for a magnified area of the set boundary, the number of times an inner-loop instruction is executed is likely to be large.

Variable Types

On some language interpreters or compilers, the use of integer variables for loop counters can have some impact on execution speed. For example, the use of the resident integer variables in BBC BASIC as loop control variables will cause the loop to be executed more quickly than using the non-resident variable types, due primarily to the BASIC interpreter knowing precisely where in memory to find these variables. Generally, the higher the numerical precision specified when setting up variables, the more processing time will be required when handling these variables, simply because of the greater amount of data that needs to be recovered from memory with the higher precision variables (because these are stored in more bytes than lower precision variables).

Calculation types

Computers are good at performing only two types of arithmetic operation quickly; addition and subtraction. Multiplication, division, etc., all take a considerable amount of processing time. In fact, the amount of time taken to multiply two numbers together can easily be 10 times as long as the time needed to add the same two numbers. Operations like the calculation of square roots, etc. are likely to take even longer. For this reason, we need to take the following points into account when using these operations:

1. Reduce the number of operations like SIN, COS, $SQRT$, etc., to the absolute minimum, and, if at all possible, *do not* carry them out inside a loop. One useful trick with SIN or COS functions, for example, is to create arrays containing the SIN or COS values over the angle range required in the program before any loops are executed. Then, within the loop, to calculate the value of $SIN(23)$ we'd simply say $x = sin\_val[23]$ where sin_val is an array of real numbers containing sines. Element 23 of the array would contain the sine of 23. You will note in Listing 7-3 that we don't calculate the real absolute value of the z term of the function; we leave it squared, and compare it to 4, rather than 2, when seeing if the function will escape to infinity. This saves us having to do a square-root operation.

2. Carry out a similar policy with operations like multiplication and division. In some algorithms, it's possible to optimise the commands to remove multiplication operations from the scene. For example, if we needed to work out $2*x$, we could actually replace this with $x+x$.

A worked example

We can apply these optimisations to the inner loop of the Mandelbrot set algorithm used in Listing 7-3:

```
x:=0
y:=0
repeat
   xsquare:=x*x
   ysquare:=y*y
   colour:=colour+1
   if (xsquare+ysquare)<=MaxSizeSquared then begin
      y:=2*x*y+(ImagMin+row*deltaimaginary)
      x:=xsquare-ysquare+(RnumMin+col*deltaRnum)
   end
until (xsquare+ysquare) > MaxSizeSquared or
      colour>MaxIterations
```

The first step we could do here is to remove the calculation of the $ImagMin$ and $RnumMin$ terms outside this loop. This would give us:

```
x:=0
y:=0
ImagInc:=ImagMin+row*deltaimaginary
RealInc:=RnumMin+col*deltaRnum
repeat
   xsquare:=x*x
```

DIFFERENT STARTING Z VALUES

```
            ysquare:=y*y
            colour:=colour+1
            if (xsquare+ysquare)<=MaxSizeSquared then begin
                y:=2*x*y+ImagInc
                x:=xsquare-ysquare+Realinc
            end
        until (xsquare+ysquare) > MaxSizeSquared or
              colour>MaxIterations
```

Of course, this generates a couple of new variables, but this isn't a problem. The next stage is to replace the multiplications with additions wherever possible. This gives us:

```
        x:=0
        y:=0
        ImagInc:=ImagMin+row*deltaimaginary
        RealInc:=RnumMin+col*deltaRnum
        repeat
            xsquare:=x*x
            ysquare:=y*y
            colour:=colour+1
            if (xsquare+ysqaure)<=MaxSizeSquared then begin
                y:=(x+x)*y+ImagInc
                x:=xsquare-ysquare+Realinc
    end
        until (xsquare+ysquare) > MaxSizeSquared or
              colour>MaxIterations
```

This has an impact on the execution speed of the program, as you'll find out if you alter Listing 7-3 to include these changes. The speed up of programs by this sort of tweaking can vary depending upon type of program and the number of times the tweaked sections of code are executed. For this reason, it's worth spending time on the sections of code executed most frequently.

The best way of getting a quicker performance, though, is to employ a faster algorithm, as already mentioned. Most of these different algorithms concentrate on reducing the amount of time spent investigating points in the set. For example, one method I have seen described in *Fractal Report* actually follows the boundary of the set, and never actually investigates the points that lie in the set!

Different starting z values

In the traditional Mandelbrot set, the function is iterated with an initial z value of 0. This is set by setting the real and imaginary parts of z to 0, as shown in Listing 7-3 when we set x (real) and y (imaginary) to 0. There's nothing to stop us using non-zero initial z values, and this can be done by replacing the two lines:

```
x:=0;
y:=0;
```

or the BASIC equivalents with lines that assign non-zero values to these variables. Considering the changes made, the resulting images are quite different to the Mandelbrot set, and again indicate the essentially chaotic nature of the iteration of the complex function $z^2 + c$. For example, you might try using $x = 0.5$ and $y = -0.5$ as starting values. One result of changing the initial x and y values is shown in the colour plates.

Mandelbroids

The Mandelbrot set is, strictly speaking, the set generated by the iteration of the function $z_{n+1} = z_n^2 + c$. However, similar images can be constructed from iterating other, similar complex functions. The easiest ones to start with are those involving higher powers of z, such as z^3 or z^4. The resultant sets are often called Mandelbroids due to their relationship to the classical Mandelbrot set. You could work out the real and complex parts of these equations from the rules for complex arithmetic given earlier in this chapter. However, to save time, here they are:

$$z = x + i*y$$

$$z^3 = x*x*x - 3*x*y*y + \quad \text{real}$$
$$i(3*x*x*y - y*y*y) \quad \text{imaginary}$$

$$z^4 = x*x*x*x + y*y*y*y - 6*x*x*y*y + \quad \text{real}$$
$$i(4*x*x*x*y - 4*x*y*y*y) \quad \text{imaginary}$$

Substituting these expressions into our Mandelbrot set listing will give different images to the Mandelbrot set than we've seen so far, but they are quite similar in many ways and will still exhibit the complexity of our original set. In addition, you could, of course, alter the term involving the constant c and see what happens; there is much experimentation to be done in these areas. One thing that you will note about patterns generated from these similar iterations is that they are often rather bumpy and angular images. However, they're quite interesting and offer a separate avenue for exploration, rather than digging deeper and deeper into the traditional Mandelbrot set. Here are a few functions you might like to try. The first one to try is the $z^3 + c$ function;. This is as follows:

MANDELBROIDS

```
x:=0
y:=0
ImagInc:=ImagMin+row*deltaimaginary
RealInc:=RnumMin+col*deltaRnum
repeat
   xsquare:=x*x
   ysquare:=y*y
   xt=x
   yt=y
   colour:=colour+1
   if (xsquare+ysquare)<=MaxSizeSquared then begin
      y:=3*xt*yt*yt-yt*yt*yt+ImagInc
      x:=xt*xt*xt-3*xt*yt*yt+Realinc
   end
until (xsquare+ysquare) > MaxSizeSquared or
      colour > MaxIterations
```

There are a couple of useful programming tips that come out of using this function. The first is to remember that the *xsquare* and *ysquare* values are computed so that we can calculate the absolute value of the complex number z. They have nothing at all to do with the power that z is raised to, so don't try calculating *xcubed* and *ycubed*! A further point to note is that the calculations are more complicated and so take longer to execute.

A further thing to try is the simple step of changing the form of the equation that is iterated away from the simple $z^n + c$ function. New functions can give very different images, but they are still derived by starting with a value of z and varying the real and imaginary parts of c. For example, you could try

```
x:=0
y:=0
ImagInc:=ImagMin+row*deltaimaginary
RealInc:=RnumMin+col*deltaRnum
repeat
   xsquare:=x*x
   ysquare:=y*y
   colour:=colour+1
   if (xsquare+ysquare)<=MaxSizeSquared then begin
      y:=2*x*y+sin(ImagInc)
      x:=xsquare-ysquare+cos(Realinc)
   end
until (xsquare+ysquare) > MaxSizeSquared or
   colour > MaxIterations
```

Here the increments of c applied in the function have been altered by the use of SIN and COS functions. This gives a different image to the standard Mandelbrot set, but one which is still worth examining in close-up. This image is offset from the origin of $(0,0)$, so you may wish to change the real minimum and maximum to -4 and 0 for your first investigations.

A further way to get different results out of the standard algorithm is to generally mess about with it until something interesting happens. This may sound rather unscientific, but a little serendipity has always been found useful in all areas of scientific endeavour! A good starting point is to alter the line:

```
x:=xsquare-ysquare+realinc
```

to read:

```
x:=ysquare-xsquare+realinc
```

where *realinc* is as described above. This sort of accident will generate occasionally interesting patterns, although the mathematical meaning of such patterns is not as clear-cut as for the proper Mandelbrot set.

Three-dimensional representations

A further way of showing the escape times of points around the Mandelbrot set is to display the set as a pseudo three-dimensional image, thus making the set look like a flat-topped mountain surrounded by ridges and hills. Personally, I prefer the flat images that we've been seeing already, and the code involved is quite complex for a true three-dimensional view of the set. However, a simplified version of this process is given in Listing 7-5. This uses a simple procedure, *ThreeD*, to convert (x, y) and colour values (colour representing the escape time of the point) into an (x, y) point that represents a three-dimensional transformation of that point.

Periodicity in the Mandelbrot set

As I've already noted, the dark area that constitutes the set is all those points which do not escape to infinity. We might expect there to be some periodicity in the set, where sequences of numbers repeat themselves in much the same way as sequences in the logistic equation repeated. In fact, if you remember back to the beginning of this chapter we examined the trajectories followed by values of z; some of these were distinctly periodic. The boundary of the set would be analogous to the chaotic region of the logistic equation. We should, therefore, be able to draw a bifurcation diagram for the Mandelbrot set, and also we could try shading the *inside* of the set different colours to represent the periodicity of different regions

PERIODICITY IN THE MANDELBROT SET

of the set. Figure 7-5 shows a bifurcation diagram for the set generated by Listing 7-6, and Listing 7-7 produces a shaded periodicity diagram of the inside of the set. Again, these techniques can be applied to any of the Mandelbroid functions mentioned above.

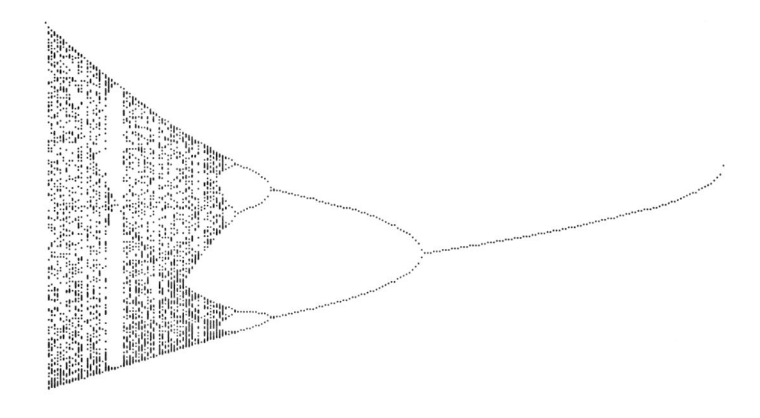

Figure 7-5. A bifurcation diagram for the Mandelbrot set.

Bifurcation diagram

The bifurcation diagram is generated by holding the imaginary part of c at 0, and varying the value of the real part of c from -2 to $+0.5$. This is analogous to the process of varying k in the logistic equation. You could, of course, create bifurcation diagrams of different cross-sections of the Mandelbrot set by varying the imaginary value of c rather than the real part, or by setting an imaginary part value to something other than 0, and so on. The techniques adopted in Chapter 2 to expand areas of the bifurcation diagram could also be applied here. Again, you will find that within this bifurcation diagram there are areas of order in chaos, just like in the logistic equation.

Shading the interior

After deciding that a point is within the set, a further set of iterations are carried out. The first absolute value of z is recorded and compared with other values subsequently generated and the colour is set depending upon the period discovered. Because you'll never get absolutely equal values due to rounding errors, etc., a check is made to see if the two numbers are within a small distance of each other; this distance can be varied between 0.1 and 0.0001, say, to get different resolution maps and in the program 0.01 is used. The smaller this value is, the more accurate the graphic will be in terms of displaying internal periodicity. The exact patterns obtained depend upon the maximum number of iterations chosen when the program is started off, as well as the periodicity inside the set. The only other point to note is the check on the absolute value of z not exceeding 6E6. This is to prevent numeric overflow errors.

There are many other ways in which you can explore the Mandelbrot set and the Mandebroids described in this chapter. You will also find many sources of inspiration in the books and software listed in the bibliography. However, there is another way of iterating the complex function $z^2 + c$, and we'll look at this in Chapter 8.

BBC BASIC listings

Listing 7-1

```
10 REM Program ComplexIterate
30 REM Start with CReal and Cimaginary=1
50 MODE 6
60 INPUT "Real Value ",CReal
70 INPUT "Imaginary Value ",CImaginary
90 REM  Initialise the Z variable to 0 by setting both the Real and Imaginary
100 REM Parts
110 REM of the number to 0.
130 ZReal=0
140 ZImaginary=0
160 REM Now loop around
180 FOR i=1 TO 10
200   REM The value of ZImaginary will be jumped all over by calculating the
      new
210   REM Zimaginary value before the new ZReal value can be calculated.  So,
      save
220   REM it away
240   temp=ZImaginary
260   REM To prevent numeric overflows, only calculate next stage if the
      absolute
280   REM value of z is less than 4E6 - this is totally arbitrary.
300   IF SQR((ZReal*ZReal)+(ZImaginary*ZImaginary))<4E6 THEN
      ZImaginary=2*ZReal*ZImaginary+CImaginary: ZReal=(ZReal*ZReal)-(temp*temp)+CReal
320   PRINT SQR((ZReal*ZReal)+(ZImaginary*ZImaginary))
330 NEXT
360 END
```

Listing 7-2

```
10 REM Program ComplexIterate
30 REM Try CReal=0.2 Cimaginary=0.3
50 MODE 6
70 INPUT "Real Value? ",CReal
80 INPUT "Imaginary Value? ",CImaginary
100 REM  Initialise the Z variable to 0 by setting both the Real and Imaginary
110 REM Parts
120 REM of the number to 0.
140 ZReal=0
150 ZImaginary=0
170 REM Now loop around
190 MODE 5
210 FOR i=1 TO 1000
230   REM The value of ZImaginary will be jumped all over by calculating the
      new
240   REM Zimaginary value before the new ZReal value can be calculated.  So,
      save
250   REM it away
270   temp=ZImaginary
290   REM To prevent numeric overflows, only calculate next stage if the
      absolute
300   REM value
310   REM of z is less than 4E6 - this is totally arbitrary.
330   IF SQR((ZReal*ZReal)+(ZImaginary*ZImaginary))>1000 THEN GOTO 400
340   ZImaginary=2*ZReal*ZImaginary+CImaginary :
350   ZReal=(ZReal*ZReal)-(temp*temp)+CReal
400   PLOT69,ZReal*500+500,ZImaginary*500+500
410 NEXT
440 END
```

Listing 7-3

```
10 REM Program MandelbrotSet
30 REM Program for Chaos Book, draws the Mandelbrot set
60 ON ERROR RUN
100 REM Initial parameters are for the full set are RMax=0.5, RMin=-2
110 REM IMax=1.25, IMaxs=-1.25, Numits=256
140 REPEAT
160   REM    Get all parameters in from the user.
190   MODE 6
210   INPUT "Minimum Real? ",Rnummin
220   INPUT "Maximum Real? ",Rnummax
230   INPUT "Minimum Imaginary? ",ImagMin
240   INPUT "Maximum Imaginary? ",ImagMax
250   INPUT "Iterations? ",Max_Iterations
270   REM See text for value of max_size details
290   max_size=2
310   MODE 5
330   REM Work out increments of Real and Imaginary from max and min. real and
340   REM imaginary numbers, and numbers of maximum rows and columns.
360   maxsizeSquared=max_size*max_size
370   maxcol=250
380   maxrow=160
390   max_colours=4
400   deltaRnum=(Rnummax-Rnummin)/(maxcol-1)
410   deltaImaginary=(ImagMax-ImagMin)/(maxrow-1)
420   col=0
430   REPEAT
440     row=0
450     REPEAT
470       REM Initialise X and Y values to 0 for each point.  X is teh real
          part of each
480       REM point and Y represents the imaginary part of each number.
500       X=0.0
510       Y=0.0
520       colour=0
530       REPEAT
540         colour=colour+1
550         Xsquare=X*X
560         Ysquare=Y*Y
570         IF (Xsquare+Ysquare)> (maxsizeSquared) THEN GOTO 810
600         REM Recalculate X and Y values the Imagmin and Rnummin terms of
            each X and Y
610         REM expression is the 'C' term discussed in the text, arrived at
            by calculating
620         REM the current value of C number from the original value and the
            current
630         REM column or row position.
650         Y=2*X*Y+(ImagMin+row*deltaImaginary)
670         REM The next bit is why we stored xsquared and ysquared away, as
            the
680         REM calculation
690         REM of the Imaginary Part of the number, Y, wrecks the previous
            value of Y
700         REM that we need for correctly calculating the Real part!
720         X=Xsquare-Ysquare+(Rnummin+col*deltaRnum)
740         REM Exit the loop when the colour pointer has exceeded the
            maximum iteration
750         REM number or
760         REM when the square root of the sum of X**2 and Y**2 exceeds
            max_size
810       UNTIL (colour>=Max_Iterations) OR
          ((Xsquare+Ysquare)>(maxsizeSquared))
840       REM   If max iteration number has been exceeded then set colour to 0
860       IF colour>=Max_Iterations THEN colour=0
```

BBC BASIC LISTINGS   209

```
 880      REM    Put a pixel on the screen at col,row
 900      GCOL0,(colour MOD max_colours)
 910      PLOT69,col*4,row*4
 930      REM Increment row value until either maximum number of row values or
 940      REM keypressed
 960      row=row+1
 970    UNTIL (row>=maxrow)
 990    REM Increment column until max columns done, or a key pressed
1010     col=col+1
1020  UNTIL (col>=maxcol)
1040  REM Now save screen - use *SAVE and change file name when needed
1060  *SAVE 5800  8000
1080  REM Wait until a key is pressed...
1100  PRINT CHR$(7) : G$=GET$
1120  REM If key pressed is 'X', then finish tidily.
1140 UNTIL (G$="x") OR (G$="X")
1150 END
```

Listing 7-4

```
  10 REM Program MandelbrotSet
  30 REM Program for Chaos Book, draws the Mandelbrot set
  60 ON ERROR RUN
 220 REM Default of Rmax=0.5, Rmin=-2, ImagMax=1.25, ImagMin=-1.25, Numits=256
 240 REPEAT
 250    MODE 6
 270    REM Get all parameters in from the user.
 290    INPUT "Maximum Real? ",Rnummax
 300    INPUT "Minimum Real? ",Rnummin
 310    INPUT "Minimum Imaginary? ",ImagMin
 320    INPUT "Maximum Imaginary? ",ImagMax
 330    INPUT "Iterations? ",Max_Iterations
 350    REM See text for value of max_size details
 370    max_size=2
 390    MODE 5
 391    REM Use VDU 19 to set palette up.
 392    VDU19,1,7,0,0,0
 393    VDU19,2,4,0,0,0
 394    VDU19,3,6,0,0,0
 410    REM Work out increments of Real and Imaginary from max and min. real and
 420    REM imaginary numbers, and numbers of maximum rows and columns.
 440    maxsizeSquared=max_size*max_size
 450    maxcol=250
 460    maxrow=160
 470    max_colours=4
 480    deltaRnum=(Rnummax-Rnummin)/(maxcol-1)
 490    deltaImaginary=(ImagMax-ImagMin)/(maxrow-1)
 500    col=0
 510    REPEAT
 520      row=0
 530      REPEAT
 550        REM   Initialise X and Y values to 0 for each point. X is teh real
                  part of
 570        REM each point and Y represents the imaginary part of each number.
 590        x=0.0
 600        y=0.0
 610        colour=0
 620        REPEAT
 630          colour=colour+1
 640          Xsquare=x*x
 650          Ysquare=y*y
 660          IF (Xsquare+Ysquare)> (maxsizeSquared) THEN GOTO 860
 680          REM Recalculate X and Y values the Imagmin and Rnummin terms of
```

```
                each X and Y
690             REM expression is the 'C' term discussed in the text, arrived at
                by calculating
700             REM the current value of C number from the original value and the
                current
710             REM column or row position.
730             y=2*x*y+(ImagMin+row*deltaImaginary)
750             REM The next bit is why we stored xsquared and ysquared away, as
                the
760             REM calculation
770             REM of the Imaginary Part of the number, Y, wrecks the previous
                value of Y
780             REM that we need for correctly calculating the Real part!
800             x=Xsquare-Ysquare+(Rnummin+col*deltaRnum)
820             REM Exit the loop when the colour pointer has exceeded the
                maximum iteration
830             REM number or
840             REM when the square root of the sum of X**2 and Y**2 exceeds
                max_size
860            UNTIL ((colour>=Max_Iterations) OR
                ((Xsquare+Ysquare)>(maxsizeSquared)))
890             REM If max iteration number has been exceeded then set colour to 0
910             IF colour>=Max_Iterations THEN colour=0
990             GCOL0,(colour / 3)
1000            PLOT69,col*4,row*4
1020            REM Increment row value until either maximum number of row values
                or keypressed
1040               row=row+1
1050          UNTIL (row>=maxrow)
1070          REM Increment column until max columns done, or a key pressed
1090            col=col+1
1100       UNTIL (col>=maxcol)
1120       REM Now save screen - change name if needed
1140       *SAVE mandel2 5800 8000
1170       REM Wait until a key is pressed...
1190       G$=GET$
1200       REM    If key pressed is 'X', then finish tidily.
1220  UNTIL (G$="X") OR (G$="x")
1230  END
```

Listing 7-5

```
10  REM Program MandelbrotPsuedo3D
30  REM Program for Chaos Book, draws the Mandelbrot set in Pseudo 3d
70  REM Default of Rmax=0.5, Rmin=-2, ImagMax=1.25, ImagMin=-1.25, Numits=256
90  REPEAT
110    MODE 6
130    REM Get all parameters in from the user.
150    INPUT "Maximum Real? ",Rnummax
160    INPUT "Minimum Real? ",Rnummin
170    INPUT "Minimum Imaginary? ",ImagMin
180    INPUT "Maximum Imaginary? ",ImagMax
190    INPUT "Iterations? ",Max_Iterations
210    REM See text for value of max_size details
250    max_size=2
270    MODE 5
290    REM Work out increments of Real and Imaginary from max and min. real and
300    REM imaginary numbers, and numbers of maximum rows and columns.
320    maxsizeSquared=max_size*max_size
330    maxcol=250
340    maxrow=160
350    max_colours=4
360    deltaRnum=(Rnummax-Rnummin)/(maxcol-1)
```

BBC BASIC LISTINGS

```
 370    deltaImaginary=(ImagMax-ImagMin)/(maxrow-1)
 380    row=maxrow
 400    REPEAT
 420      REM Do the iterations from TOP to BOTTOM - this way, the pseudo 3-d
          image
 430      REM is built up correctly.
 460      col=0
 470      REPEAT
 490        REM Initialise x and Y values to 0 for each point.  x is teh real
            part of each
 500        REM point and Y represents the imaginary part of each number.
 520        x=0.0
 530        y=0.0
 540        colour=0
 550        REPEAT
 560          colour=colour+1
 570          Xsquare=x*x
 580          Ysquare=y*y
 590          IF (Xsquare+Ysquare)> (maxsizeSquared) THEN GOTO 820
 610          REM Recalculate X and Y values the ImagMin and Rnummin terms of
              each X and Y
 620          REM expression is the 'C' term discussed in the text, arrived at
              by calculating
 630          REM the current value of C number from the original value and the
              current
 640          REM column or row position.
 660          y=2*x*y+(ImagMin+row*deltaImaginary)
 680          REM The next bit is why we stored xsquared and ysquared away, as
              the
 690          REM calculation
 700          REM of the Imaginary Part of the number, Y, wrecks the previous
              value of Y
 710          REM that we need for correctly calculating the Real part!
 730          x=Xsquare-Ysquare+(Rnummin+col*deltaRnum)
 750          REM Exit the loop when the colour pointer has exceeded the
              maximum iteration
 760          REM number or
 770          REM when the square root of the sum of X**2 and Y**2 exceeds
              max_size
 820        UNTIL ((colour>=Max_Iterations) OR
            ((Xsquare+Ysquare)>(maxsizeSquared)))
 850        REM Put a pixel on the screen at x3d,y3d after modifying the
            plotted y value
 860        REM using the ThreeD procedure described above.  Colour is selected
            by modding
 880        REM it with the max_colours variable.
 900        PROCThreeD(col,row,colour)
 910        x3d=xout
 920        y3d=yout
 930        GCOL0,colour MOD 4
 940        PLOT69,2*x3d,4*y3d
 960        REM Increment column value until either maximum number of column
            values or
 970        REM keypressed
 990        col=col+1
1000      UNTIL (col>=maxcol)
1020      REM Decrement row until all rows done, or a key pressed
1040      row=row-1
1050    UNTIL (row<=0)
1090    *SAVE mandel3 5800 8000
1110    REM Wait until a key is pressed...
1130    G$=GET$
1150    REM If key pressed is 'X', then finish tidily.
1170 UNTIL (G$="X") or (G$="x")
1180 END
```

212   CHAPTER 7. THE MANDELBROT SET

```
1240 DEFPROCThreeD(xtemp,ytemp,ztemp)
1260 REM x coordinate calculated by using cos(angle of view) * ycoordinate
     passed.
1270 REM y coordinate calculated from colour, representing height, and the
     sin(angle
1280 REM of view).   Here, angle is 30', sin(30) is 0.5, cos(30) is 0.866
1300 xout=INT(xtemp+0.866*ytemp)
1310 yout=INT(ztemp+0.5*ytemp)
1330 REM You might like to try other angles.
1350 ENDPROC
```

Listing 7-6

```
10 REM Program MandelbrotBifurcation
20 REM For Chaos Programming Book, Joe Pritchard December 1990
30 REM Allows drawing of Bifurcation diagram for the Mandelbrot Set
50 REM Try Y scale of 200, Y offset of 500 to start with
80 MODE 6
90 INPUT "Y Scale? ",YScale
100 INPUT "Y Offset? ",YOffset
120 MODE 5
130 i=1
140 RealInc=-2
160 REM Start value of Realinc set up
180 max_iterations=100
220 REPEAT
240   x=0.0
250   y=0.0
260   colour=0
280   FOR k=1 TO 300
290     REPEAT
300       colour=colour+1
310       Xsquare=x*x
320       Ysquare=y*y
330       IF (Xsquare+Ysquare)>4 THEN GOTO 600
350       REM Recalculate X and Y values When calculating the imaginary part
          we
360       REM don't have to add a C imaginary term as we're keeping that at 0
380       y=2*x*y
400       REM The next bit is why we stored xsquared and ysquared away, as the
410       REM calculation
420       REM of the Imaginary Part of the number, Y, wrecks the previous
          value of Y
430       REM that we need for correctly calculating the Real part!
450       x=Xsquare-Ysquare+RealInc
470       REM After 200 iterations, plot the point on the graph
490       IF (k>200) THEN PLOT69,i*4,(x*YScale)+YOffset
520       REM Exit the loop when the colour pointer has exceeded the maximum
          iteration
540       REM number or when the square root of the sum of X**2 and Y**2
          exceeds max_size
600     UNTIL ((colour>=max_iterations) OR (Xsquare+Ysquare)>4)
620   NEXT
660   RealInc=RealInc+0.01
670   i=i+1
690   REM move to next horizontal graphing position.
720 UNTIL (RealInc>0.5)
740 REM repeat until all real parts of c graphed
770 REM Now beep when the graph is drawn, then press any key to go on
790 PRINT CHR$(7)
800 G=GET
810 END
```

Listing 7-7

```
 10 REM Program MandelbrotSetInterior
 30 REM Program for Chaos Book, draws the Mandelbrot set
 60 ON ERROR RUN
 80 REM Default of Rmax=0.5, Rmin=-2, ImagMax=1.25, ImagMin=-1.25, Numits=256
100 REPEAT
110   MODE 6
130   REM Get all parameters in from the user.
150   INPUT "Maximum Real? ",Rnummax
160   INPUT "Minimum Real? ",Rnummin
170   INPUT "Minimum Imaginary? ",ImagMin
180   INPUT "Maximum Imaginary? ",ImagMax
190   INPUT "Iterations? ",Max_Iterations
210   REM See text for value of max_size details
250   max_size=2
270   MODE 5
290   REM Work out increments of Real and Imaginary from max and min. real and
300   REM imaginary numbers, and numbers of maximum rows and columns.
320   maxsizeSquared=max_size*max_size
330   maxcol=250
340   maxrow=160
350   max_colours=4
360   deltaRnum=(Rnummax-Rnummin)/(maxcol-1)
370   deltaImaginary=(ImagMax-ImagMin)/(maxrow-1)
380   col=0
390   REPEAT
400     row=0
410     REPEAT
430       REM Initialise X and Y values to 0 for each point. X is teh real
          part of each
440       REM point and Y represents the imaginary part of each number.
460       x=0.0
470       y=0.0
480       colour=0
490       REPEAT
500         colour=colour+1
510         Xsquare=x*x
520         Ysquare=y*y
530         IF (Xsquare+Ysquare)>(maxsizeSquared) THEN GOTO 740
550         REM Recalculate X and Y values the ImagMin and Rnummin terms of
            each X and Y
560         REM expression is the 'C' term discussed in the text, arrived at
            by calculating
570         REM the current value of C number from the original value and the
            current
580         REM column or row position.
600         y=2*x*y+(ImagMin+row*deltaImaginary)
620         REM The next bit is why we stored xsquared and ysquared away, as
            the
630         REM calculation
640         REM of the Imaginary Part of the number, Y, wrecks the previous
            value of Y
650         REM that we need for correctly calculating the Real part!
670         x=Xsquare-Ysquare+(Rnummin+col*deltaRnum)
700         REM Exit the loop when the colour pointer has exceeded the
            maximum iteration
720         REM number or when the square root of the sum of X**2 and Y**2
            exceeds max_size
740       UNTIL ((colour>=Max_Iterations) OR
          (Xsquare+Ysquare)>(maxsizeSquared))
770       REM If max iteration number has been exceeded THEN see if a
          repeating sequence
780       REM can be detected.
800       IF colour<Max_Iterations THEN GOTO 920
```

```
810     t_square=Xsquare+Ysquare
820     colour=0
830     REPEAT
840       colour=colour+1
850       Xsquare=x*x
860       Ysquare=y*y
870       y=2*x*y+(ImagMin+row*deltaImaginary)
880       x=Xsquare-Ysquare+(Rnummin+col*deltaRnum)
900     UNTIL (ABS(t_square-(Xsquare+Ysquare))<0.01) OR
        (ABS(Xsquare+Ysquare)>6E6) OR (colour>200)
920     IF (colour>200) OR (ABS(Xsquare+Ysquare)>6E6) THEN  colour=0
940     GCOL0,colour MOD max_colours
950     PLOT69,col*4,row*4
970     REM Increment row value UNTIL either maximum number of row values
        or keypressed
990     row=row+1
1000   UNTIL (row>=maxrow)
1020    REM Increment column until max columns done, or a key pressed
1040    col=col+1
1050  UNTIL (col>=maxcol)
1080  *SAVE mandelin 5800 8000
1100  REM Wait until a key is pressed...
1120  G$=GET$
1140  REM If key pressed is 'X', THEN finish tidily.
1160 UNTIL (G$="X") OR (G$="x")
1170 END
```

This colour plate section was taken using a Canon EOS600 camera and 100 ASA slide film. The screens are all taken from the IBM PC versions of the programs listed in this book, using a VGA screen.

Plates 4, 6 and 7 were generated by magnifying a small section of the Mandelbrot Set. Some of the areas magnified were originally documented in journals such as *Scientific American* and *Fractal Report*. The parameters given are in the order Real Maximum, Real Minimum, Imaginary Maximum and Imaginary Minimum.

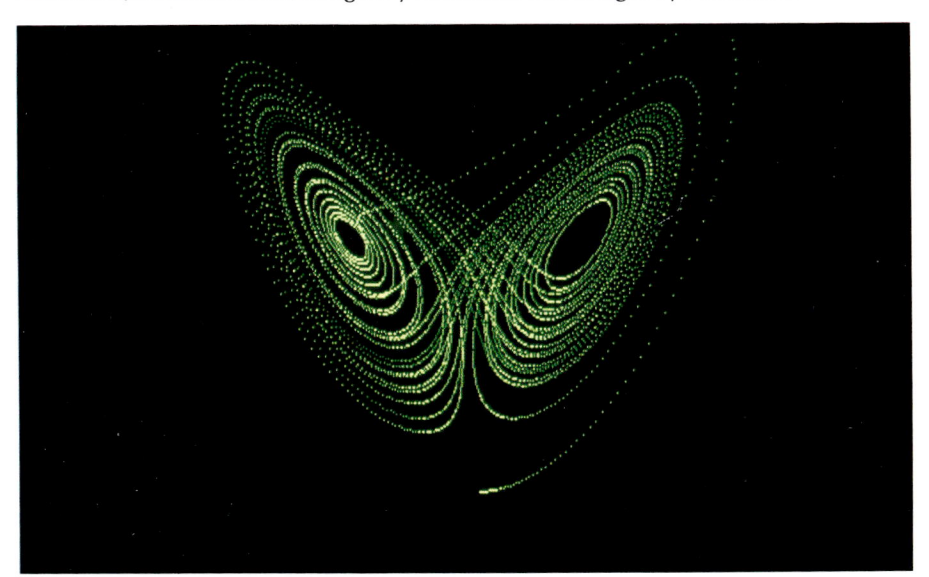

Plate 1 *View of the Lorenz Attractor*

Plate 2 *Julia Set, default parameters*

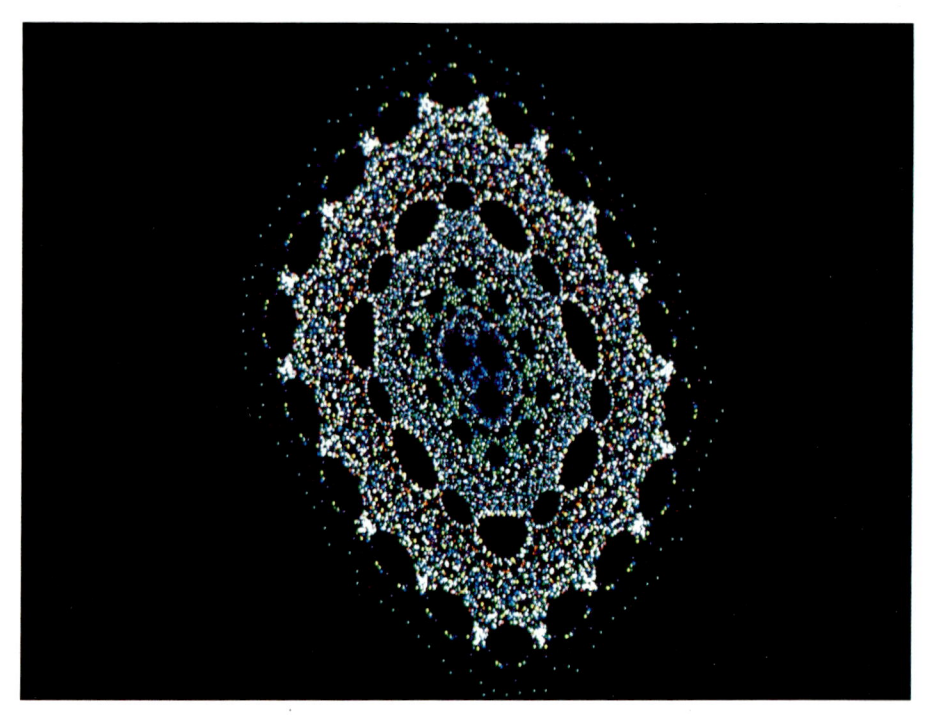

Plate 3 *Typical Martins Mapping*

Plate 4 *M Set. −0.7408, −0.75104, 0.11536, 0.10511*

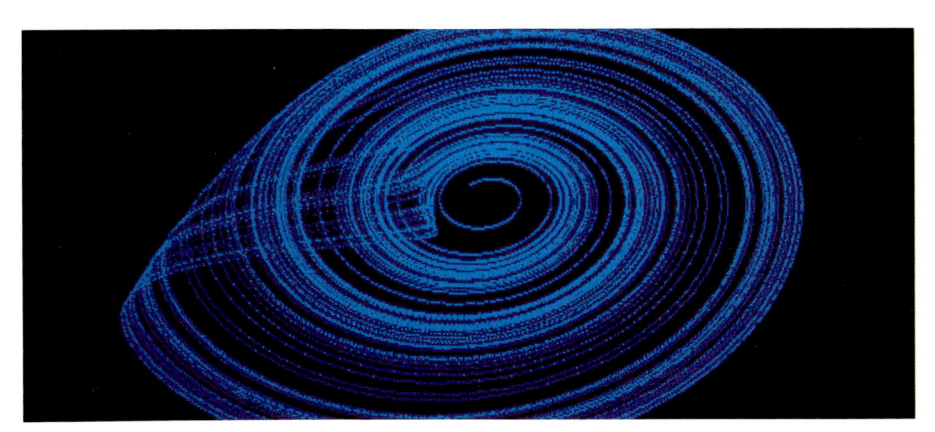

Plate 5 *Rossler Attractor*

Plate 6 *M Set. −0.690950, 691020, 0.387205, 0.387125*

Plate 7 *M Set.* −0.742, −0.744, −0.1805, −0.1825

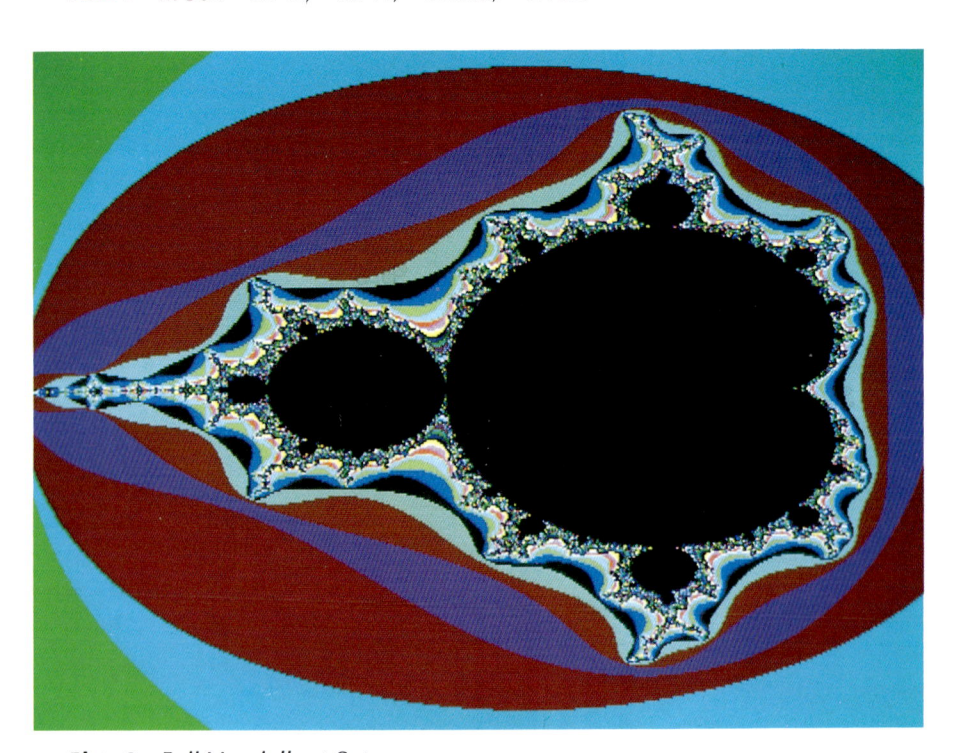

Plate 8 *Full Mandelbrot Set*

TURBO PASCAL LISTINGS   215

Turbo Pascal listings

Listing 7-1

```pascal
Program ComplexIterate;

Uses Crt,Graph,extend;

Var

zReal, ZImaginary      : Real;
CReal, CImaginary      : Real;
Counter                : Integer;
CrealString            : String;
CimaginaryString       : String;
i                      : Integer;
temp                   : Real;

Begin

CRealString:='1';
CImaginaryString:='1';

    TextBackGround(Blue);
    ClrScr;
    PrintAt(2,2,'Complex Number Iteration program. Joe Pritchard, 1989',yellow);
    PrintAt(2,4,'Real value:',Green);
    StringEdit(CRealString,20,30,4,Yellow);
    Val(CRealString,Creal,i);
    PrintAt(2,5,'Imaginary value:',Green);
    StringEdit(CImaginaryString,20,30,5,Yellow);
    Val(CImaginaryString,CImaginary,i);

    writeln;
    writeln;

{ Initialise the Z variable to 0 by setting both the Real and Imaginary Parts
  of the number to 0.   }

    zReal:=0;
    ZImaginary:=0;

{ Now loop around   }

    for i:=1 to 10 do begin

{ The value of ZImaginary will be jumped all over by calculating the new
  Zimaginary value before the new ZReal value can be calculated.  So, save
  it away    }

        temp:=ZImaginary;

{ To prevent numeric overflows, only calculate next stage if the absolute
  value
  of z is less than 4E6 - this is totally arbitrary.   }

        if sqrt((ZReal*ZReal)+(Zimaginary*ZImaginary))<4E6 then begin
            Zimaginary:=2*ZReal*ZImaginary+CImaginary;
            ZReal:=(ZReal*ZReal)-(temp*temp)+CReal;
        end;
        Writeln(sqrt((ZReal*ZReal)+(Zimaginary*ZImaginary)));
    end;
end.
```

Listing 7-2

```
Program ComplexIterate;

Uses Crt,Graph,extend;

Var

zReal, ZImaginary      : Real;
CReal, CImaginary      : Real;
Counter                : Integer;
CrealString            : String;
CimaginaryString       : String;
i                      : Integer;
temp                   : Real;
Colour                 : Integer;

Begin

CRealString:='1';
CImaginaryString:='1';

    TextBackGround(Blue);ClrScr;
    PrintAt(2,2,'Complex Number Iteration program. Joe Pritchard, 1989',yellow);
    PrintAt(2,4,'Real value:',Green);
    StringEdit(CRealString,20,30,4,Yellow);
    Val(CRealString,Creal,i);
    PrintAt(2,5,'Imaginary value:',Green);
    StringEdit(CImaginaryString,20,30,5,Yellow);
    Val(CImaginaryString,CImaginary,i);

{ Initialise the Z variable to 0 by setting both the Real and Imaginary Parts
  of the number to 0.  }

    zReal:=0;   ZImaginary:=0;

{ Now loop around  }

    ChoosePalette;
    Colour:=7;                  { Set colour for plotted picture }
    for i:=1 to 1000 do begin

{ The value of ZImaginary will be jumped all over by calculating the new
  Zimaginary value before the new ZReal value can be calculated.  So, save
  it away   }

        temp:=ZImaginary;

{ To prevent numeric overflows, only calculate next stage if the absolute
  value
  of z is less than 4E6 - this is totally arbitrary.  }

        if sqrt((ZReal*ZReal)+(Zimaginary*ZImaginary))<1000 then begin
            Zimaginary:=2*ZReal*ZImaginary+CImaginary;
            ZReal:=(ZReal*ZReal)-(temp*temp)+CReal;

{ The scaling factor here is chosen to suit VGA screens, but can be modified
  for other screens.             }

            PutPixel(trunc(ZReal*300+(GetMaxX/2)),trunc(ZImaginary*300+(GetMaxY/2)),Colour);

        end;
    end;
end.
```

Listing 7-3

```pascal
Program MandelbrotSet;

{   Program for Chaos Book, draws the Mandelbrot set }

Uses Crt,Graph,Extend;

Var
    maxcol,maxrow,max_colours,i:                            Integer;
    max_iterations,row,col,colour:                          Integer;
    Rnummax,ImagMax,Rnummin,ImagMin,Rnum,deltaRnum,deltaImaginary:  Real;
    X,Y,Xsquare,Ysquare,max_size:                           Real;
    Rmaxs,Rmins,Imaxs,Imins,Numits,Fname:                   String;
    MaxSizeSquared :                                        Real;
    ch:                                                     Char;

Begin

{ Initialise strings with the default values for full picture of set   }
{   Initial parameters are for the full set                    }

Rmaxs:='0.5';
Rmins:='-2';
Imaxs:='1.25';
Imins:='-1.25';
Numits:='256';
Fname:='Mandel';

 Repeat

{   Get all parameters in from the user. }

    TextBackGround(Blue);
    ClrScr;
    PrintAt(2,2,'Mandelbrot Set plotting program. Joe Pritchard, 1989',yellow);
    PrintAt(2,4,'Maximum real value:',Green);
    StringEdit(Rmaxs,20,30,4,Yellow);
    Val(Rmaxs,Rnummax,i);
    PrintAt(2,5,'Minimum real value:',Green);
    StringEdit(Rmins,20,30,5,Yellow);
    Val(Rmins,Rnummin,i);
    PrintAt(2,6,'Minimum Imaginary value:',Green);
    StringEdit(Imins,20,30,6,Yellow);
    Val(Imins,ImagMin,i);

    PrintAt(2,7,'Maximum Imaginary value:',Green);
    StringEdit(Imaxs,20,30,7,Yellow);
    Val(Imaxs,ImagMax,i);

    PrintAt(2,9,'Number of iterations:',Green);
    StringEdit(Numits,20,30,9,Yellow);
    Val(Numits,Max_Iterations,i);
    PrintAt(2,10,'Screen File Name: ',Green);
    StringEdit(Fname,20,30,10,Yellow);

{ See text for value of max_size details                          }

    max_size:=2;

    ChoosePalette;

{ Work out increments of Real and Imaginary from max and min. real and
  imaginary numbers, and numbers of maximum rows and columns.       }
```

CHAPTER 7. THE MANDELBROT SET

```
maxsizeSquared:=max_size*max_size;
maxcol:=GetMaxX;
maxrow:=GetMaxY;
max_colours:=GetMaxColor;
deltaRnum:=(Rnummax-Rnummin)/(maxcol-1);
deltaImaginary:=(ImagMax-ImagMin)/(maxrow-1);
col:=0;
Repeat
        row:=0;
        Repeat
```

{ Initialise X and Y values to 0 for each point. X is teh real part of each
point and Y represents the imaginary part of each number. }

```
                X:=0.0;
                Y:=0.0;
                colour:=0;
                repeat
                   colour:=colour+1;
                   Xsquare:=x*x;
                   Ysquare:=y*y;
                   if (Xsquare+Ysquare)<= (maxsizeSquared) then
                      Begin
```

{ Recalculate X and Y values; the Imagmin and Rnummin terms of each X and Y
expression is the 'C' term discussed in the text, arrived at by calculating
the current value of C number from the original value and the current
column or row position.   }

```
                         Y:=2*x*y+(Imagmin+row*deltaImaginary);
```

{ The next bit is why we stored xsquared and ysquared away, as the calculation
of the Imaginary Part of the number, Y, wrecks the previous value of Y
that we need for correctly calculating the Real part!   }

```
                         X:=Xsquare-Ysquare+(Rnummin+col*deltaRnum);
                      end;
```

{ Exit the loop when the colour pointer has exceeded the maximum iteration
number or
when the square root of the sum of X**2 and Y**2 exceeds max_size
}

```
                until ((colour>=max_iterations) or
                  ((Xsquare+Ysquare)>(maxsizeSquared)));
```

{ If max iteration number has been exceeded then set colour to 0   }

```
                if colour>=max_iterations then
                   colour:=0;
```

{ Put a pixel on the screen at col,row after modifying the plotted y value
so that the maximum y value (corresponding to max. Imaginary value) is
plotted at the top of the screen. Colour is selected by modding it with
the max_colours variable.   }

```
                PutPixel(col,trunc(GetMaxY-row),trunc(colour mod max_colours));
```

{ Increment row value until either maximum number of row values or
keypressed }

```
                row:=row+1;
           Until (row>=maxrow) or (Keypressed);
```

TURBO PASCAL LISTINGS

```
{   Increment column until max columns done, or a key pressed        }

            col:=col+1;
        Until (col>=maxcol) or (KeyPressed);

{   Now save screen - uses a library routine                         }

        SaveScreen(Fname);

{   Wait until a key is pressed...                 }

        ch:=Readkey;
        CloseGraph;
        TextBackGround(Blue);
        ClrScr;

{   If key pressed is 'X', then finish tidily.             }

    until (ch='X') or (ch='x');
end.
```

Listing 7-4

```
Program MandelbrotSet;

{   Program for Chaos Book, draws the Mandelbrot set }

Uses Crt,Graph,Extend;

Var
    maxcol,maxrow,max_colours,i:                                Integer;
    max_iterations,row,col,colour:                              Integer;
    Rnummax,ImagMax,Rnummin,ImagMin,Rnum,deltaRnum,deltaImaginary:  Real;
    X,Y,Xsquare,Ysquare,max_size:                               Real;
    Rmaxs,Rmins,Imaxs,Imins,Numits,Fname:                       String;
    MaxSizeSquared :                                            Real;
    ch:                                                         Char;
    plotcolour                                              :   array[0..10]
        of integer;

Begin

{ Initialise strings with the default values for full picture of set    }

{   Initial parameters are for the full set                  }

plotcolour[0]:=black;
plotcolour[1]:=Blue;
plotcolour[2]:=LightBlue;
plotcolour[3]:=cyan;
plotcolour[4]:=LightCyan;
plotcolour[5]:=Green;
plotcolour[6]:=LightGreen;
PlotColour[7]:=Yellow;
plotcolour[8]:=Brown;
plotcolour[9]:=LightRed;
plotcolour[10]:=Red;

Rmaxs:='0.5';
Rmins:='-2';
Imaxs:='1.25';
Imins:='-1.25';
```

```
Numits:='256';
Fname:='Mandel';

Repeat

{   Get all parameters in from the user. }

    TextBackGround(Blue);
    ClrScr;
    PrintAt(2,2,'Mandelbrot Set plotting program. Joe Pritchard, 1989',yellow);
    PrintAt(2,4,'Maximum real value:',Green);
    StringEdit(Rmaxs,20,30,4,Yellow);
    Val(Rmaxs,Rnummax,i);
    PrintAt(2,5,'Minimum real value:',Green);
    StringEdit(Rmins,20,30,5,Yellow);
    Val(Rmins,Rnummin,i);
    PrintAt(2,6,'Minimum Imaginary value:',Green);
    StringEdit(Imins,20,30,6,Yellow);
    Val(Imins,ImagMin,i);

    PrintAt(2,7,'Maximum Imaginary value:',Green);
    StringEdit(Imaxs,20,30,7,Yellow);
    Val(Imaxs,ImagMax,i);

    PrintAt(2,9,'Number of iterations:',Green);
    StringEdit(Numits,20,30,9,Yellow);
    Val(Numits,Max_Iterations,i);
    PrintAt(2,10,'Screen File Name: ',Green);
    StringEdit(Fname,20,30,10,Yellow);

{   See text for value of max_size details                                   }

    max_size:=2;

    ChoosePalette;

{   Work out increments of Real and Imaginary from max and min. real and
    imaginary numbers, and numbers of maximum rows and columns.              }

    maxsizeSquared:=max_size*max_size;
    maxcol:=GetMaxX;
    maxrow:=GetMaxY;
    max_colours:=GetMaxColor;
    deltaRnum:=(Rnummax-Rnummin)/(maxcol-1);
    deltaImaginary:=(ImagMax-ImagMin)/(maxrow-1);
    col:=0;
    Repeat
          row:=0;
          Repeat

{   Initialise X and Y values to 0 for each point. X is teh real part of each
    point and Y represents the imaginary part of each number. }

                X:=0.0;
                Y:=0.0;
                colour:=0;
                repeat
                   colour:=colour+1;
                   Xsquare:=x*x;
                   Ysquare:=y*y;
                   if (Xsquare+Ysquare)<= (maxsizeSquared) then
                      Begin

{   Recalculate X and Y values; the Imagmin and Rnummin terms of each X and Y
    expression is the 'C' term discussed in the text, arrived at by calculating
```

TURBO PASCAL LISTINGS 221

```
           the current value of C number from the original value and the current
           column or row position.                           }

                         Y:=2*x*y+(Imagmin+row*deltaImaginary);

{ The next bit is why we stored xsquared and ysquared away, as the calculation
  of the Imaginary Part of the number, Y, wrecks the previous value of Y
  that we need for correctly calculating the Real part!        }

                         X:=Xsquare-Ysquare+(Rnummin+col*deltaRnum);
                    end;

{  Exit the loop when the colour pointer has exceeded the maximum iteration
   number or
   when the square root of the sum of X**2 and Y**2 exceeds max_size
   }
              until ((colour>=max_iterations) or
                    ((Xsquare+Ysquare)>(maxsizeSquared)));

{  If max iteration number has been exceeded then set colour to 0        }

                    if colour>=max_iterations then
                       colour:=0;

{  Put a pixel on the screen at col,row after modifying the plotted y value
   so that the maximum y value (corresponding to max. Imaginary value) is
   plotted at the top of the screen.  Colour is selected by modding it with
   the max_colours variable.                                    }

                    PutPixel(col,trunc(GetMaxY-row),plotcolour[trunc(colour / 10)]);

{  Increment row value until either maximum number of row values or
   keypressed  }

                    row:=row+1;
              Until (row>=maxrow) or (Keypressed);

{  Increment column until max columns done, or a key pressed           }

                    col:=col+1;
              Until (col>=maxcol) or (KeyPressed);

{  Now save screen - uses a library routine                            }

              SaveScreen(Fname);

{  Wait until a key is pressed...                   }

           ch:=Readkey;
           CloseGraph;
           TextBackGround(Blue);
           ClrScr;

{  If key pressed is 'X', then finish tidily.       }

       until (ch='X') or (ch='x');
end.
```

Listing 7-5

```
Program MandelbrotPsuedo3D;

{  Program for Chaos Book, draws the Mandelbrot set in Pseudo 3d }

Uses Crt,Graph,Extend;

Var
    maxcol,maxrow,max_colours,i:                            Integer;
    max_iterations,row,col,colour                           Integer;
    Rnummax,ImagMax,Rnummin,ImagM:    n,deltaRnum,deltaImaginary:  Real;
    X,Y,Xsquare,Ysquare,max_size:                           Real;
    Rmaxs,Rmins,Imaxs,Imins,Numits F ame:                   String;
    MaxSizeSquared :                                        Real;
    ch:                                                     Char;
    x3d,y3d                                               : integer;

Procedure ThreeD(xtemp,ytemp ztemp : integer; var xout,yout : integer);
begin

{ x coordinate calculated by using cos(angle of view) * ycoordinate passed.
  y coordinate calculated from colour, representing height, and the sin(angle
  of view).   Here, angle is 30', sin(30) is 0.5, cos(30) is 0.866    }

    xout:=trunc(xtemp+0.866*ytemp);
    yout:=trunc(ztemp+0.5*ytemp);

{ You might like to try other angles.                                 }

end;

Begin

{ Initialise strings with the default values for full picture of set  }

{   Initial parameters are for the full set                }

Rmaxs:='0.5';
Rmins:='-2';
Imaxs:='1.25';
Imins:='-1.25';

{ Note that the 'height' of the Mandelbrot set above the 'ground' depends upon
  the maximum iterations    }

Numits:='256';
Fname:='Mandel';

Repeat

{   Get all parameters in from the user. }

    TextBackGround(Blue);
    ClrScr;
    PrintAt(2,2,'Mandelbrot Set plotting program. Joe Pritchard, 1989',yellow);
    PrintAt(2,4,'Maximum real value:',Green);
    StringEdit(Rmaxs,20,30,4,Yellow);
    Val(Rmaxs,Rnummax,i);
    PrintAt(2,5,'Minimum real value:',Green);
    StringEdit(Rmins,20,30,5,Yellow);
    Val(Rmins,Rnummin,i);
    PrintAt(2,6,'Minimum Imaginary value:',Green);
    StringEdit(Imins,20,30,6,Yellow);
    Val(Imins,ImagMin,i);
```

TURBO PASCAL LISTINGS

```
        PrintAt(2,7,'Maximum Imaginary value:',Green);
        StringEdit(Imaxs,20,30,7,Yellow);
        Val(Imaxs,ImagMax,i);

        PrintAt(2,9,'Number of iterations:',Green);
        StringEdit(Numits,20,30,9,Yellow);
        Val(Numits,Max_Iterations,i);
        PrintAt(2,10,'Screen File Name: ',Green);
        StringEdit(Fname,20,30,10,Yellow);

{ See text for value of max_size details                            }

        max_size:=2;

        ChoosePalette;

{ Work out increments of Real and Imaginary from max and min. real and
  imaginary numbers, and numbers of maximum rows and columns.       }

        maxsizeSquared:=max_size*max_size;
        maxcol:=GetMaxX;
        maxrow:=GetMaxY;
        max_colours:=GetMaxColor;
        deltaRnum:=(Rnummax-Rnummin)/(maxcol-1);
        deltaImaginary:=(ImagMax-ImagMin)/(maxrow-1);
        row:=maxrow;
        Repeat

{   Do the iterations from TOP to BOTTOM - this way, the pseudo 3-d image
    is built up correctly.         }

                col:=0;
                Repeat

{   Initialise X and Y values to 0 for each point. X is teh real part of each
    point and Y represents the imaginary part of each number. }

                        X:=0.0;
                        Y:=0.0;
                        colour:=0;
                        repeat
                           colour:=colour+1;
                           Xsquare:=x*x;
                           Ysquare:=y*y;
                           if (Xsquare+Ysquare)<= (maxsizeSquared) then
                                Begin

{   Recalculate X and Y values; the Imagmin and Rnummin terms of each X and Y
    expression is the 'C' term discussed in the text, arrived at by calculating
    the current value of C number from the original value and the current
    column or row position.                       }

                        Y:=2*x*y+(Imagmin+row*deltaImaginary);

{   The next bit is why we stored xsquared and ysquared away, as the calculation
    of the Imaginary Part of the number, Y, wrecks the previous value of Y
    that we need for correctly calculating the Real part!           }

                        X:=Xsquare-Ysquare+(Rnummin+col*deltaRnum);
                                end;

{   Exit the loop when the colour pointer has exceeded the maximum iteration
    number or
    when the square root of the sum of X**2 and Y**2 exceeds max_size
```

224   CHAPTER 7. THE MANDELBROT SET

```
      }
                      until ((colour>=max_iterations) or
                      ((Xsquare+Ysquare)>(maxsizeSquared)));

{    Put a pixel on the screen at x3d,y3d after modifying the plotted y value
     using the ThreeD procedure described above.  Colour is selected by modding
     it with
     the max_colours variable.                                                }

                      ThreeD(col,row,colour,x3d,y3d);
                      PutPixel(x3d,trunc(GetMaxY-y3d),trunc(colour mod 15));

{    Increment column value until either maximum number of column values or
     keypressed  }

                      col:=col+1;
                 Until (col>=maxcol) or (Keypressed);

{    Decrement row until all rows done, or a key pressed                      }

                  row:=row-1;
             Until (row<=0) or (KeyPressed);

{    Now save screen - uses a library routine                                 }

             SaveScreen(Fname);

{    Wait until a key is pressed...                            }

             ch:=Readkey;
             CloseGraph;
             TextBackGround(Blue);
             ClrScr;

{    If key pressed is 'X', then finish tidily.                }

     until (ch='X') or (ch='x');
end.
```

Listing 7-6

```
Program MandelbrotBifurcation;
{ For Chaos Programming Book, Joe Pritchard December 1990
  Allows drawing of Bifurcation diagram for the Mandelbrot Set  }

Uses Crt,Graph,Extend;

Var

    i,yscale,yoffset              :              integer;
    x,y                           :              real;
    RealInc                                      Real;
    xsquare,ysquare                              Real;
    colour                        :              Integer;
    max_iterations                :              Integer;
    k                             :              integer;

{ Get the start parameters in from the user           }

{ Try Y scale of 50, Y offset of 100 to start with    }
```

TURBO PASCAL LISTINGS   225

```pascal
Begin
  Write('Y Scale          : ');
  Readln(YScale);
  Write('Y Offset         : ');
  Readln(YOffset);

  ChoosePalette;              { Use Extend function to set up graphics mode }
  i:=1;                       { Left hand edge of screen. }
  RealInc:=-2;                { Start value of Realinc set up }
  max_iterations:=100;

  Repeat
              X:=0.0;
              Y:=0.0;
              colour:=0;

              for k:=1 to 300 do begin
              repeat
                 colour:=colour+1;
                 Xsquare:=x*x;
                 Ysquare:=y*y;
                 if (Xsquare+Ysquare)<=4 then
                     Begin

{  Recalculate X and Y values; When calculating the imaginary part we
   don't have to add a C imaginary term as we're keeping that at 0  }

                         Y:=2*x*y;

{ The next bit is why we stored xsquared and ysquared away, as the calculation
  of the Imaginary Part of the number, Y, wrecks the previous value of Y
  that we need for correctly calculating the Real part!              }

                         X:=Xsquare-Ysquare+RealInc;

{  After 200 iterations, plot the point on the graph                 }

                         if (k>200) then
                              putpixel(i*2,(GetMaxY-trunc(x*YScale)-Yoffset),3);

                  end;

{  Exit the loop when the colour pointer has exceeded the maximum iteration
   number or
   when the square root of the sum of X**2 and Y**2 exceeds max_size
   }
                  until ((colour>=max_iterations) or ((Xsquare+Ysquare)>4));

         end;

  RealInc:=RealInc+0.01;
  i:=i+1;                { move to next horizontal graphing position. }
until (Realinc>0.5);   { repeat until all real parts of c graphed    }

{ Now beep when the graph is drawn, then press any key to go on }

Sound(220);Delay(250);NoSound;
Readln;
CloseGraph;

end.
```

Listing 7-7

```
Program MandelbrotSetInterior;

{   Program for Chaos Book, draws the Mandelbrot set }

Uses Crt,Graph,Extend;

Var
    maxcol,maxrow,max_colours,i:                              Integer;
    max_iterations,row,col,colour:                            Integer;
    Rnummax,ImagMax,Rnummin,ImagMin,Rnum,deltaRnum,deltaImaginary:   Real;
    X,Y,Xsquare,Ysquare,max_size:                             Real;
    Rmaxs,Rmins,Imaxs,Imins,Numits,Fname:                     String;
    MaxSizeSquared,t_square :                                 Real;
    ch:                                                       Char;

Begin

{ Initialise strings with the default values for full picture of set    }
{   Initial parameters are for the full set               `            }

Rmaxs:='0.5';
Rmins:='-2';
Imaxs:='1.25';
Imins:='-1.25';
Numits:='256';
Fname:='Mandel';

  Repeat

{ Get all parameters in from the user. }

    TextBackGround(Blue);
    ClrScr;
    PrintAt(2,2,'Mandelbrot Set plotting program. Joe Pritchard, 1989',yellow);
    PrintAt(2,4,'Maximum real value:',Green);
    StringEdit(Rmaxs,20,30,4,Yellow);
    Val(Rmaxs,Rnummax,i);
    PrintAt(2,5,'Minimum real value:',Green);
    StringEdit(Rmins,20,30,5,Yellow);
    Val(Rmins,Rnummin,i);
    PrintAt(2,6,'Minimum Imaginary value:',Green);
    StringEdit(Imins,20,30,6,Yellow);
    Val(Imins,ImagMin,i);

    PrintAt(2,7,'Maximum Imaginary value:',Green);
    StringEdit(Imaxs,20,30,7,Yellow);
    Val(Imaxs,ImagMax,i);

    PrintAt(2,9,'Number of iterations:',Green);
    StringEdit(Numits,20,30,9,Yellow);
    Val(Numits,Max_Iterations,i);
    PrintAt(2,10,'Screen File Name: ',Green);
    StringEdit(Fname,20,30,10,Yellow);

{ See text for value of max_size details                          }

    max_size:=2;

    ChoosePalette;

{ Work out increments of Real and Imaginary from max and min. real and
  imaginary numbers, and numbers of maximum rows and columns.          }
```

TURBO PASCAL LISTINGS

```
maxsizeSquared:=max_size*max_size;
maxcol:=GetMaxX;
maxrow:=GetMaxY;
max_colours:=GetMaxColor;
deltaRnum:=(Rnummax-Rnummin)/(maxcol-1);
deltaImaginary:=(ImagMax-ImagMin)/(maxrow-1);
col:=0;
Repeat
        row:=0;
        Repeat
```

{ Initialise X and Y values to 0 for each point. X is teh real part of each
  point and Y represents the imaginary part of each number. }

```
                X:=0.0;
                Y:=0.0;
                colour:=0;
                repeat
                    colour:=colour+1;
                    Xsquare:=x*x;
                    Ysquare:=y*y;
                    if (Xsquare+Ysquare)<= (maxsizeSquared) then
                        Begin
```

{ Recalculate X and Y values; the Imagmin and Rnummin terms of each X and Y
  expression is the 'C' term discussed in the text, arrived at by calculating
  the current value of C number from the original value and the current
  column or row position.   }

```
                        Y:=2*x*y+(Imagmin+row*deltaImaginary);
```

{ The next bit is why we stored xsquared and ysquared away, as the calculation
  of the Imaginary Part of the number, Y, wrecks the previous value of Y
  that we need for correctly calculating the Real part!   }

```
                        X:=Xsquare-Ysquare+(Rnummin+col*deltaRnum);
                    end;
```

{ Exit the loop when the colour pointer has exceeded the maximum iteration
  number or
  when the square root of the sum of X**2 and Y**2 exceeds max_size
}

```
                until ((colour>=max_iterations) or
                ((Xsquare+Ysquare)>(maxsizeSquared)));
```

{ If max iteration number has been exceeded then see if a repeating sequence
  can be detected.   }

```
                if colour>=max_iterations then begin
                    t_square:=xsquare+ysquare;
                    colour:=0;
                    repeat
                    colour:=colour+1;
                    Xsquare:=x*x;
                    Ysquare:=y*y;
                        Y:=2*x*y+(Imagmin+row*deltaImaginary);
                        X:=Xsquare-Ysquare+(Rnummin+col*deltaRnum);

                until (abs(t_square-(xsquare+ysquare))<0.01) or
                (abs(xsquare+ysquare)>6E6) or (colour>200);
                if (colour>200) or (abs(xsquare+ysquare)>6E6) then
                    colour:=0;
```

```
                    end;
{   Put a pixel on the screen at col,row after modifying the plotted y value
    so that the maximum y value (corresponding to max. Imaginary value) is
    plotted at the top of the screen.  Colour is selected by modding it with
    the max_colours variable.                                               }

                PutPixel(col,trunc(GetMaxY-row),trunc(colour mod max_colours));

{   Increment row value until either maximum number of row values or
    keypressed  }

                row:=row+1;
            Until (row>=maxrow) or (Keypressed);

{   Increment column until max columns done, or a key pressed               }

                col:=col+1;
            Until (col>=maxcol) or (KeyPressed);

{   Now save screen - uses a library routine                    }

            SaveScreen(Fname);

{   Wait until a key is pressed...                 }

            ch:=Readkey;
            CloseGraph;
            TextBackGround(Blue);
            ClrScr;

{   If key pressed is 'X', then finish tidily.             }

        until (ch='X') or (ch='x');
end.
```

Chapter 8

Julia sets

We've just seen how the iteration of a complex function, $z^2 + c$, can produce some rather interesting graphical images. If you recollect, the Mandelbrot set was drawn by taking a start value for the complex number z of 0 and then the real and imaginary parts of c were varied and the escape time for each resulting value of c was computed for values of c corresponding to positions on the complex plane. The position of c in the complex plane, in terms of real and imaginary parts, was used as a coordinate pair and a pixel plotted in a colour that corresponded to the escape time for that value of c. Iterating a complex function in this way, with varying values of c, is called generating a c plane, as the complex plane represents variations in the value of c.

There is another way in which we can investigate the iteration of the function $z^2 + c$; that is to hold the real and imaginary parts of c constant and vary the initial real and imaginary parts of the complex number z. In this instance, the x coordinate on the complex plane would correspond to the real part of z and the y coordinate would be the imaginary part of z. Again, a colour is assigned to the pixel at these coordinates depending upon the escape time for that particular initial value of z. This process is called generating a z plane, as the complex plane represents variations in the value of z in the function being iterated.

This process of generating a z plane gives rise to what are known as Julia sets, after the French mathematician Gaston Julia who first investigated these iterations in the early years of this century. As you can probably imagine from the fact that they are generated from the same function as the Mandelbrot set, Julia sets are complicated, beautiful structures that can be generated very simply—provided that you have a computer to do all the donkey work. It is a great tribute to Julia and rivals like Pierre Fatou that they managed to draw even crude images of the shapes that these iterations led to; don't forget, these people were doing the iteration of complex functions using pencil and paper, with at most a slide rule and a set of logarithm tables to help them!

Fortunately for us, we can quite easily modify the Mandelbrot set program given in Chapter 7 to generate Julia sets. The algorithm for creating a Julia set is as follows:

1. Select upper and lower limits for the real and imaginary parts of z that are going to be examined. This defines a portion of the complex plane, with the x-axis representing real numbers and the y-axis representing imaginary numbers.

2. Select a real and imaginary value for the number c.

3. Select a maximum number of iterations to perform for each z value iterated, and select a threshold value for z beyond which we will assume that the function will go off to an attractor at infinity.

4. For each pair of real and imaginary parts of z, iterate the function as we did for the Mandelbrot set.

5. If the value of the function after the number of iterations specified is still less than the threshold value selected at stage 3, then assume that this point in the plane is within the Julia set, and so colour it black.

6. If the function value exceeds the threshold, then colour the point depending upon the number of iterations it has taken to exceed the threshold value. If this colour would be black, then set this point to another colour than black.

7. Repeat for each pair of z real and imaginary values in your desired section of the complex plane.

If we adopt this colouring scheme, then any coloured areas of the resultant graphical image are points that we've determined are going to escape to the attractor at infinity. Any black areas are those z values which, when iterated, do not go to the attractor at infinity but stay within the limits of the Julia set. The black areas are, therefore, the points that lay in the Julia set for a particular c value and the coloured areas are those points outside the set. Again, as we'll later see, the boundaries of the Julia sets are the most interesting areas in terms of patterns. The Julia sets are, not surprisingly, fractal, and can be magnified in a similar way to the Mandelbrot set in order to create close-up views of particular regions of a set.

Trajectories for the Julia set

You may recall that our first investigations into the Mandelbrot set were done using a simple program to explore the trajectories followed by the absolute value of z for a specific value of c. The only difference in tracking z trajectories for the Julia set is that the initial value of z does not have to be 0. You may like to try plotting a few trajectories before using Listing 8-1 to plot some Julia sets.

Plotting the Julia set

Listing 8-1 shows a simple program to plot Julia sets. It uses the same sort of algorithm as the Mandelbrot set generating program, and also suffers from the disadvantages of that program in that it is a little slow compared to other algorithms. However, it works and again the mechanism by which the sets are generated is self-explanatory from the listing. The only differences between the algorithm used here and that in Listing 7-3 are as follows:

1. The initial values of x and y (start real and imaginary values of z) are set to a value corresponding to the position of the number in the complex plane, not to 0.

2. A higher value of the value that the function iterates to before we assume that it's going to go to infinity is used (100 rather than 4). You may like to vary this.

3. Only points that do not escape to infinity are coloured in black in this program. Any points that are 'accidentally' black are coloured with one of the other colours.

If you run the program with the default values given, you will obtain a colourful image like that shown in the colour plates of this book. You may like to experiment with altering some of the parameters of the system using this program, before discussing the theory behind why the Julia sets behave in the way that they do. Before we do that, it will be useful to describe the nomenclature that you may encounter when examining Julia sets.

The attractors

Like in the Mandelbrot set, Julia sets two attractors. The values of z for which iteration does not cause an escape to infinity constitute the finite attractor. You will often find that this is described as the *basin* of the finite attractor. This is analogous to the situation in geography where a river has a basin around it which drains in to the river. The points that go off to infinity consitute the basin of the attractor at infinity.

Boundary

The *boundary* of the Julia set is, as in the case of the Mandelbrot set, the area immediately surrounding the points that do belong to the Julia set, that is, it's the boundary between the basins of the two attractors. As with the Mandelbrot set, the boundary is the most interesting area of the Julia sets.

Contour bands

Points that iterate off to infinity at the same rate outside the boundary appear as bands of colour. Within the boundary region, *contour bands* may also appear in the set under suitable magnification.

Connectedness

Connectedness is a mathematical term that describes the Julia set under consideration in terms of whether you can reach every point in the finite attractor basin without crossing the basin of the attractor at infinity. In terms of our program, can we visit all the black points on our image without crossing a coloured point. A set is said to be connected if this is the case. With Julia sets that have a complex internal arrangement, it can be difficult to see whether a set is connected or not from graphical images without considerable magnification, as we did with the Mandelbrot set. In fact, a set that may appear to be connected may be seen to be unconnected at a high enough magnification.

Pinch points

A *pinch point* in the Julia set (or in the Mandelbrot set) is a point where two or more separate parts of the image touch each other. Point 'A' in Figure 8-1 is a pinch point. Let's now look at the main parameters we can alter to change the Julia set image created.

Iteration number

For a specific Julia set (i.e. a set created from a specific pair of real and imaginary parts of c) a small number of iterations will give an outline of the set with very little internal detail. You will, however, see the banded nature of the area outside the set that is similar to the Mandelbrot set. As you increase the number of iterations, the fractal nature of the internal structure of the set becomes more and more apparent, and a large number of iterations leads to a very complex internal structure that can be magnified like the boundary region of the Mandelbrot set. An excessively large

THE VALUE OF C

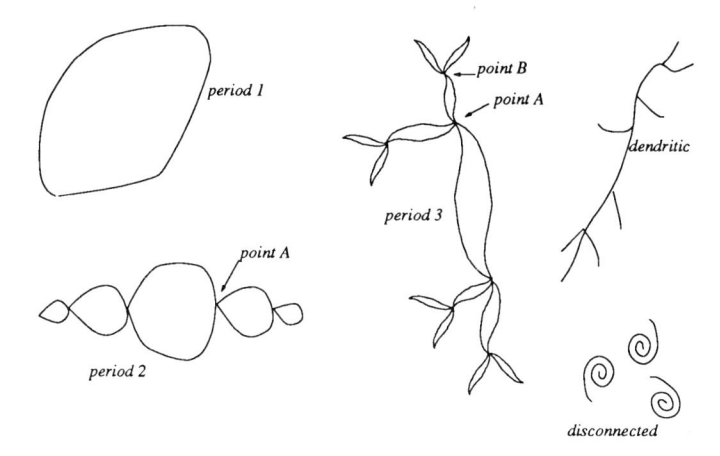

Figure 8-1. Examples of different classes of Julia sets.

number of iterations will lead to a lack of detail in the image, as the Julia set begins to look like a jumble of points. In addition, the more iterations that are carried out, the slower the drawing of the full image will be as points in the image that never go off to infinity will have to be iterated the full number of times. When experimenting with the iteration number in Julia set evaluations, the following rules can be usefully applied.

1. Start with a number of iterations around 50–60, and run the program.

2. Inspect the image; if the display appears to lack detail, and consists of blobs rather than any well-defined structures within the boundary of the set, increase the number of iterations.

3. If the image appears to consist of a jumble of coloured points inside the boundary of the image, then you may need to reduce the number of iterations carried out.

4. If the image appears like a *dust* (a totally disconnected set of points), you might like to try reducing the number of iterations.

The value of c

Of course, the real parameter that affects the shape of the Julia set produced is the value of c that is used. Each different value of c, made up of a real and an imaginary part, will create a different image. To start with a varied series of images, here are a few values of c for you to try in the program:

Real	Imaginary
0	−1
−0.1	0.1
−1	0
−0.22	−0.74
−0.5	0.55
0	0.7

There is a more convenient way of writing down these values:

$$c = real + imag * i$$

and so the first entry in this table would be written as:

$$c = 0 + -1 * i$$
$$\text{or} \quad c = -1 * i$$
$$\text{or} \quad c = -i$$

This is simply a shorthand form of writing down the c value for a Julia set. You also need to remember, of course, which function is being iterated to generate the Julia set from these c values. In this case, it's $z^2 + c$, as we've already discussed.

Try these values of c; you may wish to experiment with different numbers of iterations. In addition, you might like to magnify a specific part of the Julia set being generated by reducing the area of the complex plane being mapped. Enter different values when prompted for minimum and maximum real and complex values to vary the area of the plane mapped. You will notice that these patterns are very different, but we can categorise all Julia sets in terms of their *connectedness* and *period*. We've already discussed the concept of a set being connected, and this is illustrated in Figure 8-1 and also by $c = 0 + 0.7i$. However, what about the period of a Julia set?

Periodicity and link to the Mandelbrot set

If you examine the diagrammatic representations in Figure 8-1, you'll see that for shapes with period 2 and 3 I've labelled a point 'A'. At this point, two or more closed loops meet each other; in a period 2 set ($c = -1.1 + 0.04 * i$, for example) there are two loops which contact each other at this point. In a period 3 set there are three loops, and so on. If there are no points of contact, then the set is unconnected. Some sets, like that given by $c = -1 * i$, are spidery, branch-like things; these are called *dendrites*.

PERIODICITY AND LINK TO THE MANDELBROT SET

An examination of sets with period of 2 or 3 will also indicate the fractal nature of the shapes. For example, point 'B' in the period 3 set shown in Figure 8-1 is similar to point 'A' but is on a smaller scale. This self similarity is repeated at smaller and smaller scales if you magnify the appropriate areas of the set by scanning a smaller area of the complex plane than the default area offered by the program, and, as we've already seen, is a characteristic of a fractal system.

The actual shape of a Julia set obviously depends upon the value of c that is used. At first glance, there is no apparent order to the way in which different values of c give rise to Julia sets of different values. However, a closer look shows that the precise type of pattern obtained from a value of c is determined where that value of c lies in the Mandelbrot set. Remember that each point in or around the Mandelbrot set is a particular value of c, represented as real and imaginary parts. In fact, if you select a particular value of c, draw the Julia set for that value of c and then draw an enlarged section of the Mandelbrot set around the corresponding value of c on the Mandelbrot set, you will see some similarities. It can be shown that the Mandelbrot set, for the function $z = z^2 + c$, acts as a map for all possible Julia sets that can be derived from the same function. The actual shape of a particular Julia set is determined by the position of c within the Mandelbrot set in the following manner.

Values of c outside the set

Values of c outside the Mandelbrot set will give rise to unconnected Julia sets. The area outside the Mandelbrot set is, don't forget, the area in which the only attractor is at infinity. A typical value of c to give an unconnected set would be $c = 0 + 0.7 * i$. Values of c within the main cardioid of the Mandelbrot set

A value of c within the main body of the set—the large, heart shape region to the right of centre—will give a set that is a distorted circle of some sort. Note that the term 'circle' is used very loosely here, as some of the shapes obtained are very ragged indeed, especially if the value of c chosen lies near the boundary of the Mandelbrot set. Julia sets obtained from these locations are said to have a period of 1, and are particularly uninteresting! However, the boundaries of such sets are still fractal.

Values of c within the buds of the Mandelbrot set

If we move to the second largest circle on the Mandelbrot set, to the left of the central 'pinch', then we see Julia sets with a period of 2. A typical value of c to give rise to this type of set is $c = -1 + 0 * i$ (or, $c = -1$). Moving to the last of the large buds on the real axis, around $c = -1.3$, we see Julia sets generated with a period of 4. Areas within the set in other buds, like the two buds at the top and bottom of the main cardioid of the Mandelbrot set, will generate Julia sets of a given period; I'll leave it to you to select suitable values of c to work out the periodicity of Julia sets in these regions.

Values of c on the boundary

Values of c on the boundary of the Mandelbrot set give quite complex Julia sets, with a great deal of internal detail. These sets are well worth magnifying.

Values of c on a filament

A value of c on a filament will give rise to a dendritic structure, very similar to the filament of the Mandelbrot set.

Summary

Before moving away from Julia sets generated from the function $z = z^2 + c$ it's worth making a few comments about what you'll find in the depths of these sets if you magnify them. Like magnifications of the Mandelbrot set, you will need to be prepared to play around with the numbers of iterations used to get attractive images. However, you will find the following structures:

1. The fractal nature of Julia sets will be evident from the self-similarity at many levels of magnification.

2. Within many sets, magnification will show spiral structures that split into finer spirals by bifurcation. This branching continues at greater magnifiucations until a fine network of spirals is generatd. This is reminiscent of the period doubling route to chaos that we investigated in Chapter 2.

3. Within Julia sets, there is a great deal of sensitivity to initial conditions, again the hallmark of a chaotic system. This can be seen by the changes noted in the fine detail (obtained by magnification) of Julia sets generated from similar value of c. Although the shapes obtained may appear to be similar at first glance, the magnified views will be different.

4. The boundaries of Julia sets, like those of the Mandelbrot set, are fractal.

The Julia sets are all mapped by the Mandelbrot set; indeed, the Mandelbrot set can be treated as a one-page atlas of the Julia sets. In addition, as already noted, the Julia set for a particular value of c looks similar to a magnified plot of the Mandelbrot set around the same value of c.

Generating the Julia set in this way, by mapping the complex plane and seeing if a point lies within a set or not, is prone to the principal disadvantage that we saw with the same technique when it was used to draw the Mandelbrot set. It is rather slow. There are other mechanisms for drawing Julia sets, but I will not be exploring them here. The most commonly used alternative is based upon a technique called *backward iteration*. Here, recursive programming methods are used to quickly define the boundary of the set.

The technique of iterating complex functions to generate pictures of the boundaries of attractor basins, as we've done to generate the Julia sets, can be applied to any complex function. The results are still called Julia sets, although they no longer refer to the function $z = z^2 + c$. Let's now look at iterating some other functions in the complex plane and see what Julia sets they generate.

Newton's method

Sir Isaac Newton keeps cropping up in this book; here, I'm going to look at how chaos, in the form of Julia sets, arises from using Newton's method to solve equations such as:

$$z^3 - 1 = 0$$

We are clearly looking for values of a complex number z such that $z*z*z = 1$. An obvious answer is the real number 1, with no imaginary part, but there are other roots to this equation in the complex plane. Figure 8-2 shows the positions of these roots on the Gaussian plane. Missing out the mathematics, it can be shown that you can iterate a function given starting real and imaginary parts of z and that you'll eventually come to rest at one of these three attractors. Obviously, if there were more than three attractors, as happens in the cases of more complex functions, then the iterative function for that case would lead to a different number of attractors. For the $z^3 - 1$ case, the function is:

$$z_{n+1} = 0.66666666 * z_n + 1/(3 * z_n * z_n)$$

Repeated iteration of this function for a start value of the complex number z would lead to one of the three attractors in Figure 8-2. The equation can be rewritten to give separate functions for the real and imaginary parts of z as:

$$\begin{aligned} real_{n+1} &= 0.666666 * real_n + (real_n^2 - imag_n^2)/(3 * (real_n^2 + imag_n^2)^2) \\ imag_{n+1} &= 0.666666 * (imag_n - ((real_n * imag_n)/((real_n^2 + imag_n^2)^2))) \end{aligned}$$

```
                    imaginary
                        ↑
(-0.5, 0.86)            │
      •                 │
                        │
                        │
                        │
←───────────────────────┼──────•──────────→
              (0,0)     │    (1,0)   real
                        │
                        │
(-0.5, -0.86)           │
      •                 │
                        │
                        ↓
```

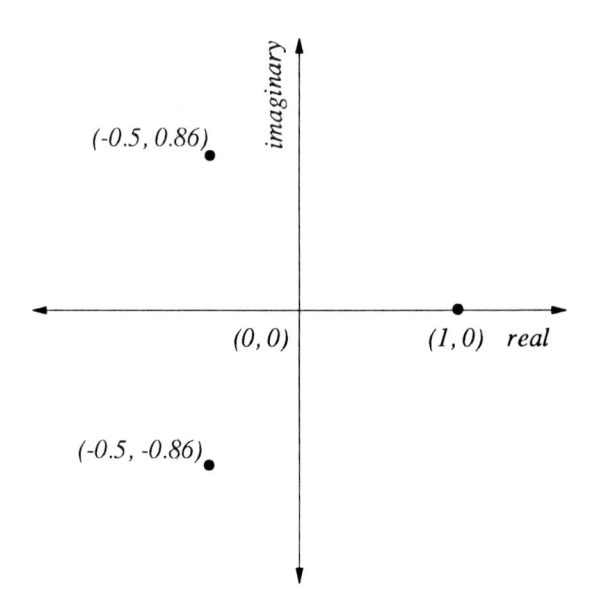

Figure 8-2. The three solutions to $z^3 = 1$.

Armed with these equations, we can map the Julia set for $z * z * z = 1$ and a program to do this is shown in Listing 8-2.

Run the program with an iteration number of 100 or so and use the default limits for the area to be mapped. You will notice the following about the image created:

1. The three attractors are easily visible, and the three basins of attraction are separated from each other by a complicated boundary. A complex number within one of these basins of attraction will iterate to that attractor.

2. A close-up examination of the boundary will show that it is fractal, exhibiting self-similarity and great detail. A complex number that lies on the boundary will iterate to one of the attractors, but precisely which attractor it will go to depends upon the precise value of the number.

3. A few minutes thought will indicate that this last point indicates sensitivity to initial conditions—this is yet another chaotic system.

In programming terms, this listing is fairly straightforward. The iteration loop is exited under two conditions. The first is when the number of iterations exceeds the pre-defined limit. The second is when the value

NEWTON'S METHOD

of the complex number being iterated settles down to a steady value—this indicates that the function has iterated to an attractor. This is detected by the simple process of checking whether the current value of the function is the same as the last value. If it is, then we assume that an attractor has been reached. Due to rounding errors, we're never going to get the two values to be absolutely the same, even at the attractor, so we just check to see whether the two values are within 1E-6 of each other. There is the chance that a divide-by-zero error may arise in this routine if care is not taken, as the function will, at some point, iterate the function with real and imaginary parts of the start complex number equal to 0, and this would lead to a division by 0. This is gotten around by simply checking with an *if* statement.

A further way of looking at this image is to change the way in which colours are assigned to the various points on the screen. In Listing 8-2, a pixel is coloured in depending on how long it takes the function to get to an attractor when using the complex number represented by that point. This gives us no feeling for *which* attractor a particular point iterates towards, only an idea of how rapidly it does so. You might care to alter the colouring of this image along the lines suggested in Chapter 7. A more useful method of colouring the pixels is to colour them dependant upon which of the three attractors the point will iterate to. Of course, for us to do this, we need the values of the three attractors. In this particular case, the three attractors are:

$$z = 1$$
$$z = -0.5 + 0.866 * i$$
$$z = -0.5 + -0.866 * i$$

Listing 8-3 gives a program to do this, colouring in a point depending upon which attractor a point goes to. Magnifying the boundary area using this program is very interesting, as it shows up very clearly the self-similarity within the boundary—each area that iterates to one particular attractor is bordered by areas that iterate to the other two attractors. This is rather a strange arrangement, but it is possible—magnify a section of the boundary and you'll see what I mean! Again, in programming terms when checking to see if a point has iterated to an attractor a comparison is made between the attractor value and the current function value and if the difference is less than 1E-6 then it is assumed that the attractor has been reached. In this program, whether we've reached an attractor or not is determined by the use of Pythagoras' equation, again. By the way, in my version of this program running on a VGA monitor I noted some spots of grey in the attractor areas—these are probably artifacts of the program.

You may like to experiment with this program by generally altering the equations that are iterated. The resultant maps won't have as much mathematical significance as the original listing, but can generate attractive images.

Quaternions

Quaternions are a special form of complex number which has one real part and three imaginary parts. They are written as follows.

$$Q = a + b*i + c*j + d*k$$

where a, b, c and d are different real numbers and i, j and k are imaginary numbers. Quaternions can be worked on like normal complex numbers, so it is possible to generate a function like:

$$Q_{n+1} = Q_n^2 + c$$

which is analogous to the function that generates the Mandelbrot set and the Julia sets with which we started this chapter. Iterating a Quaternion function will give rise to Julia sets, but of a rather different type to those we saw previously. They are much more 'fussy' and the symmetry that we saw with the previous Julia sets is absent. It's important to realise that Quaternions represent a four-dimensional complex number, and when we draw a Julia set based upon iteration of Quaternions we're actually looking at a slice through a much more complex shape.

With 'normal' complex numbers, we have just two parts to the number to think about—a real and a complex part. This is rather convenient for us, as we can plot the real part of the complex plane on one axis and the imaginary part on the other axis, and use the position on the screen as initial values of the complex number to iterate. With Quaternions, we've got a more complicated situation in that we have one real part but three imaginary parts, but we've still only got two dimensions on the computer screen. We can thus only initialise two of these four numbers with real and imaginary values corresponding to positions on the screen. So we provide initial values for the iteration for the real part of the Quaternion (corresponding to a position on the x-axis of the screen) and one of the imaginary numbers (corresponding to the y value on the screen). The initial values of the other two imaginary numbers are given as constants in the program, and these effectively define the 'slice' of the Quaternion that is drawn on the screen.

As to the constant that is added to Q, it consists of four parts—a real number and three imaginary parts. These can be entered into the program, just as we entered values for the usual Julia set program. The values used here can be within the range used to generate non-Quaternion Julia sets,

INVERTED JULIA SETS

but the image will be quite different. The Quaternion can be split into four parts as follows:

$$Q_{real} = a*a - b*b - c*c - d*d + C_{real}$$
$$Q_{imag1} = 2*a*b + C_{imag1}$$
$$Q_{imag2} = 2*a*c + C_{imag2}$$
$$Q_{imag3} = 2*a*d + C_{imag3}$$

You will see that these equations are analogous to those we've already seen for the non-Quaternion Julia sets. The arithmetic involved in iterating these equations is also similar, as can be seen by examining Listing 8-4. However, the programs to generate these Quaternion Julia sets will run more slowly than those that generate normal Julia sets due to the incresed number of multiplications that is involved.

Inverted Julia sets

What, I hear you saying, is an inverted Julia set? Well, let's look at a mathematical process that you might have come across in day-to-day life—finding the *reciprocal* of a number. The reciprocal of a number is the number that, when multiplied by the original number, gives the answer 1. We can write:

$$n * n_r = 1$$

where n_r is the reciprocal of n. You obtain a reciprocal by dividing 1 by n. For a complex number, a similar process exists. This is called *inversion* and involves the following steps:

$$invert = real*real + imaginary*imaginary$$
$$invert\_real = real/invert$$
$$invert\_imaginary = imaginary/invert$$

If we apply this method to our programs for generating Julia sets, then we obtain a listing like Listing 8-5. A little extra code is employed to trap the situations when the value of *invert* is 0. If you were to try dividing by this a divide-by-zero error will be generated in most languages.

The images obtained by inverting the complex number used as the seed for the iteration process is quite different to that from the non-inverted Julia set, and is well worth investigating. When creating an inverted Julia set, as we are here, we invert the value of z corresponding to the point on the complex plane being iterated. You can apply the technique to the Mandelbrot set as well, to create an inverted Mandelbrot set. I'll leave this as an excercise for you to try, but the main thing to remember is that to invert the Mandelbrot set you need to invert the value of the c term that is added to z^2 during iteration.

Some final comments

Before leaving Julia sets, here are a few pointers and ideas for further experiments.

1. Try experimenting with different colour schemes. You might like to try the systems of colouring that were applied in Chapter 7 to the Mandelbrot set. This can give a whole new world of images.

2. If you feel that your maths is up to it, try Julia sets based upon different functions than $z^2 + c$. For example, try higher powers of z, or throw in a few functions like sin. In the latter case, the mathematical significance of the images will not be obvious, but the images can be quite strange!

3. Try magnifying various areas of Julia sets of $z^2 + c$, and match the images resulting from this against the corresponding areas of the Mandelbrot set.

BBC BASIC listings

Listing 8-1

```
 10 REM Program JuliaSets
 30 REM For Chaos Book. Joe Pritchard
 50 REM Try initial values as follows MaxX=2, MaxY=1, MinX=-2, MinY=-1
 60 REM CompReal=-0.6, CompImagin=-0.6, Max_Iterations=256
 80 ON ERROR RUN
100 REPEAT
110   MODE 6
120   PRINT "Plot Julia Set"
130   INPUT "Maximum X Value ",MaxX
140   INPUT "Minimum X Value ",MinX
150   INPUT "Maximum Y Value ",MaxY
160   INPUT "Minimum Y Value ",MinY
170   INPUT "Real part of c ",CompReal
180   INPUT "Imaginary part of c ",CompImagin
190   INPUT "Number of iterations ",Max_Iterations
210   REM Set up limit for iterations at which its decided that this point
      goes off
220   REM to infinity or not.
240   max_size=100
250   MODE 5
260   MaxCol=250
270   MaxRow=160
280   max_colours=4
290   deltaX=(MaxX-MinX)/(MaxCol-1)
300   deltaY=(MaxY-MinY)/(MaxRow-1)
310   col=0
350   REM the next stage of the program solves the equation:
370   REM            Zn+1 = Zn*Zn+c
390   REM where c is a complex number.
440   REPEAT
460     row=0
470     REPEAT
490       REM Now initialise the values of X any Y (real and imaginary parts
          of Z)
500       REM to the current position in the Complex plane.
520       X=MinX+col*deltaX
540       Y=MinY+row*deltaY
550       colour=0
560       REPEAT
570         colour=colour+1
580         Xsquare=X*X
590         Ysquare=Y*Y
600         IF (Xsquare+Ysquare)> max_size THEN GOTO 640
610         Y=2*X*Y + CompImagin
620         X=Xsquare-Ysquare+CompReal
640       UNTIL ((colour>=Max_Iterations) OR ((Xsquare+Ysquare)>max_size))
650       IF (colour<Max_Iterations) AND (colour=0) THEN colour=1
660       IF colour>=Max_Iterations THEN colour=0
680       GCOL0,colour MOD max_colours
690       PLOT69,4*col,4*row
700       row=row+1
710     UNTIL (row>=MaxRow)
720     col=col+1
730   UNTIL (col>=MaxCol)
740   *SAVE JULIA1 5800 8000
750   G$=GET$
760 UNTIL (G$="X") OR (G$="x")
770 END
```

244   CHAPTER 8. JULIA SETS

Listing 8-2

```
    10 REM Program JuliaNewtonSet
    30 REM Try MaxX=1, MaxY=1, MinX=-1, MinY=-1, Max_Iterations=100
    60 ON ERROR RUN
→   80 REPEAT            LINE 880 / WHILE (G$="X") OR (G$="x")
    90   MODE 6
   100   PRINT "Plot Newton Julia Set"
   110   INPUT "Maximum X Value ",MaxX
   120   INPUT "Minimum X Value ",MinX
   130   INPUT "Maximum Y Value ",MaxY
   140   INPUT "Minimum Y Value ",MinY
   150   INPUT "Number of iterations ",Max_Iterations
   170   REM Set up limit for iterations at which its decided that this point
         goes off
   180   REM to infinity or not.
   200   max_size=100
   210   MODE 5
   220   MaxCol=250
   230   MaxRow=160
   240   max_colours=4
   250   deltaX=(MaxX-MinX)/(MaxCol-1)
   260   deltaY=(MaxY-MinY)/(MaxRow-1)
   270   col=0
   290   Current_Z=0
→  330   REPEAT  LINE 850 / WHILE (COL >= MAXCOL)
   350     row=0
   360     REPEAT  LINE 830 / WHILE (ROW >= MAXROW)
   380       REM Now initialise the values of X any Y (real and imaginary parts
           of Z)
   390       REM to the current position in the Complex plane.
   110       x=MinX+col*deltaX
   430       y=MinY+row*deltaY
   460       colour=0
   470       REPEAT  LINE 710 / WHILE ((COLOUR) = MAX_ITER) OR (ABS ETC
   480         colour=colour+1
   500         REM Now calculate the x and y values.  Again, we take the current
             x and y
   510         REM values
   520         REM before we start calculations so as not to lose any data!
   540         xsquare=x*x
   550         ysquare=y*y
   560         xold=x
   570         yold=y
   580         IF (xsquare+ysquare)*(xsquare+ysquare)=0 THEN GOTO 660
   600         x=0.6666*xold+(xsquare-ysquare)/(3*(xsquare+ysquare)*(xsquare+
             ysquare))
   620         y=0.6666*(yold-((xold*yold)/((xsquare+ysquare)*(xsquare+
             ysquare))))
   660         REM IF, after a number of iterations, we've not arrived at an
             attractor, finish
   670         REM any way.
   680         REM Otherwise, say we've arrived at an attractor if the new z
             value is the same
   690         REM as the last z value.
→  710       UNTIL ((colour>=Max_Iterations) OR (ABS(Current_Z-(xsquare+  / WEND
           ysquare))<1E-16))
   730       IF colour>=Max_Iterations THEN colour=0
   740       GCOL0,colour MOD max_colours
   750       PLOT69,4*col,4*row
   760       row=row+1
   780       REM Now re-evaluate Current_z to get the current value oz z*z use
           this to
   790       REM save time - we don't have to calculate a square root.
   810       Current_Z=x*x+y*y
```

BBC BASIC LISTINGS

```
→ 830       UNTIL (row>=MaxRow)     /WEND
  840       col=col+1
→ 850     UNTIL (col>=MaxCol)       /WEND
  860     *SAVE NEWTON 5800 8000
  870     G$=GET$
→ 880 UNTIL (G$="X") OR (G$="x")    /WEND
  890 END
```

Listing 8-3

```
10  REM Program JuliaNewtonAttractor
30  REM For Chaos Book.  Joe Pritchard
50  REM Try MaxRnum=1, MinRnum=-1, MinImag=-1, MaxImag=1,Max_Iterations=100
80  REPEAT
90    MODE 6
100   PRINT "Newton Attractor Plotting"
110   INPUT "Maximum Real Value ",MaxRnum
120   INPUT "Minimum Real Value ",MinRnum
130   INPUT "Maximum Imaginary Value ",MaxImag
140   INPUT "Minimum Imaginary Value ",MinImag
150   INPUT "Number of Iterations ",Max_Iterations
170   REM Set up limit for iterations at which its decided that this point
      isn't
180   REM going to an attractor
220   max_size=100
240   MODE 5
250   maxcol=250
260   maxrow=160
270   max_colours=4
280   deltaRnum=(MaxRnum-MinRnum)/(maxcol-1)
290   deltaImag=(MaxImag-MinImag)/(maxrow-1)
300   col=0
310   Current_z=0
350   REPEAT
370     row=0
380     REPEAT
400       REM Now initialise the values of real any Imag (real and imaginary
          parts of Z)
410       REM to the current position in the Complex plane.
430       Rnum=MinRnum+col*deltaRnum
450       Imag=MinImag+row*deltaImag
480       colour=0
490       REPEAT
500         colour=colour+1
520         REM Now calculate the x and y values.  Again, we take the current
            real and y
530         REM values
540         REM before we start calculations so as not to lose any data!
560         Rnumsquare=Rnum*Rnum
570         Imagsquare=Imag*Imag
580         Rnumold=Rnum
590         Imagold=Imag
600         IF ((Rnumsquare+Imagsquare)*(Rnumsquare+Imagsquare))=0 THEN GOTO
            750
610         Rnum=0.6666*Rnumold+(Rnumsquare-Imagsquare) / (3*(Rnumsquare+
            Imagsquare)*(Rnumsquare+Imagsquare))
620         Imag=0.6666*Imagold-(Rnumold*Imagold)/((Rnumsquare+Imagsquare)*
            (Rnumsquare+Imagsquare))
700         REM If, after a number of iterations, we've not arrived at an
            attractor, finish
710         REM any way.
720         REM Otherwise, say we've arrived at an attractor if the new z
            value is the same
```

```
730         REM as the last z value.
750         UNTIL ((colour>=Max_Iterations) OR (ABS(Current_z-(Rnumsquare+
            Imagsquare))<1E-2))
751
790         t_colour=colour
800         colour=0
810         IF (t_colour>=Max_Iterations) THEN GOTO 870
820         IF (SQR(ABS(Rnum--0.5)) + SQR(ABS(Imag-0.866))) < 0.6 THEN colour=1
830         IF (SQR(ABS(Rnum--0.5)) + SQR(ABS(Imag--0.866))) < 0.6 THEN colour=2
840         IF (SQR(ABS(Rnum-1)) + SQR(ABS(Imag-0))) < 0.6 THEN colour=3
870         GCOL0,colour
880         PLOT69,col*4,row*4
890         row=row+1
910         REM Now re-evaluate Current_z to get the current value oz z*z use
            this to
920         REM save time - we don't have to calculate a square root.
940         Current_z=Rnum*Rnum+Imag*Imag
960      UNTIL (row>=maxrow)
970      col=col+1
980   UNTIL (col>=maxcol)
990   *SAVE NEWTATT 5800 8000
1000  G$=GET$
1010 UNTIL (G$="X") OR (G$="x")
1020 END
```

Listing 8-4

```
10  REM Program QuaternionJuliaSets
30  REM For Chaos Book.  Joe Pritchard
50  REM Try MaxX=2, MaxY=1, MinX=-2, MinY= 1
60  REM and CompReal=-0.192, CompImagin=0.05, CompImagin2=0.66, CompImagin3=-
    0.05
70  REM Max_Iterations=125
100 REPEAT
110   MODE 6
120   PRINT "Quaternion Julia Set "
130   INPUT "Maximum X Value ",MaxX
140   INPUT "Minimum X Value ",MinX
150   INPUT "Minimum Y Value ",MinY
160   INPUT "Maximum Y Value ",MaxY
170   INPUT "Real part of Q ",CompReal
180   INPUT "i part of Q ",CompImagin
190   INPUT "j part of Q ",CompImagin2
200   INPUT "k part of Q ",CompImagin3
210   INPUT "Number of Iterations ",Max_Iterations
230   REM Set up limit for iterations at which its decided that this point
      goes off
240   REM to infinity or not.
260   max_size=4
280   MODE 5
290   MaxCol=250
300   MaxRow=160
310   max_colours=4
320   deltaX=(MaxX-MinX)/(MaxCol-1)
330   deltaY=(MaxY-MinY)/(MaxRow-1)
340   col=0
390   REPEAT
410     row=0
420     REPEAT
440       REM Now initialise the values of X any Y (real and imaginary parts
          of Z)
450       REM to the current position in the Complex plane.
470       Q1=MinX+col*deltaX
```

```
480     QImag3=MinY+row*deltaY
500     REM Because we're plotting on a 2d surface, we can only iterate
        two of the
510     REM parts
520     REM of the Quaternion. So, set up the other two parts with a dummy
        initil
530     REM value
540     REM of 0.05
560     QImag1=0.05
570     QImag2=0.05
580     colour=0
590     REPEAT
600       colour=colour+1
620       REM Now evaluate bits that we need for the calculation of the
          Quaternion
630       REM function
650       Q1square=Q1*Q1
660       QImag1square=QImag1*QImag1
670       QImag2square=QImag2*QImag2
680       QImag3square=QImag3*QImag3
690       xtemp=Q1
710       IF (Q1square+QImag1square+QImag2square+QImag3square)>max_size
          THEN GOTO 790
740       Q1=Q1square-QImag1square-QImag2square-QImag3square+CompReal
750       QImag1=2*xtemp*QImag1 + CompImagin
760       QImag2=2*QImag2*xtemp + CompImagin2
770       QImag3=2*QImag3*xtemp + CompImagin3
790     UNTIL ((colour>=Max_Iterations) OR ((Q1square+QImag1square+
        QImag2square+QImag3square)>max_size))
810     IF (colour<Max_Iterations) AND (colour=0) THEN colour=1
820     IF colour>=Max_Iterations THEN colour=0
840     GCOL0,colour MOD max_colours
850     PLOT69,4*col,4*row
860     row=row+1
870   UNTIL (row>=MaxRow)
880   col=col+1
890 UNTIL (col>=MaxCol)
900 *SAVE QUAT 5800 8000
910 G$=GET$
930 UNTIL (G$="X") OR (G$="x")
940 END
```

Listing 8-5

```
 10 REM Program JuliaSets
 30 REM For Chaos Book.  Joe Pritchard
 60 REM Try initial values as follows MaxX=2, MaxY=1, MinX=-2, MinY=-1
 70 REM CompReal=-0.6, CompImagin=-0.6, Max_Iterations=256
 90 REM ON ERROR RUN
110 REPEAT
120   MODE 6
130   PRINT "Plot Inverse Julia Set"
140   INPUT "Maximum X Value ",MaxX
150   INPUT "Minimum X Value ",MinX
160   INPUT "Maximum Y Value ",MaxY
170   INPUT "Minimum Y Value ",MinY
180   INPUT "Real part of c ",CompReal
190   INPUT "Imaginary part of c ",CompImagin
200   INPUT "Number of iterations ",Max_Iterations
220   REM Set up limit for iterations at which its decided that this point
      goes off
230   REM to infinity or not.
250   max_size=100
```

CHAPTER 8. JULIA SETS

```
260   MODE 5
270   MaxCol=250
280   MaxRow=160
290   max_colours=4
300   deltaX=(MaxX-MinX)/(MaxCol-1)
310   deltaY=(MaxY-MinY)/(MaxRow-1)
320   col=0
350   REM the next stage of the program solves the equation
370   REM                    Zn+1 = Zn*Zn+c
390   REM where c is a complex number.
440   REPEAT
460     row=0
470     REPEAT
490       REM Now initialise the values of X any Y (real and imaginary parts
          of Z)
500       REM to the current position in the Complex plane.
520       X=MinX+col*deltaX
540       Y=MinY+row*deltaY
560       REM The next bit inverts the complex number represented by X and Y
580       invert=X*X + Y*Y
590       IF (invert<>0) THEN X=X/invert : Y=Y/invert ELSE X=1E6 : Y=1E6
610       colour=0
620       REPEAT
630         colour=colour+1
640         Xsquare=X*X
650         Ysquare=Y*Y
660         IF (Xsquare+Ysquare)>max_size THEN GOTO 710
670         Y=2*X*Y + CompImagin
680         X=Xsquare-Ysquare+CompReal
710       UNTIL ((colour>=Max_Iterations) OR ((Xsquare+Ysquare)>max_size))
730       IF (colour<Max_Iterations) AND (colour=0) THEN colour=1
740       IF colour>=Max_Iterations THEN colour=0
750       GCOL0,colour MOD max_colours
760       PLOT69,4*col,4*row
770       row=row+1
780     UNTIL (row>=MaxRow)
790     col=col+1
800   UNTIL (col>=MaxCol)
810   *SAVE INVERT 5800 8000
820   G$=GET$
830 UNTIL (G$="X") OR (G$="x")
840 END
```

Turbo Pascal listings

Listing 8-1

```pascal
Program JuliaSets;

{   For Chaos Book.   Joe Pritchard        }

Uses Crt,Graph,Extend;

Var
   maxcol,maxrow,max_colours,i:                              Integer;
   max_iterations,row,col,colour:                            Integer;
   Compreal,Compimagin,MinX,MaxX,MinY,MaxY,deltaX,deltaY:    Real;
   X,Y,Xsquare,Ysquare,max_size:                             Real;
   Yvalues:                                                  array[0..400]
   of real;
   MaxXs,MaxYs,MinXs,MinYs,ComprealS,CompImaginS,Numits:     String;
   ch:                                                       Char;
   Fname:                                                    String;

Begin

{ Initialise strings with the default values for full picture of set     }

MaxXs:='2';
MinXs:='-2';
MaxYs:='1';
MinYs:='-1';
CompRealS:='-0.6';
CompImaginS:='-0.6';
Numits:='125';
Fname:='Julia';

 Repeat
    TextBackGround(Blue);
    ClrScr;
    PrintAt(2,2,'Julia Set plotting program. Joe Pritchard, 1989',yellow);
    PrintAt(2,4,'Maximum X value:',Green);
    StringEdit(MaxXs,20,30,4,Yellow);
    Val(MaxXs,MaxX,i);
    PrintAt(2,5,'Minimum X value:',Green);
    StringEdit(MinXs,20,30,5,Yellow);
    Val(MinXs,MinX,i);
    PrintAt(2,6,'Minimum Y value:',Green);
    StringEdit(MinYs,20,30,6,Yellow);
    Val(MinYs,MinY,i);
    PrintAt(2,7,'Maximum Y value:',Green);
    StringEdit(MaxYs,20,30,7,Yellow);
    Val(MaxYs,MaxY,i);
    PrintAt(2,8,'Real part of c:',Green);
    StringEdit(CompRealS,20,30,8,Yellow);
    Val(CompRealS,CompReal,i);
    PrintAt(2,9,'Imaginary part of c:',Green);
    StringEdit(CompImaginS,20,30,9,Yellow);
    Val(CompImaginS,CompImagin,i);
    PrintAt(2,11,'Number of iterations:',Green);
    StringEdit(Numits,20,30,11,Yellow);
    Val(Numits,Max_Iterations,i);
    PrintAt(2,12,'File Name:',Green);
    StringEdit(Fname,20,30,12,Yellow);

{ Set up limit for iterations at which its decided that this point goes off
  to infinity or not.                                                     }
```

```
   max_size:=100;

   ChoosePalette;
   maxcol:=GetMaxX;
   maxrow:=GetMaxY;
   max_colours:=GetMaxColor;
   deltaX:=(MaxX-MinX)/(maxcol-1);
   deltaY:=(MaxY-MinY)/(maxrow-1);
   col:=0;

{ the next stage of the program solves the equation:

              Zn+1 = Zn*Zn+c

  where c is a complex number.

}

   Repeat

         row:=0;
         Repeat
{  Now initialise the values of X any Y (real and imaginary parts of Z)
   to the current position in the Complex plane.                      }

               X:=MinX+col*deltaX;

               Y:=MinY+row*deltaY;
               colour:=0;
               repeat
                  colour:=colour+1;
                  Xsquare:=x*x;
                  Ysquare:=y*y;
                  if (Xsquare+Ysquare)<= max_size then
                     Begin
                        Y:=2*x*y + Compimagin;
                        X:=Xsquare-Ysquare+CompReal;
                     end;

               until ((colour>=max_iterations) or
               ((Xsquare+Ysquare)>max_size));
               if (colour<max_iterations) and (colour=0) then
                     colour:=1;
               if colour>=max_iterations then
                     colour:=0;

               PutPixel(col,(MaxRow-row),trunc(colour mod max_colours));
               row:=row+1;
           Until (row>=MaxRow) or (Keypressed);
           col:=col+1;
       Until (col>=maxcol) or (KeyPressed);
       ch:=Readkey;
       SaveScreen(Fname);
       CloseGraph;
       TextBackGround(Blue);
       ClrScr;
   until (ch='X') or (ch='x');
end.
```

TURBO PASCAL LISTINGS   251

Listing 8-2

Program JuliaNewtonSet;

{   For Chaos Book.   Joe Pritchard   }

Uses Crt,Graph,Extend;

Var
    maxcol,maxrow,max_colours,i: Integer;
    max_iterations,row,col,colour: Integer;
    Compreal,Compimagin,MinX,MaxX,MinY,MaxY,deltaX,deltaY: Real;
    X,Y,Xsquare,Ysquare,max_size,yold,xold,Current_Z: Real;
    MaxXs,MaxYs,MinXs,MinYs,ComprealS,CompImaginS,Numits: String;
    ch: Char;
    Fname: String;

Begin

{ Initialise strings with the default values for full picture of set   }

MaxXs:='1';
MinXs:='-1';
MaxYs:='1';
MinYs:='-1';
Numits:='100';
Fname:='Newton';

  Repeat
    TextBackGround(Blue);
    ClrScr;
    PrintAt(2,2,'Newton Julia Set plotting program. Joe Pritchard, 1990',yellow);
    PrintAt(2,4,'Maximum X value:',Green);
    StringEdit(MaxXs,20,30,4,Yellow);
    Val(MaxXs,MaxX,i);
    PrintAt(2,5,'Minimum X value:',Green);
    StringEdit(MinXs,20,30,5,Yellow);
    Val(MinXs,MinX,i);
    PrintAt(2,6,'Minimum Y value:',Green);
    StringEdit(MinYs,20,30,6,Yellow);
    Val(MinYs,MinY,i);
    PrintAt(2,7,'Maximum Y value:',Green);
    StringEdit(MaxYs,20,30,7,Yellow);
    Val(MaxYs,MaxY,i);

{ Set up limit for iterations at which its decided that this point isn't going
  to an
  attractor }

    PrintAt(2,11,'Number of iterations:',Green);
    StringEdit(Numits,20,30,11,Yellow);
    Val(Numits,Max_Iterations,i);
    PrintAt(2,12,'File Name:',Green);
    StringEdit(Fname,20,30,12,Yellow);

    max_size:=100;

    ChoosePalette;
    maxcol:=GetMaxX;
    maxrow:=GetMaxY;
    max_colours:=GetMaxColor;
    deltaX:=(MaxX-MinX)/(maxcol-1);
    deltaY:=(MaxY-MinY)/(maxrow-1);
    col:=0;
    Current_z:=0;

```
        Repeat

            row:=0;
            Repeat
{   Now initialise the values of X any Y (real and imaginary parts of Z)
    to the current position in the Complex plane.                      }

                X:=MinX+col*deltaX;

                Y:=MinY+row*deltaY;

                colour:=0;
                repeat
                    colour:=colour+1;
{   Now calculate the x and y values.  Again, we take the current x and y values
    before we start calculations so as not to lose any data!           }

                    Xsquare:=x*x;
                    Ysquare:=y*y;
                    xold:=x;
                    yold:=y;
                    if (xsquare+ysquare)*(xsquare+ysquare)<>0 then begin

                      x:=0.6666*xold+(xsquare-ysquare)/(3*(xsquare+ysquare)*(xsquare+ysq

                      y:=0.6666*(yold-((xold*yold)/((Xsquare+ysquare)*(xsquare+ysquare))
                    end;

[   If, after a number of iterations, we've not arrived at an attractor, finish
    any way.
    Otherwise, say we've arrived at an attractor if the new z value is the same
    as the last z value.                }

                until ((colour>=max_iterations) or
                  (abs(Current_Z-(Xsquare+Ysquare))<1e-16));
                if colour>=max_iterations then
                    colour:=0;

                PutPixel(col,(MaxRow-row),trunc(colour mod max_colours));
                row:=row+1;
{   Now re-evaluate Current_z to get the current value oz z*z; use this to
    save time - we don't have to calculate a square root.              }

                Current_Z:=x*x+y*y;

            Until (row>=MaxRow) or (Keypressed);
            col:=col+1;
        Until (col>=maxcol) or (KeyPressed);
        ch:=Readkey;
        SaveScreen(Fname);
        CloseGraph;
        TextBackGround(Blue);
        ClrScr;
    until (ch='X') or (ch='x');
end.
```

Listing 8-3

```pascal
Program JuliaNewtonAttractor;

{   For Chaos Book.   Joe Pritchard        }

Uses Crt,Graph,Extend;

Var
   maxcol,maxrow,max_colours,i,t_colour:              Integer;
   max_iterations,row,col,colour:                     Integer;
   Compreal,Compimagin,MinRnum,MaxRnum,MinImag,MaxImag,deltaRnum,deltaImag:
   Real;
   Rnum,Imag,Rnumsquare,Imagsquare,max_size,Imagold,Rnumold,Current_z:
   Real;
   MaxRnums,MaxImags,MinRnums,MinImags,ComprealS,CompImaginS,Numits:
   String;
   ch:                                                Char;
   Fname:                                             String;

Begin

{ Initialise strings with the default values for full picture of set        }
MaxRnums:='1';
MinRnums:='-1';
MaxImags:='1';
MinImags:='-1';
Numits:='100';
Fname:='Newtatt';

 Repeat
   TextBackGround(Blue);
   ClrScr;
   PrintAt(2,2,'Newton Attractor plotting program. Joe Pritchard, 1990',yellow);
   PrintAt(2,4,'Maximum Rnum value:',Green);
   StringEdit(MaxRnums,20,30,4,Yellow);
   Val(MaxRnums,MaxRnum,i);
   PrintAt(2,5,'Minimum Rnum value:',Green);
   StringEdit(MinRnums,20,30,5,Yellow);
   Val(MinRnums,MinRnum,i);
   PrintAt(2,6,'Minimum Imag value:',Green);
   StringEdit(MinImags,20,30,6,Yellow);
   Val(MinImags,MinImag,i);
   PrintAt(2,7,'Maximum Imag value:',Green);
   StringEdit(MaxImags,20,30,7,Yellow);
   Val(MaxImags,MaxImag,i);

{ Set up limit for iterations at which its decided that this point isn't going
  to an
  attractor }

   PrintAt(2,11,'Number of iterations:',Green);
   StringEdit(Numits,20,30,11,Yellow);
   Val(Numits,Max_Iterations,i);
   PrintAt(2,12,'File Name:',Green);
   StringEdit(Fname,20,30,12,Yellow);

   max_size:=100;

   ChoosePalette;
   maxcol:=GetMaxX;
   maxrow:=GetMaxY;
   max_colours:=GetMaxColor;
   deltaRnum:=(MaxRnum-MinRnum)/(maxcol-1);
```

```
        deltaImag:=(MaxImag-MinImag)/(maxrow-1);
        col:=0;
        Current_z:=0;

        Repeat

                row:=0;
                Repeat
{       Now initialise the values of real any Imag (real and imaginary parts of Z)
        to the current position in the Complex plane.                            }

                        Rnum:=MinRnum+col*deltaRnum;

                        Imag:=MinImag+row*deltaImag;

                        colour:=0;
                        repeat
                            colour:=colour+1;
{       Now calculate the x and y values. Again, we take the current real and y
        values
        before we start calculations so as not to lose any data!                 }

                            Rnumsquare:=Rnum*Rnum;
                            Imagsquare:=Imag*Imag;
                            Rnumold:=Rnum;
                            Imagold:=Imag;
                            if (Rnumsquare+Imagsquare)*(Rnumsquare+Imagsquare)<>0 then
                            begin
                                Rnum:=0.6666*Rnumold+(Rnumsquare-Imagsquare) /
        (3*(Rnumsquare+Imagsquare)*(Rnumsquare+Imagsquare));

                                Imag:=0.6666*(Imagold-((Rnumold*Imagold) /
        ((Rnumsquare+Imagsquare)*(Rnumsquare+Imagsquare))));
                            end;

{       If, after a number of iterations, we've not arrived at an attractor, finish
        any way.
        Otherwise, say we've arrived at an attractor if the new z value is the same
        as the last z value.                      }

                        until ((colour>=max_iterations) or
                            (abs(Current_Z-(Rnumsquare+Imagsquare))<1e-16));

{       Now get current value of colour (holding iterations performed) and store.
        Then assign colour to gray, and process t_colour to see if we're at an
        attractor. NOTE - these colours are for EGA / VGA. Change for CGA        }

                        t_colour:=colour;
                        colour:=DarkGray;
                        if (t_colour<max_iterations) then begin
                            if (sqr(Rnum--0.5) + sqr(Imag-0.866)) < 1e-6 then
                                colour:=Red;
                            if (sqr(Rnum--0.5) + sqr(Imag--0.866)) < 1e-6 then
                                colour:=Green;
                            if (sqr(Rnum-1) + sqr(Imag-0)) < 1e-6 then
                                colour:=Blue;

                        end;
                        PutPixel(col,(MaxRow-row),colour);
                        row:=row+1;

{       Now re-evaluate Current_z to get the current value oz z*z; use this to
        save time - we don't have to calculate a square root.                    }
```

TURBO PASCAL LISTINGS

```
                    Current_Z:=Rnum*Rnum+Imag*Imag;

               Until (row>=MaxRow) or (Keypressed);
               col:=col+1;
           Until (col>=maxcol) or (KeyPressed);
           ch:=Readkey;
           SaveScreen(Fname);
           CloseGraph;
           TextBackGround(Blue);
           ClrScr;
     until (ch='X') or (ch='x');
end.
```

Listing 8-4

```
Program QuaternionJuliaSets;

{    For Chaos Book.  Joe Pritchard        }

Uses Crt,Graph,Extend;

Var
   maxcol,maxrow,max_colours,i:                              Integer;
   max_iterations,row,col,colour:                            Integer;
   Compreal,Compimagin,MinX,MaxX,MinY,MaxY,deltaX,deltaY:    Real;
   q1,qimag1,q1square,qimag1square,max_size:                            Real;
   QImag2,QImag3,QImag2square,QImag3square,xtemp:
   Real;
   MaxXs,MaxYs,MinXs,MinYs,ComprealS,CompImaginS,Numits:     String;
   CompImagin3,CompImagin2 :                                 Real;
   ch:                                                       Char;
   Fname,CompImagin3s,CompImagin2s:                          String;

Begin

{ Initialise strings with the default values for full picture of set       }

MaxXs:='2';
MinXs:='-2';
MaxYs:='1';
MinYs:='-1';
CompRealS:='-0.193';
CompImaginS:='0.05';
CompImagin2s:='0.66';
CompImagin3s:='-0.05';
Numits:='125';
Fname:='quat';

 Repeat
   TextBackGround(Blue);
   ClrScr;
   PrintAt(2,2,'Quaternion Julia Set plotting program. Joe Pritchard,
   1989',yellow);
   PrintAt(2,4,'Maximum X value:',Green);
   StringEdit(MaxXs,20,30,4,Yellow);
   Val(MaxXs,MaxX,i);
   PrintAt(2,5,'Minimum X value:',Green);
   StringEdit(MinXs,20,30,5,Yellow);
   Val(MinXs,MinX,i);
   PrintAt(2,6,'Minimum Y value:',Green);
   StringEdit(MinYs,20,30,6,Yellow);
```

CHAPTER 8. JULIA SETS

```
          Val(MinYs,MinY,i);
          PrintAt(2,7,'Maximum Y value:',Green);
          StringEdit(MaxYs,20,30,7,Yellow);
          Val(MaxYs,MaxY,i);
          PrintAt(2,8,'Real part of Q:',Green);
          StringEdit(CompRealS,20,30,8,Yellow);
          Val(CompRealS,CompReal,i);
          PrintAt(2,9,'i part of Q:',Green);
          StringEdit(CompImaginS,20,30,9,Yellow);
          Val(CompImaginS,CompImagin,i);
          PrintAt(2,10,'j part of Q:',Green);
          StringEdit(CompImagin2S,20,30,10,Yellow);
          Val(CompImagin2S,CompImagin2,i);
          PrintAt(2,11,'k part of Q:',Green);
          StringEdit(CompImagin3S,20,30,11,Yellow);
          Val(CompImagin3S,CompImagin3,i);

          PrintAt(2,12,'Number of iterations:',Green);
          StringEdit(Numits,20,30,12,Yellow);
          Val(Numits,Max_Iterations,i);
          PrintAt(2,13,'File Name:',Green);
          StringEdit(Fname,20,30,13,Yellow);

    { Set up limit for iterations at which its decided that this point goes off
      to infinity or not.                                                     }

          max_size:=4;

          ChoosePalette;
          maxcol:=GetMaxX;
          maxrow:=GetMaxY;
          max_colours:=GetMaxColor;
          deltaX:=(MaxX-MinX)/(maxcol-1);
          deltaY:=(MaxY-MinY)/(maxrow-1);
          col:=0;

          Repeat

                row:=0;
                Repeat
    {   Now initialise the values of X any Y (real and imaginary parts of Z)
        to the current position in the Complex plane.                        }

                    q1:=MinX+col*deltaX;
                    QImag3:=MinY+row*deltaY;

    {  Because we're plotting on a 2d surface, we can only  iterate two of the parts
       of the Quaternion.  So, set up the other two parts with a dummy initil value
       of 0.05    }

                    qimag1:=0.05;
                    QImag2:=0.05;
                    colour:=0;
                    repeat
                       colour:=colour+1;

    {  Now evaluate bits that we need for the calculation of the Quaternion
       function }

                           q1square:=q1*q1;
                           qimag1square:=qimag1*qimag1;
                           QImag2square:=QImag2*QImag2;
                           QImag3square:=QImag3*QImag3;
                           xtemp:=q1;
```

```
            if (q1square+qimag1square+QImag2square+QImag3square)<=
            max_size then
               Begin

                    q1:=q1square-qimag1square-QImag2square-QImag3square+CompReal;
                    qimag1:=2*xtemp*qimag1 + CompImagin;
                    QImag2:=2*QImag2*xtemp + CompImagin2;
                    QImag3:=2*QImag3*xtemp + CompImagin3;

               end;

            until ((colour>=max_iterations) or
            ((q1square+qimag1square+QImag2square+QImag3square)>max_size));
            if (colour<max_iterations) and (colour=0) then
               colour:=1;
            if colour>=max_iterations then
               colour:=0;

            PutPixel(col,(MaxRow-row),trunc(colour mod max_colours));
            row:=row+1;
         Until (row>=MaxRow) or (Keypressed);
         col:=col+1;
      Until (col>=maxcol) or (KeyPressed);
      ch:=Readkey;
      SaveScreen(Fname);
      CloseGraph;
      TextBackGround(Blue);
      ClrScr;
   until (ch='X') or (ch='x');
end.
```

Listing 8-5

```
Program JuliaSets;

{   For Chaos Book.  Joe Pritchard       }

Uses Crt,Graph,Extend;

Var
   maxcol,maxrow,max_colours,i:                         Integer;
   max_iterations,row,col,colour:                       Integer;
   Compreal,Compimagin,MinX,MaxX,MinY,MaxY,deltaX,deltaY:   Real;
   X,Y,Xsquare,Ysquare,max_size:                        Real;
   invert:                                              Real;
   MaxXs,MaxYs,MinXs,MinYs,CoprealS,CompImaginS,Numits: String;
   ch:                                                  Char;
   Fname:                                               String;

Begin

{ Initialise strings with the default values for full picture of set     }

MaxXs:='2';
MinXs:='-2';
MaxYs:='1';
MinYs:='-1';
CompRealS:='-0.6';
CompImaginS:='-0.6';
Numits:='125';
Fname:='Julia';
```

```
Repeat
  TextBackGround(Blue);
  ClrScr;
  PrintAt(2,2,'Inverse Julia Set plotting program. Joe Pritchard,
  1990',yellow);
  PrintAt(2,4,'Maximum X value:',Green);
  StringEdit(MaxXs,20,30,4,Yellow);
  Val(MaxXs,MaxX,i);
  PrintAt(2,5,'Minimum X value:',Green);
  StringEdit(MinXs,20,30,5,Yellow);
  Val(MinXs,MinX,i);
  PrintAt(2,6,'Minimum Y value:',Green);
  StringEdit(MinYs,20,30,6,Yellow);
  Val(MinYs,MinY,i);
  PrintAt(2,7,'Maximum Y value:',Green);
  StringEdit(MaxYs,20,30,7,Yellow);
  Val(MaxYs,MaxY,i);
  PrintAt(2,8,'Real part of c:',Green);
  StringEdit(CompRealS,20,30,8,Yellow);
  Val(CompRealS,CompReal,i);
  PrintAt(2,9,'Imaginary part of c:',Green);
  StringEdit(CompImaginS,20,30,9,Yellow);
  Val(CompImaginS,CompImagin,i);
  PrintAt(2,11,'Number of iterations:',Green);
  StringEdit(Numits,20,30,11,Yellow);
  Val(Numits,Max_Iterations,i);
  PrintAt(2,12,'File Name:',Green);
  StringEdit(Fname,20,30,12,Yellow);

{ Set up limit for iterations at which its decided that this point goes off
  to infinity or not.                                                      }

  max_size:=100;

  ChoosePalette;
  maxcol:=GetMaxX;
  maxrow:=GetMaxY;
  max_colours:=GetMaxColor;
  deltaX:=(MaxX-MinX)/(maxcol-1);
  deltaY:=(MaxY-MinY)/(maxrow-1);
  col:=0;

{ the next stage of the program solves the equation:

               Zn+1 = Zn*Zn+c

  where c is a complex number.

}

  Repeat

        row:=0;
        Repeat
{   Now initialise the values of X any Y (real and imaginary parts of Z)
    to the current position in the Complex plane.                        }

              X:=MinX+col*deltaX;

              Y:=MinY+row*deltaY;

{   The next bit inverts the complex number represented by X and Y       }
```

TURBO PASCAL LISTINGS

```pascal
                invert:=x*x + y*y;
                if (invert<>0) then begin
                   x:=x/invert;
                   y:=y/invert;
                end
                else begin
                   x:=1E6;
                   y:=1E6;

                end;

                colour:=0;
                repeat
                   colour:=colour+1;
                   Xsquare:=x*x;
                   Ysquare:=y*y;
                   if (Xsquare+Ysquare)<= max_size then
                      Begin
                          Y:=2*x*y + Compimagin;
                          X:=Xsquare-Ysquare+CompReal;
                      end;

                until ((colour>=max_iterations) or
                ((Xsquare+Ysquare)>max_size));
                if (colour<max_iterations) and (colour=0) then
                   colour:=1;
                if colour>=max_iterations then
                   colour:=0;

                PutPixel(col,(MaxRow-row),trunc(colour mod max_colours));
                row:=row+1;
             Until (row>=MaxRow) or (Keypressed);
             col:=col+1;
        Until (col>=maxcol) or (KeyPressed);
        ch:=Readkey;
        SaveScreen(Fname);
        CloseGraph;
        TextBackGround(Blue);
        ClrScr;
    until (ch='X') or (ch='x');
end.
```

Chapter 9

Other fractal systems

In this chapter, I'll briefly look at other fractal systems that you may like to play with on your computer. If you touch nothing else in this chapter, do try out the 'chaos game' programs. The game still impresses me after a couple of years of playing!

Fractal landscapes

One area in which fractal geometry has received some commercial attention has been in the simulation of landscapes in films or computer games. In the latter case, storage is at a premium so a procedure for generating large amounts of landscape for you to fly over in your computer-simulated aircraft that does not involve lots of disc or memory storage is very useful. In the cinema, fractals are often used to create planets and landscapes in science-fiction films, the most famous probably being the Genesis planet in *Star Trek II*. Well, I'm not going to show you how to do those sorts of graphics. To start with, they're fairly complex and need a considerable amount of processing power; in addition the quality of graphics available on most home-computer systems isn't really up to the job of really realistic landscapes. However, I will explore the basic principles behind one such technique, and provide a program that you can experiment with to get some idea of how these systems work.

There are a variety of ways in which two- and three-dimensional representations of coastlines, hills, etc., can be obtained using fractal techniques. Let's start by looking briefly at the basic methods used, and then we'll examine a program to generate two-dimensional 'maps'.

Mid-point displacement

You may remember that we used this technique for simulating Brownian motion using a fractal. Here a grid of squares of triangles is set up and the middle of randomly selected squares or triangles are displaced either above or below the plane of the grid a random number of times. This gives areas of the grid, represented in the computer by a two-dimensional array, which are above and below the initial plane of the grid, which represents 'sea level' in the system. The amount of displacement applied to each individual mid-point on the grid is randomly selected, and this corresponds to the height of the landscape. The resultant 'map' can then be coloured to give a contour map or it can be converted into a perspective view to give a pseudo three-dimensional view.

Fault line modelling

A further way of generating landscapes is shown in Listing 9.1. This is based upon the technique of generating random 'faults' across a landscape and then raising or lowering the 'land' either side of the faults by a random amount. In practical terms, the area to be modelled is represented by a two-dimensional array. A fault line is represented in this system by using the following method. A straight line has a gradient and this gradient is given by:

$$gradient = (y - y_1)/(x - x_1)$$

where (x, y) and (x_1, y_1) are two points through which the line passes. We can rearrange this as:

$$(y - y_1) = gradient * (x - x_1)$$
$$y = gradient * (x - x_1) + y_1$$

This means that, if we were to generate a random fault across the landscape we could work out which side of the fault line a particular value of y in the landscape would be. A brief examination of the program will show how this is implemented. The program generates realistic maps when a fair number of faults are introduced into the system. The value of 250 chosen in the program is quite good but takes rather a long time to run. I suggest that when you are trying this program out for the first time, replace this 250 with something like 25, and try a map of 70×70 squares rather than 170×170. This will execute faster, but will still be very slow on 8086 machines or on the BBC Micro; it took several minutes to run for the full 250 faults and 170×170 map on my 386SX PC, so you have been warned.

FAULT LINE MODELLING

Once the displacement values have been applied to the map, each point on the map is plotted and given a colour according to how 'high' on the map that point is. The procedure adopted for this is:

1. Calculate the minimum and maximum 'heights' of squares on the map. Let's call the difference between these two values *difference*.

2. Scan the grid of squares, and assign a colour to each square according to the following rules:

    (a) If the square is less than 0.25**difference*, then colour it blue; this is the sea.

    (b) If between 0.25 and 0.5**difference*, then call it a plain and colour this square green.

    (c) If between 0.5 and 0.75**difference*, then call it a hill and colour it brown.

    (d) If over 0.75**difference*, it's a mountain; colour this white.

You will find that some of these maps are better than others, so be prepared to experiment with the program. Here are a few experiments to try:

1. The relative proportion of mountain to hill to plain to sea can be altered by changing the steps 2(a) to 2(d) above. For example, to create a more watery map you would increase the fraction of the total difference between maximum and minimum from 0.25 to, for example, 0.4, and then modify the plain, hill and mountain values accordingly.

2. A further enhancement to this program would be to increase the number of colours used. This will require that the splitting up of the grid done in steps 2(a) to 2(d) above would have to be modified to introduce more levels.

3. You might like to try rendering the grid produced by this program into a pseudo three-dimensional display. Look at the Mandelbrot set pseudo three-dimensional code in Chapter 7 for some inspiration, but rather than plotting points, as we did there, plot solid, coloured columns starting from the back of the screen (top) and working forward (down).

There are a couple of defects (if you'll pardon the pun) with this and the mid point displacement method of drawing landscapes. The first is that a lot of the edges are so very straight! This is an artifact of the method of drawing used, but can be troublesome. The other problem is that such landscapes are often very fiddly and have too much irregularity for them to ring true as landscapes. The closest real landscapes to some of the maps generated are probably areas like Finland, containing many small lakes, or coastlines like Norway, with all the fjords. In his book *The Fractal geometry of nature* Benoit Mandelbrot describes fractal 'forgeries' of landscapes in some detail.

There are other, more complex methods of generating fractal landscapes that I won't go in to here. These involve complex mathematical methods such as Fourier synthesis. However, these methods do have the advantage of producing images that are a little closer to what we expect from a landscape.

Fractal plants and trees

As has already been mentioned, trees and plants in nature often exhibit fractal properties, and so it's not surprising that we can generate reasonable models of trees and plants using fractal graphics techniques. One method was mentioned when the L-language was described. A further method is to use what are called *graftals*; these work in a similar way to the L-language, in that a tree shape is defined as a string of symbols. You'll find reference to these in Becker and Doerfler. A more straightforward way to generate a tree is to use recursion. To understand how this works, we simply have to think about trees in a slightly different way than we usually do. What does a tree consist of?

1. A trunk—OK, nothing complicated there.

2. Branches, which lead to smaller branches, which lead to twigs.

Now, for the purposes of our model we could say that a branch is actually a small tree, with its own branches, each of which is again a small tree, with branches, right down to the twig level. In real trees, there are some differences at different scales, but using this method we can get a reasonable tree by repeatedly calling a routine to actually draw a branch, and have each call to this routine call itself again a set number of times. This is the basis of the program shown in Listing 9-2 to a draw a tree. Again, you may care to experiment with varying the different parameters of this program, although the default values listed will give a good starting place. A further method of drawing plant-like structures is to use *iterated function systems*, as will be given later in this chapter.

Iterated function systems

If there's anything that approximates to mathematical magic in this book, this is it. The area of iterated function systems in fractal geometry and chaos science is likely to be the first that makes a really big impact in the

ITERATED FUNCTION SYSTEMS

real world, particularly in the fields of data compression and communications. However, before we see how it all works, let's play a game. This game, first described by Michael Barnsley, is called the *chaos game*.

The chaos game

The rules of this game are very simple. You could, if you have great patience, play the game with a sheet of paper and a die, but we'll use a computer; after all, life's short!

1. Select three points on the computer graphics grid. Call these *heads*, *tails* and *side*. The original game used a rather remarkable coin which can fall on its side and stay there as well as falling heads or tails. In the true spirit of experimental mathematics, we'll assume we have such a coin, and label our points accordingly. Each point has an x and y coordinate to identify that point on the grid.

2. Select a random point anywhere on the grid. Mark this initial point.

3. Toss the coin. Let's assume it falls heads. Mark a point exactly halfway between the initial point and the point labelled *heads*. Call this new point p_1.

4. Toss the coin again. Let's assume it falls tails, this time. Now, draw in another point exactly between p_1 and *tails*. This process is then repeated a few thousand (or hundreds of thousands) of times. Figure 9-1 shows the now familiar shape of the Sierpinski carpet that emerges from this process.

Listing 9-3 demonstrates how this can be done on a computer. Again, it's well worth watching how this shape is generated as the program runs. The outline of the shape appears fairly quickly, but the detail is filled in over time. The implications of this are as follows:

1. A small number of iterations will give the outline of the shape but will not present us with any great internal detail.

2. The fine detail of the pattern will become apparent when more points are plotted, but this will take time.

As to the technical details of this program, the only point to note is the use of the two variables i and k to control the number of points that are plotted. Integers usually have a limited range, so if we wished to exceed this range we might have to consider using real numbers. Here, however, an alternative is used. An outer and an inner loop gives the number of points plotted as $i * k$, where i and k are the highest values reached by the two variables in the counting loop. So, in the default conditions in the

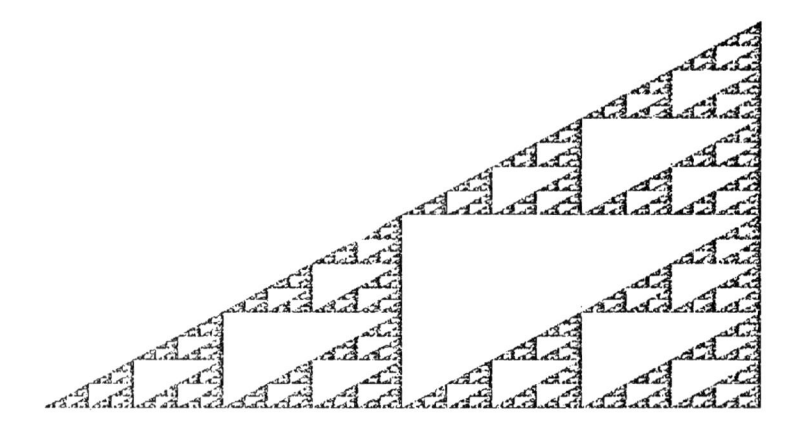

Figure 9-1. The familiar shape of the Sierpinski carpet.

program 30,000 points would be plotted. If you increased k from 1 to 10, then 300,000 points would be plotted. Be warned—this can take time! The only other thing to note is that the first 20 points that are calculated are not plotted. This gives the program time to settle down—plotting these 20 points would give a few points that do not fall into the desired shape.

It is interesting to note that we feed this program from a random-number generator, so that the plotting of points depends purely on probabilities. There should be a 1/3 chance of a head, tail or side being chosen. In the short term, therefore, the shape may appear distorted if the weighting of the numbers generated is away from one or more of these values. However, for a good random-number generator over a period of time the weighting offered to each of the three values should be the same and so a full image will appear. A bad random number generator will generate the image in a rather odd way. To see this, try the following experiment. Change the k loop to run from 1 to 4, so that a total of 120,000 points will be plotted, and then change the line (in the PC Version):

```
j:=random(3);
```

to read:

```
j:=random(4);
```

On running the program you'll find the Sierpinski carpet disappears into a more raggedy image. Here we've simply modified the random number generator so that, whilst the digits 0, 1 and 2 are still generated (and used by the program) the digit 3 is also generated. There's no rule for how to deal with 3 (we're only using head, tail and side, remember) so this number is

ITERATED FUNCTION SYSTEMS

ignored by the system. However, this leaves a gap in the random numbers; we can process three random digits, but a possible four can be generated. Thus, we're only dealing with 75% of all the possible probabilities and the system produces a rather odd image.

If we leave head, tail and side for the time being, you can play the game with more than three fixed points. Listing 9-4 shows a four-point version of the game. Note the other changes needed:

1. Random number generator altered.

2. More points plotted.

In addition, I've made the part of the program dealing with the rules of the game more general purpose by using arrays. You might like to carry out some further experiments with this simple program, bearing the following in mind:

1. Varying the number of fixed points will produce different images.

2. Although the programs, as listed, plot a point half-way between the randomly selected x, y point and a specific fixed point, there is no reason why this should be so.

In fact, the half-way rule is quite interesting for three fixed points, but other numbers of fixed points give better results if you change this. To do this in Listing 9-4 requires a change to the lines:

```
t_x:=trunc(abs(fixed_x[j]-x) / 2)
t_y:=trunc(abs(fixed_y[j]-y) / 2)
```

which specify that the next point will be half-way between the two existing points. The altered lines read:

```
t_x:=trunc(abs(fixed_x[j]-x) * n)
t_y:=trunc(abs(fixed_y[j]-y) * n)
```

where n is a number between 0 and 1 specifying the distance between the two existing points that the new point will be. For example, to make it half-way, $n=0.5$. If you allow the value of n to exceed 1 then there is a great likelihood that, although a pattern will be generated, the picture will grow rather uncontrollably! A useful starting point is to use Listing 9-4 but alter these two lines to give an n value of 0.7. This produces a pattern of squares on the screen very reminiscent of Cantor dusts.

If you play with this program for a while, you might get the impression that we've dealt with something similar to this before. You'd be correct. The system behaves very much like the mappings we looked at earlier in the book, and the final image created is analogous to the attractor for the system. For the Hénon mapping, for example, we iterated the function repeatedly to draw the attractor. Here, generate the map by repeatedly applying a simple process to randomly generated numbers. The principles are very similar. The big difference is that there is a probabilistic element in the chaos game. The image generated for a set of fixed points and a rule set will always follow the same attractor, given a reasonable random-number generator.

A natural extension of this simple idea would be for us to be able to have some predictable control over the image created. For example, we might like the idea of generating, say, a fractal plant or spiral. Could we do this? After all, with the different images you can generate from this program by altering the distances between fixed and varying points, the spread of the random numbers, and the probability of a rule being applied, surely it should be possible to get some images of other objects.

Transformations

It is possible to produce images of other objects, but to see how the technique works we need to go back to a few first principles in graphics. Take any shape drawn on a plane. It's made up of a series of points, and each point has an x and y coordinate. Let's assume that it is centred on the origin, $x = 0$, $y = 0$. Now, if you think about it, we can modify several parameters of the image. To start with, we could shift it away from the origin. Then, we might enlarge or shrink the image. Finally, we could consider rotating it around the origin or with respect to any other point in the plane.

Each of these alterations is called a *transformation* of the shape, and transformations consisting of a rotation, scaling and a linear shift of this type are called *affine transformations*. When a transformation is applied to a shape, the transformation is applied to each point in the shape. An affine transformation is described in terms of a matrix:

$$\begin{pmatrix} x_{n+1} \\ y_{n+1} \end{pmatrix} = \begin{bmatrix} a & b \\ c & d \end{bmatrix} (x_n, y_n) + (e, f)$$

where x_{n+1}, y_{n+1} are the resulting coordinates of applying the affine transformation to points x_n, y_n. The variables a, b, c and d define a rotation to be applied to the coordinate point being transformed. The size of these values also gives rise to scaling effects. If the overall effect of this part of the transformation is to give a scaling factor of greater than one, then the transformed image will be larger than the original, and if this is the case then repeated applications of the transformation can lead to the shape 'growing like Topsy'. If the overall scaling factor is less than one, then the transformed image will be smaller than the original image, and repeated applications of the transformation will lead to a shrinking image. Such a

ITERATED FUNCTION SYSTEMS

transformation is called a *contractive transformation*, because it causes the elements of the shape to shrink together whilst retaining their positions relative to one another. By altering the scaling factor applied to the x and y parts of the shape, the shape can be slightly distorted; for example, a different x and y scaling factor when applied to a circle results in an ellipse. The terms e and f refer to a linear shift of the shape, not altering its orientation or size but simply moving the shape in the plane. The matrix operation given above can be converted into two equations:

$$x_{n+1} = a*x + b*y + e$$
$$y_{n+1} = c*x + d*y + f$$

If we know the position of each point on a shape, then actually applying the transformation to the shape is rather easy; we simply use each point on the shape as (x, y) and generate, from these equations, the corresponding (x_{n+1}, y_{n+1}) pairs which can then be plotted. However, what if we don't know the position of all points on the shape? What can we do then? Well, the simplest way of handling this is to randomly select a point that lies somewhere in the vicinity of the shape that is having the transformation applied to it. Note the colour of the pixel on the screen at that location, and apply the transformation to the point selected. Now, plot the resultant (x, y) pair in the colour of the original point. If the original point was part of the shape—i.e. had a colour that was not that of the background—then the transformed version of this point will be plotted in the same colour. Background colour points will be transformed and plotted in the background colour, and so will not show up. There is a one-to-one correspondence between points in the original image and points in the transformed image. After a large number of iterations of this process, there will be a transformed version as well as the original shape in the plane.

Now comes the clever bit. Let's take away the original shape, and see what happens then. The first thing that we need is more than one affine transformation to apply. We'll come on to how these transformations are defined later, but suffice to say for now that we need at least two different affine transformations. Associated with each transformation is a measure of how often the transformation needs to be applied to generate a desired final image. This is usually specified as a probability, between 0 and 1.

In a system with several affine transformations, the probabilities associated with each affine transformation must add up to 1.0. Whether a particular affine transformation is applied or not is specified by a random-number generator; for example, a value of between 0 and 0.33 might cause transformation 1 to be applied, between 0.34 and 0.67 might cause transformation 2 to be applied, and above 0.67 will cause transformation 3 to be applied.

CHAPTER 9. OTHER FRACTAL SYSTEMS

Well, have you noticed what's missing here? Yes, we're not actually applying the transformation to anything! We've already disposed of the original shape, so what do we apply the transformation to? Well, we start with a randomly selected point and plot this. We then assume that this is effectively part of the original shape, and apply a randomly selected transformation to this point. The new point is now effectively part of the transformed image, and is itself subject to being transformed again. After a few iterations are made to allow things to settle down, the resultant pattern of dots will be a transformed image started off from the first randomly positioned point. To generate a complete image requires that the random-number generator used is a good one; that is, it produces truly random numbers. In theory, better results would be obtained from a hardware random-bit-sequence generator than with the *RANDOM* function of computer languages which actually uses a deterministic algorithm to create pseudo random numbers. In addition, many thousands or even hundreds of thousands of points need to be plotted to get a detailed drawing, although a rough outline of the shape will be drawn fairly rapidly. A final point to note about this technique is that it is extremely well suited to generating fractal shapes by the very nature of the process; after all, a point is transformed repeatedly to generate the image, and so it would be surprising if this *didn't* produce self-similar images somewhere along the way.

This technique is quite fascinating to work with; in fact, when demonstrating it to people I often feel it's the computing equivalent of pulling half a dozen rabbits and a flock of doves from a small top hat! Seriously though, for this technique of rendering the iterated function system to work we need the following:

1. A set of affine transformations to apply to the points to create the desired image. These are usually contractive, to prevent the image blowing up.

2. A set of probabilities to decide which transformation is applied.

3. A good random number generator.

The principal difficulty is actually defining the transformations to be used and the probabilities. However, as an example of the technique in action try Listing 9-5. This uses a transformation matrix to generate a fern-leaf-type pattern. The image improves with the number of points plotted, so a detailed image requires a considerable amount of time to plot, especially on slower computers. The problem with this system is the need for a set of matrices that holds the details of different images. Some starting points for you are shown below.

Tree

a	b	c	d	e	f	probability
0.42	0.42	−0.42	0.42	0	0.2	0.4
0.42	−0.42	0.42	0.42	0	0.2	0.4
0.1	0	0	0.1	0	0.2	0.15
0	0	0	0.5	0	0	0.05

Square

a	b	c	d	e	f	probability
0.5	0	0	0.5	0.5	0	0.25
0.5	0	0	0.5	0	0.5	0.25
0.5	0	0	0.5	0.5	0.5	0.25
0.5	0	0	0.5	0	0	0.25

Sierpinski triangle

a	b	c	d	e	f	probability
0.5	0	0	0.5	0	0	0.33
0.5	0	0	0.5	0.25	0.5	0.33
0.5	0	0	0.5	0.5	0	0.34

Don't forget to sum the probabilities when substituting these values into the arrays in Listing 9-3. The probabilities for the square would be entered as 0.25, 0.5, 0.75 and 1.0.

You'll find sample iterated function system matrices in some of the books listed in the Bibliography. How do we go about constructing our own matrices? Well, it's not that hard, but requires a great deal of rather tedious work for complex shapes. To give you an idea, I'll go through the working out needed for the square given above.

The Collage Theorem

The technique is called the *Collage Theorem*, and works by taking the final image you wish to attain—say a square, or a fern leaf—and several smaller copies of itself. Now, try and fit the smaller copies into the large original in such a way that:

1. The overlap of smaller copies is as small as possible.

2. As much of the original image is covered by the small copies as is possible.

Although you can, if you wish, alter the size of the smaller copies, and shift them around, you can't hack chunks off the smaller copies or use copies of different sizes to tile the larger copy. Now, if we look at our square, we have a simple proposition, as shown in Figure 9-2. We can tile the large

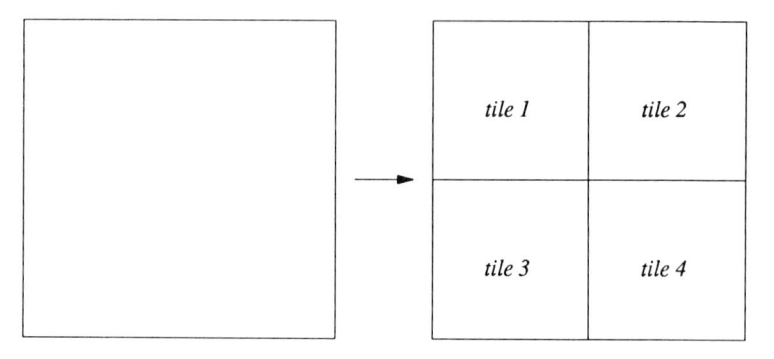

```
+-------------+        +------+------+
|             |        |      |      |
|             |        |tile 1|tile 2|
|             |   -->  |      |      |
|             |        +------+------+
|             |        |      |      |
|             |        |tile 3|tile 4|
|             |        |      |      |
+-------------+        +------+------+
```

Figure 9-2. Tiling a surface.

square with four smaller ones. In this instance, the tiling is perfect, as we might expect for such a relatively simple image. Right, now we come to the pencil and paper bit. Mark a little dot in the centre of the lower left tile of the square. We'll use this as the starting point for our transformations. What we have to do is to create a set of transformations that ensures that points are plotted in all parts of the image in which we're interested, but nowhere else. Now, by using a random-number generator, we'll get a spread of points; we simply need to apply the transformations to ensure that each of the four tiles in the square is visited by the plotting process. Assume that the full square has sides of 1 unit; this will make the sums easier. Finally, don't forget that in order to prevent the whole thing 'blowing up' we need to ensure that the scaling aspects of the transformations are less than 1.

Visiting the lower left tile

Visiting the lower left tile is the least obvious part of the process, so how do we generate a transformation to visit this tile? Well, let's assume that we're anywhere at all in the large square. The lower left tile occupies the space of the square from $0 <= x <= 0.5$ and $0 <= y <= 0.5$. A few seconds though will show that to get back into the lower left tile from anywhere on the square, we simply divide the current x and y values by 2. We can thus write our first set of transformations as:

$$x = 0.5 * x$$
$$y = 0.5 * y$$

THE COLLAGE THEOREM

Visiting the lower right tile

Starting from the lower left tile, visiting the lower right tile is straightforward enough; we keep the y value constant and add 0.5 to x. Now, remembering the comments about scaling made above, we can write a transformation for this operation of:

$$x = 0.5 * x + 0.5$$
$$y = 0.5 * y$$

This will ensure that the values of x and y for this process will never exceed 1, and so will ensure that the points plotted are kept within the boundaries of the square that we're interested in.

Visiting the upper left tile

Visiting the upper left tile we again start from the lower left tile and can use:

$$x = 0.5 * x$$
$$y = 0.5 * y + 0.5$$

Visiting the upper right tile

Visiting the upper right tile we start from the lower left and can have:

$$x = 0.5 * x + 0.5$$
$$y = 0.5 * y + 0.5$$

Now we can take these sets of equations and rewrite them in the form needed for use in the iterated function system. These can be written as:

lower left:
$$x = 0.5 * x + 0 * y + 0 \quad = \quad 0.5x$$
$$y = 0 * x + 0.5 * y + 0 \quad = \quad 0.5y$$
transformation parameters: $0.5, 0, 0, 0.5, 0, 0$

lower right:
$$x = 0.5 * x + 0 * y + 0.5 \quad = \quad 0.5x + 0.5$$
$$y = 0 * x + 0.5 * y + 0 \quad = \quad 0.5y$$
transformation parameters: $0.5, 0, 0, 0.5, 0.5, 0$

upper left:
$$x = 0.5*x + 0*y + 0 = 0.5x$$
$$y = 0*x + 0.5*y + 0.5 = 0.5y + 0.5$$
transformation parameters: 0.5, 0, 0, 0.5, 0, 0.5

upper right:
$$x = 0.5*x + 0*y + 0.5 = 0.5x + 0.5$$
$$y = 0*x + 0.5*y + 0.5 = 0.5y + 0.5$$
transformation parameters: 0.5, 0, 0, 0.5, 0.5, 0.5

We now have four transformations, and to generate our square from these we need to ensure that each transformation is applied the same number of times. They each should have an equal probability of being applied, so we can assign a probability of 0.25 to each of these occurring. Do not forget that the program Listing 9-3 needs these probabilities to be summed.

The probabilities used in affine transformation routines must add up to 1, as otherwise the image produced will not be correct. In addition, transformations with a very small probability attached to them will, clearly, not be executed as often as transformations with larger probabilities attached to them. So, small probabilities will need more points to be plotted for the transformation to be executed frequently enough to fill in the appropriate part of the image. The corollary of this is that transformations with a large probability attached will be executed more times than needed to fill in a particular part of the screen, so these extra transformations will effectively be wasted. You can also run out of probability in some cases, where parts of the image will not be drawn in because you've not specified a high enough probability for those image parts. In these cases, you may need to adjust the other probabilities downwards before you can adjust the probability of interest upwards, as the probabilities must sum up to a value of 1 exactly.

A few minutes' thought about the nature of these transformations will make it clear that because of the linear nature of the transformation process, curves can't be generated in this way. In some of the images given, areas of the image may look curved but these are created by transformations that generate short line segments that give the appearance of a curve. This might appear to be a limitation to this technique, but it needn't be. To add non-linearity into the iterated function system, we use some non-linear equations rather than the transformations. In practical terms, we have two equations, one representing the next x coordinate and one the next y coordinate. These would normally be generated by the transformations applied to the previous point, but in this instance the x and y coordinates of the previous point are used in the equations. A probability is attached to this equation pair, and in the software a change is made to render the iterated function system, so that when that probability occurs the equations are evaluated, rather than using any transformation.

THE COLLAGE THEOREM

Michael Barnsley and his co-workers are already making use of the practical applications of iterated function systems. Once you can work out the affine transformations needed for a particular image, you can generate a set of transformations and the probabilities needed. This data can be stored in half a dozen real numbers per transformation. Now, if we add a few more numbers for colour information, we might be in a position to encode a fairly complex picture—say from a TV camera—in just a few thousand bytes of memory. This is of major significance, as it will allow high-resolution graphics images, TV pictures, etc., to be transmitted over telephone lines in a very short space of time. Or, it could allow the storage of vast amounts of graphical information on single floppy discs. We've got a very good form of data compression here, with compression ratios of hundreds or thousands to one. Of course, the set of transformations takes some time to render back to a screen image, but with advances in computing power, such as parallel processing, we might soon have real-time iterated function systems. The major technical problem, which Barnsley and his colleagues have overcome, is the generation of the transformations from the image to be processed. This is done with a process that is similar to the Collage Theorem.

276   CHAPTER 9. OTHER FRACTAL SYSTEMS

BBC BASIC listings

Listing 9-1

```
 10 REM Program FractalLandscape
 30 REM Generate a fractal map.   Inspired by an algorithm in Fractal Report
 40 REM Newsletter, algorithm modified.
 60 DIM map%(30,30)
 90 PROCInitialise
100 PROCProduceFaults
110 PROCAssignLevels
120 MODE 5
130 PROCPlotTheLandscape
140 END
170 DEFPROCCalculateDisplacement
200 REM This procedure applies the displacement to each position on the
210 REM grid.  It also computes tha lowest and highest points in the grid
220 REM so that the sea level, etc. can be calculated.
240 FOR col=0 TO 30
250   FOR row=0 TO 30
260     IF (row<(gradient*(col-col1)+row1)) THEN
           map%(row,col)=map%(row,col)+displacement
280     IF map%(row,col)>max THEN max=map%(row,col)
290     IF map%(row,col)<min THEN min=map%(row,col)
300   NEXT
310 NEXT
320 ENDPROC
360 DEFPROCInitialise
370 T=RND(-TIME)
380 REM Initialise array to all zeroes this can cause problems if missed out!!
390 FOR col=0 TO 30
400   FOR row=0 TO 30
410     map%(col,row)=0
420   NEXT
430 NEXT
440 max=-10000000
450 min=10000000
470 ENDPROC
510 DEFPROCPlotTheLandscape
540 REM plot a big pixel for each location in the map grid.  Colour it depending
550 REM upon the 'height' of the location. Use VDU commands to change palette
560 REM if you want to.
580 FOR row=0 TO 30
590   y=row*8
600   FOR col=0 TO 30
610     x=col*8
620     IF map%(row,col)<=sea_diff THEN colour=0
630     IF (map%(row,col)>sea_diff) AND (map%(row,col)<plain_diff) THEN colour=1
650     IF (map%(row,col)>=plain_diff) AND (map%(row,col)<hill_diff) THEN colour=2
670     IF (map%(row,col)>=hill_diff) THEN colour=3
680     GCOL0,colour
690     PLOT69,x,y
700     PLOT69,x+4,y
710     PLOT69,x,y+4
720     PLOT69,x+4,y+4
730   NEXT
740 NEXT
750 ENDPROC
790 DEFPROCProduceFaults
800 FOR iteration=1 TO 250
810   IF RND(10)<5 THEN displacement=1 ELSE displacement=-1
```

BBC BASIC LISTINGS   277

```
 830     gradient=10-RND(20)
 840     col1=RND(31)-1
 850     row1=RND(31)-1
 860     PROCCalculateDisplacement
 870   NEXT
 880 ENDPROC
 910 DEFPROCAssignLevels
 940 sea_diff=INT((0.25*(max-min))+min)
 950 plain_diff=INT((0.5*(max-min))+min)
 960 hill_diff=INT((0.75*(max-min))+min)
 990 ENDPROC
```

Listing 9-2

```
 10 REM Program FractalPlants
 30 REM Generate a fractal plant
 70 T=RND(-TIME)
 80 MODE 5
 90 PROCDrawTree(300,200,50)
100 END
130 DEFPROCBranch(density,startx,starty)
150 LOCAL num_branches, branching_factor, bushiness
170 REM Variable num_branches controls number of branches drawn
180 REM Low value of branching_factor bushiness will give
190 REM bushy trees.
210 branching_factor=3
220 bushiness=10
240 IF density<=branching_factor THEN GOTO 420
250 FOR num_branches=0 TO 12
260    IF density<=bushiness THEN GOTO 380
270    branch_angle=RND(180)
280    MOVE startx,starty
310    GCOL0,1
320    DRAW (startx+density*COS(branch_angle)),
  (starty+density*-SIN(branch_angle))
350    PROCBranch((density/2),(startx+density*COS(branch_angle)),
  (starty+density*-SIN(branch_angle)))
380 NEXT
420 ENDPROC
450 DEFPROCDrawTree(X_Tree,Y_Tree,Branches)
460 GCOL0,1
470 MOVE X_Tree,Y_Tree-200
480 DRAW X_Tree,Y_Tree
490 PROCBranch(Branches,X_Tree,Y_Tree)
500 ENDPROC
```

Listing 9-3

```
 10 REM Program ChaosGame
 30 REM Generate Sierpinski Triangle by a rather odd method!!
 60 MODE 5
 80 REM Initialise the three reference points of the triangle
100 headx=250
110 heady=160
120 tailx=250
130 taily=10
140 sidex=0
150 sidey=160
170 REM Initialise the start position
190 T=RND(-TIME)
```

278   CHAPTER 9. OTHER FRACTAL SYSTEMS

```
200 x=RND(250)
210 y=RND(160)
230 REM Plot more points for a more filled in picture, but it will take longer
250 FOR I%=1 TO 30000
270   REM Use the inner 'k' loop TO plot more than 32000 odd points due TO
280   REM use of an integer. Setting k TOrun from 1 to 2 will plot 60000
      points,
290   REM etc.
320   FOR K%=1 TO 1
340     REM Only plot after more than 20 points plotted
360     IF I%*K%<20 THEN GOTO 510
380     j=RND(3)-1
390     IF j=0 THEN PROCJ0
400     IF j=1 THEN PROCJ1
410     IF j=2 THEN PROCJ2
440     REM Update x and y coordinates and plot the point
460     x=x+t_x
470     y=y+t_y
480     PLOT69,x*4,y*4
510   NEXT
530 NEXT
550 END
570 DEFPROCJ2
580 t_x=INT(ABS(sidex-x)/2)
590 t_y=INT(ABS(sidey-y)/2)
600 IF x>sidex THEN t_x=-t_x
610 IF x<sidex THEN t_x=t_x
620 IF y>sidey THEN t_y=-t_y
630 IF y<sidey THEN t_y=t_y
640 ENDPROC
660 DEFPROCJ1
670 t_x=INT(ABS(tailx-x)/?)
680 t_y=INT(ABS(taily-y)/2)
690 IF x>tailx THEN t_x=-t_x
700 IF x<tailx THEN t_x=t_x
710 IF y>taily THEN t_y=-t_y
720 IF y<taily THEN t_y=t_y
730 ENDPROC
750 DEFPROCJ0
760 t_x=INT(ABS(headx-x)/2)
770 t_y=INT(ABS(heady-y)/2)
780 IF x>headx THEN t_x=-t_x
790 IF x<headx THEN t_x=t_x
800 IF y>heady THEN t_y=-t_y
810 IF y<heady THEN t_y=t_y
820 ENDPROC
```

Listing 9-4

```
10 REM Program ChaosGame4
30 MODE 5
40 DIM fixed_x(10),fixed_y(10)
80 fixed_x(0)=250
90 fixed_y(0)=160
100 fixed_x(1)=250
110 fixed_y(1)=10
120 fixed_x(2)=10
130 fixed_y(2)=160
140 fixed_x(3)=10
150 fixed_y(3)=10
170 REM Initialise the start position
190 x=RND(250)
200 y=RND(160)
```

BBC BASIC LISTINGS

```
220 REM Plot more points for a more filled in picture, but it will take longer
240 FOR I%=1 TO 30000
260   REM Use the inner 'k' loop to plot more than 32000 odd points due to
270   REM use of an integer. Setting k torun from 1 to 2 will plot 60000
      points,
280   REM etc.
310   FOR K%=1 TO 20
330     REM Only plot after more than 20 points plotted
350     IF I%*K%<20 THEN GOTO 520
370     j=RND(4)-1
380     t_x=INT(ABS(fixed_x(j)-x)/2)
390     t_y=INT(ABS(fixed_y(j)-y)/2)
400     IF x>fixed_x(j) THEN t_x=-t_x
410     IF x<fixed_x(j) THEN t_x=t_x
420     IF y>fixed_y(j) THEN t_y=-t_y
430     IF y<fixed_y(j) THEN t_y=t_y
450     REM Update x and y coordinates and plot the point
470     x=x+t_x
480     y=y+t_y
490     PLOT69,x*4,y*4
520   NEXT
530 NEXT
550 END
```

Listing 9-5

```
10 REM Program SimpleIFSProgram
30 REM IFS Program based on the work of Michael Barnsley    }
50 DIM a(6),b(6),c(6),d(6),e(6),f(6),p(6)
70 MODE 5
90 oldx=0
100 oldy=0
120 a(1)=0.85
130 b(1)=0.04
140 c(1)=-0.04
150 d(1)=0.85
160 e(1)=0
170 f(1)=1.6
180 p(1)=0.85
200 a(2)=-0.15
210 b(2)=0.28
220 c(2)=0.26
230 d(2)=0.24
240 e(2)=0.0
250 f(2)=00.44
260 p(2)=0.92
280 a(3)=0.2
290 b(3)=-0.26
300 c(3)=0.23
310 d(3)=0.22
320 e(3)=0.0
330 f(3)=1.6
340 p(3)=0.99
360 a(4)=0
370 b(4)=0
380 c(4)=0
390 d(4)=0.16
400 e(4)=0.0
410 f(4)=0
420 p(4)=1.0
450 FOR i=0 TO 30000
460   FOR m=1 TO 40
480     j=RND(1)
```

```
490     IF j<=p(4) THEN l=4
510     IF j<=p(3) THEN l=3
520     IF j<=p(2) THEN l=2
530     IF j<=p(1) THEN l=1
570     x=a(l)*oldx + b(l)*oldy + e(l)
580     y=c(l)*oldx + d(l)*oldy + f(l)
590     oldx=x
600     oldy=y
610     PLOT69,(200*x)+200,(100*y)+100
630   NEXT
640 NEXT
650 END
```

Turbo Pascal listings

Listing 9-1

```
Program FractalLandscape;

{   Generate a fractal map.   Inspired by an algorithm in Fractal Report
    Newsletter, algorithm modified.          }

Uses crt,graph,extend;

Var
        map                             : array[0..170,0..170] of integer;
        iteration,row,col               : integer;
        sea_diff,plain_diff,hill_diff   : integer;
        gradient,col1,row1,displacement : integer;
        colour                          : integer;
        max,min                         : real;
        x,y                             : integer;

Procedure CalculateDisplacement;
Begin

{   This procedure applies the displacement to each position on the
    grid.  It also computes tha lowest and highest points in the grid
    so that the sea level, etc. can be calculated.        }

  for col:=0 to 170 do begin
     for row:=0 to 170 do begin

         if (row<(gradient*(col-col1)+row1)) then
             map[row,col]:=map[row,col]+displacement;

         if map[row,col]>max then
             max:=map[row,col];
         if map[row,col]<min then
             min:=map[row,col];
         end;
   end;
end;

Procedure Initialise;
Begin
randomize;

{   Initialise array to all zeroes; this can cause problems if missed out!! }

for col:=0 to 170 do begin
   for row:=0 to 170 do begin
      map[col,row]:=0;
   end;
end;
max:=-10000000;
min:=10000000;

end;

Procedure PlotTheLandscape;
Begin

{   plot a big pixel for each location in the map grid.  Colour it depending
    upon the 'height' of the location.        }

   for row:=0 to 170 do begin
```

```
      y:=row*2;
      for col:=0 to 170 do begin
         x:=col*2;
         if map[row,col]<=sea_diff then
            colour:=Blue;
         if (map[row,col]>sea_diff) and (map[row,col]<plain_diff) then
            colour:=Green;
         if (map[row,col]>=plain_diff) and (map[row,col]<hill_diff) then
            colour:=Brown;
         if (map[row,col]>=hill_diff) then
            colour:=White;
         PutPixel(x,y,colour);
         PutPixel(x+1,y,colour);
         PutPixel(x,y+1,colour);
         PutPixel(x+1,y+1,colour);
   end;
 end;
end;

Procedure ProduceFaults;
Begin

   for iteration:=1 to 250 do begin
         if random<0.5 then
            displacement:=1
         else
            displacement:=-1;

         gradient:=10-random(20);
         col1:=random(171);
         row1:=random(171);
         CalculateDisplacement;
      end;
end;

Procedure AssignLevels;
Begin

sea_diff:=trunc((0.25*(max-min))+min);
plain_diff:=trunc((0.5*(max-min))+min);
hill_diff:=trunc((0.75*(max-min))+min);

end;

Begin
Initialise;
ProduceFaults;
AssignLevels;
ChoosePalette;
PlotTheLandscape;
end.
```

TURBO PASCAL LISTINGS   283
Listing 9-2
```pascal
Program FractalPlants;

{   Generate a fractal plant    }

Uses crt,graph,extend;

Var

branch_angle          : real;
startx,starty         : integer;
density               : integer;

Procedure Branch(density,startx,starty : integer);

Var     num_branches            : integer;
        branching_factor        : integer;
        bushiness               : integer;

{   Variable num_branches controls number of branches drawn    }

Begin

{   Low value of branching_factor bushiness will give
    bushy trees.    }

branching_factor:=3;
bushiness:=10;

   if density>branching_factor then begin
     for num_branches:=0 to 12 do begin
       if density>bushiness then begin
         branch_angle:=random(180);
         MoveTo(startx,starty);

{   Colour chosen for EGA / VGA. You can vary this, or might like to try
    introducing different colours at different points in the plot    }

         SetColor(Brown);

         LineTo(trunc(startx+density*cos(branch_angle)),
         trunc(starty-density*sin(branch_angle)));

         Branch(trunc(density/2),trunc(startx+density*cos(branch_angle)),
         trunc(starty-density*sin(branch_angle)));
       end;
      end;
    end;
end;

Procedure DrawTree(X_tree,Y_Tree,Branches: integer);
Begin
   SetColor(Brown);
   MoveTo(X_Tree,Y_Tree+200);
   LineTo(X_Tree,Y_Tree);
   Branch(Branches,X_Tree,Y_Tree);
end;

Begin
randomize;
ChoosePalette;
DrawTree(300,200,50);
end.
```

Listing 9-3

```
Program ChaosGame;

{  Generate Sierpinski Triangle by a rather odd method!! }

Uses crt,graph,extend;

Var x,y,z          : integer;
    i,j,k          : integer;
    headx,heady    : integer;
    tailx,taily    : integer;
    sidex,sidey    : integer;
    t_x,t_y        : integer;

Begin
ChoosePalette;

{  Initialise the three reference points of the triangle   }

headx:=GetMaxX;
heady:=GetMaxY;
tailx:=GetMaxX;
taily:=10;
sidex:=0;
sidey:=GetMaxY;

{  Initialise the start position   }

x:=random(GetMaxX);
y:=random(GetMaxY);

{  Plot more points for a more filled in picture, but it will take longer   }

for i:=1 to 30000 do begin

{  Use the inner 'k' loop to plot more than 32000 odd points due to
   use of an integer. Setting k torun from 1 to 2 will plot 60000 points,
   etc. }

     for k:=1 to 1 do begin

{  Only plot after more than 20 points plotted   }

       if i*k>20 then begin

          j:=random(3);
          if j=0 then begin
             t_x:=trunc(abs(headx-x)/2);
             t_y:=trunc(abs(heady-y)/2);
             if x>headx then
                t_x:=-t_x;
             if x<headx then
                t_x:=t_x;
             if y>heady then
                t_y:=-t_y;
             if y<heady then
                t_y:=t_y;
          end;
          if j=1 then begin
             t_x:=trunc(abs(tailx-x)/2);
             t_y:=trunc(abs(taily-y)/2);
             if x>tailx then
                t_x:=-t_x;
             if x<tailx then
```

TURBO PASCAL LISTINGS

```
                t_x:=t_x;
            if y>taily then
                t_y:=-t_y;
            if y<taily then
                t_y:=t_y;
        end;
        if j=2 then begin
            t_x:=trunc(abs(sidex-x)/2);
            t_y:=trunc(abs(sidey-y)/2);
            if x>sidex then
                t_x:=-t_x;
            if x<sidex then
                t_x:=t_x;
            if y>sidey then
                t_y:=-t_y;
            if y<sidey then
                t_y:=t_y;
        end;
{   Update x and y coordinates and plot the point   }

        x:=x+t_x;
        y:=y+t_y;
        PutPixel(x,y,3);
    end;

    end;
end;

end.
```

Listing 9-4

```
Program ChaosGame4;

Uses crt,graph,extend;

Var x,y,z           : integer;
    i,j,k           : integer;
    fixed_x         : array[0..6] of integer;
    fixed_y         : array[0..6] of integer;
    t_x,t_y         : integer;

Begin
ChoosePalette;

{   Initialise the three reference points of the triangle   }

fixed_x[0]:=GetMaxX;
fixed_y[0]:=GetMaxY;
fixed_x[1]:=GetMaxX;
fixed_y[1]:=10;
fixed_x[2]:=10;
fixed_y[2]:=GetMaxY;
fixed_x[3]:=10;
fixed_y[3]:=10;

{   Initialise the start position   }

x:=random(GetMaxX);
y:=random(GetMaxY);

{   Plot more points for a more filled in picture, but it will take longer   }
```

286   CHAPTER 9. OTHER FRACTAL SYSTEMS

```
for i:=1 to 30000 do begin

{   Use the inner 'k' loop to plot more than 32000 odd points due to
    use of an integer. Setting k torun from 1 to 2 will plot 60000 points,
    etc. }

    for k:=1 to 20 do begin

{   Only plot after more than 20 points plotted   }

    if i*k>20 then begin

        j:=random(4);
            t_x:=trunc(abs(fixed_x[j]-x)/2);
            t_y:=trunc(abs(fixed_y[j]-y)/2);
            if x>fixed_x[j] then
                t_x:=-t_x;
            if x<fixed_x[j] then
                t_x:=t_x;
            if y>fixed_y[j] then
                t_y:=-t_y;
            if y<fixed_y[j] then
                t_y:=t_y;

{   Update x and y coordinates and plot the point   }

        x:=x+t_x;
        y:=y+t_y;
        PutPixel(x,y,3);
    end;

    end;
end;

end.
```

Listing 9-5

```
Program SimpleIFSProgram;

{   IFS Program based on the work of Michael Barnsley   }

Uses CRT,Graph,Extend;

Var     i,k,l,m    :   integer;
        a,b,c,d,e,f,p :  array[1..6] of real;
        oldx,oldy,j,x,y   :   real;

Begin
ChoosePalette;

oldx:=0;
oldy:=0;

a[1]:=0.85;
b[1]:=0.04;
c[1]:=-0.04;
d[1]:=0.85;
e[1]:=0;
f[1]:=1.6;
p[1]:=0.85;
```

TURBO PASCAL LISTINGS

```
a[2]:=-0.15;
b[2]:=0.28;
c[2]:=0.26;
d[2]:=0.24;
e[2]:=0.0;
f[2]:=00.44;
p[2]:=0.92;

a[3]:=0.2;
b[3]:=-0.26;
c[3]:=0.23;
d[3]:=0.22;
e[3]:=0.0;
f[3]:=1.6;
p[3]:=0.99;

a[4]:=0;
b[4]:=0;
c[4]:=0;
d[4]:=0.16;
e[4]:=0.0;
f[4]:=0;
p[4]:=1.0;

for i:=0 to 30000 do begin
   for m:=1 to 40 do begin

      j:=random;
      if j<=p[4] then
         l:=4;

      if j<=p[3] then
         l:=3;
      if j<=p[2] then
         l:=2;
      if j<=p[1] then
         l:=1;

      x:=a[l]*oldx + b[l]*oldy + e[l];
      y:=c[l]*oldx + d[l]*oldy + f[l];
      oldx:=x;
      oldy:=y;
      PutPixel(trunc(200*x)+200,trunc(100*y)+100,3);

      end;
   end;
end.
```

Chapter 10

Cellular automata

From fractals, we'll now go on to look at another class of mathematical systems that, at first glance, seems to have little to do with chaos or fractal systems. These systems are *cellular automata*, which were originally developed by the mathematicians John Von Neuman and Stanislaw Ulam as a model of reproduction in biological systems. Cellular automata are also known by different names, such as *tessellation automata, cellular structures* or *iterative arrays*.

In the simplest form, a cellular automaton consists of a grid or line of cells which can be in one of two states, alive or dead. If we start out with generation n, then the state of the cells in the grid or line at generation $n+1$ will depend upon the state of cells at generation 0 and a set of rules that dictates whether a cell in generation $n+1$ is alive or dead. There are more complex versions of this, but the essence of any cellular automaton is that the state of a cell in generation $n+1$ depends upon a production rule and the state of cells in generation n. The generation of cells in state $n+1$ is dependent purely upon the state of the cells in generation n and the rules used; the state of existing cells in generation $n+1$ (i.e., those just calculated) *has no effect* on the creation of generation $n+1$. In a theoretical cellular automaton, the grid of cells is infinite in extent, and the changes of state in cells in the grid are synchronised to take place simultaneously, as if driven by an imaginary clock. In practical terms, we can implement cellular automata using anything from plates (representing cells) on a squared floor (that's how the game of Life was developed) to computer graphics displays. We get around the requirement for an infinite grid by saying that the left-hand edge of the grid wraps around to meet the right-hand edge of the grid, and the top edge of the grid wraps around, if necessary, to meet the bottom edge of the grid. This uses a finite area, but cells cannot run off the edge of it. This is often called a *wrap-round* grid.

Dimensions in automata

Like the other systems explored in this book, cellular automata can exist in different dimensional variants. For example, one-dimensional cellular automata consist of a single line of cells in different states. In line automata, the left and right edges of the line are joined together to produce a continuous line. Two-dimensional automata will consist of a grid like a chess board, occupied by cells, with the left and right edges meeting and the top and bottom edges meeting. Three-dimensional cellular automata will consist of a cubic array of cells—a bit like a Rubic cube. Again, like the other systems we've explored the general principles of cellular automata can be explored by examining the simplest case—the one-dimensional cellular automata. So, let's start there.

One-dimensional cellular automata

The recipe for creating a one-dimensional cellular automaton is as follows:

1. Create a single line of cells, each cell having a value within a specified range. For example, let's say that cells can have values between 0 and 3. Call this line of cells *seed*, generation n. Set up an empty grid to receive the cells of the next generation, $n + 1$; call this *progeny*.

2. Create some production rules to define how to create the next generation of cells. If we number cells from left to right 0 to x, then the simplest production rule might be to make cell x in generation $n + 1$ the same as cell x in generation n. This would, of course, be incredibly boring—nothing will change. A more usual approach is to do something like:

    (a) Define a rule array: $rules[0..9]$ containing values 0 to 3.

    (b) In generation n, add together the values of cells x, $x - 1$, $x + 1$ for each cell on the grid in turn. Call this *oldcells*. If $x - 1$ is less than the left edge of the screen, use the cell of generation n on the right edge of the screen. If the cell referenced by $x + 1$ is off the right edge of the screen, use the cell of generation n on the left edge. This type of rule set, where the cells of the next generation are defined by adding up the values of the cells in the current generation, is often called a *totalistic* rule set.

    (c) Assign a value to the cell x in *progeny* at generation $n + 1$ of $rules[oldcells]$.

3. Draw up the contents of *progeny* on the screen, and copy the values of these cells into *seed*. Repeat the process *ad infinitum*.

ONE-DIMENSIONAL CELLULAR AUTOMATA

Although this sounds rather simple, it produces some exceptionally fascinating displays. Listing 10-1 is a simple one-dimensional cellular automaton in which the values of *seed* and the *rules* array are randomly defined. Have a play with this program, and let's see what we get. The first thing to note is that there are a lot of different images created here and so many runs of the program allow you to see the possibilities of this simple system. In general, the following types of image will be seen:

1. A general mish-mash of colours and shapes, none of which persist for any length of time. However, there are areas of order within the apparent chaos.

2. Regular forms appear and grow for a short time, then stop dead or die away. These are often reminiscent of plants or the buildings which, according to science-fiction writers, will populate our cities in forty years time!

3. Regular forms which appear to persist through many variations.

So, we start with a randomly selected start position and a randomly generated set of rules, which are then applied throughout the program run. This gives us a deterministic system—after all, the state of the system in the next generation is determined by the current state and the application of the rules. This system can give rise to chaos or order, and, as in other chaotic systems we have explored in this book, there are areas of order within the chaos. You will also see self-similarity, one of the hallmarks of a fractal system, in the images created.

A wider implication is this: we had a chaotic set of start points, with a randomly defined set of rules. This gave rise, in some cases, to a persistent, ordered system. The rules are local in action, that is, they only affect the cells that they define. They do not say to the system 'in 340 generations time, make sure that you're in an ordered pattern', this simply drops out of the rule set. In general terms, therefore, we have a means of creating long-term order out of short-term chaos. It is for this reason that many people have studied cellular automata as models of other systems, such as biological systems, turbulence, crystal growth and so on.

To progress further with our studies of cellular automata, we need to cut down the variables somewhat. To experiment with these systems we need to be able to reproduce a specific cellular automaton on demand and with this simple program we can't do that. A one-dimensional cellular automaton is defined by two things; the start conditions and the production rules, so we need to be able to define those. Listing 10-2 shows the simplest way in which this can be done. Here, I simply set all the start conditions to 0 with the exception of a single cell in the middle, and I set up the values in the rule set array deliberately. Run this program, and see what happens.

Yes, it's amazing how often variations on the Sierpinski carpet crawl out from these mathematical systems, isn't it! Now that we've got a method of defining a cellular automaton from the start, we can experiment with the rule set and start conditions to explore these systems in greater detail. I'll suggest some experiments to start with, but after that you're on your own!

The rule set

Apart from the values in the rule-set array, the rule set also consists of how that rule-set array is addressed based upon the values of cells in generation n. In this listing, you can see that the index to the array is created from:

$$left\_one + centre\_one + right\_one$$

where these refer to cells $x-1$, x and $x+1$ as defined above. What happens if we alter this function?

Well, if you use $left\_one + centre\_one$ instead, and remove the reference to the right-hand cell, then the resultant image is biased over to the right. Similarly, removing reference to the left-hand cell biases the image to the left. Removing the reference to the centre cell generates a different image again.

You may like to try applying a *weighting* to one or other of these three parameters, so that the value of one cell has more impact than the value of the other two. Of course, you have to be careful here not to exceed the bounds of the rule set array, or, you could extend the rule-set array.

As to the rule set itself, then clearly changes here will affect the image obtained. I will shortly give a list of interesting rule sets and starting positions for this simple automata, but for now you might like to try altering the values of each rule slightly. Are there rules where alterations have a larger impact on the shape than others? Do the changes get larger with time, or do their effects stay constant? Is there a Butterfly Effect for cellular automata, where a change in the rules does not have an immediately apparent effect but does after several hundred generations? With this particular rule set, try the following changes. Between each, reset the rule set to its original values.

$rule\_set[6] := 2$ This gives a rather different pattern to the Sierpinski carpet generated previously. It's a relatively small change but it has quite an impact on the image obtained.

$rule\_set[3] := 2$ Some order is still visible but this change gives rise to a less structured image than the above change to $rule\_set[6]$.

$rule\_set[3] := 1$ This gives rise to a more chaotic image than the last change, although there are still areas of order within the apparent chaos.

THE RULE SET

$rule\_set[3] := 0$ We might expect this to follow the trend of the above two changes, but in fact more order is apparent here, certainly in the short term, than is apparent in the last rule change.

Altering a single parameter in a rule set like this shows some of the more chaotic aspects of the behaviour of cellular automata. Although an alteration to a value in a rule set will cause some change in most cases, it can be difficult to predict in advance exactly what the change will be. This is reminiscent of other chaotic systems that we've encountered.

What about other rule sets? Well, here is one to try:

$rule\_set[0] := 2 \quad rule\_set[1] := 0 \quad rule\_set[2] := 3$
$rule\_set[3] := 2 \quad rule\_set[4] := 2 \quad rule\_set[5] := 1$
$rule\_set[6] := 2 \quad rule\_set[7] := 3 \quad rule\_set[8] := 2$
$rule\_set[9] := 2$

It is interesting to execute this rule set with a randomly generated start position, i.e. like that seen in Listing 10-1 where we set up *seed* to contain random numbers between 0 and 3 *and* with a start position consisting of just one point. The single point gives a pseudo-Sierpinski carpet, but the random start position gives rise to a structure containing vertical bands which get wider, get narrower, and, in some cases, die out. Varying the value of $rule\_set[9]$ will vary the widths of the 'growths' quite considerably. This rule set introduces us to some useful ideas and interesting concepts. The first of these is in how we define the rule sets.

Nomenclature

Clearly, the method given above is rather long-winded if you intend doing a lot of work on cellular automata. A more convenient way of writing down this particular rule set would be:

$$2032212322_4$$

where the subscripted 4 indicates that there are four possible states for cells to adopt in this particular automaton. We label the rules from the left 0, 1, 2...9. In our case we need to add a further rule that describes how the states of generation n are combined to create a reference to the rule we wish to use in generation $n+1$. This extra rule can be written as:

$$index_{n+1} = state_n^{x-1} + state_n^x + state_n^{x+1}$$

where x corresponds to the position of the cell in generation n. The status of the new cell in generation $n+1$ is thus given by the entry in the rule set at *index*. So, if *index* was equal to 6 the new cell would have a value of 1, this being the sixth entry in the rule set.

Initial states

So far, we've examined two types of initial case; the simple case of one cell in the centre of *seed* having a non-zero value (single-cell initiator) and the complex case of each entry in *seed* being given a random value (random initiator). There is another case to consider. This is where we have several cells each with a known value, rather than just one cell in the centre of *seed* that has a defined value. Thus, we set off the automaton with a start position consisting of a line of several cells, like 10012300_1, in the centre of *seed*. The rules are applied and the automaton evolves. Such a line of cells we can call a multiple-cell initiator. This can be valuable, as many interesting patterns can only be seen when we start the automaton with initial settings of more than one cell set to a non-zero state. You may like to experiment with Listing 10-2, by adding a few more statements after $seed[200] := 1$, to put start values in the cells around this one. Don't forget that the start values don't have to be 1, and you don't have to put the start values in the middle of the seed line.

Computational irreducibility

Some automata give rise to very simple repeating structures, for example, a series of cells in the sequence 1, 2, 3, 1, 2, 3, 1, 2, 3 ... may appear down the screen. If we assume that this isn't going to change for any reason in the future, than we can compute the value of a cell in this line at any future generation we choose provided that we know the status of the cell at the current generation. In effect, we can find out what the automaton is going to do via a mathematical formula. This is similar to the situation for the logistic equation before the threshold into chaos is passed. However, some systems are such that the only way we can see what state the system will be in at a particular generation is to go there by iterating each generation between the current one and the generation of interest. A system that exhibits this sort of behaviour is showing *computational irreducibility*—there is no way in which its behaviour can be simplified to a simple mathematical formula into which we just stick a generation number to get the state of a cell or of the whole automaton at that generation. The only way to explore such a system is to set it up and play with it. The 20322123224_4 automaton exhibits computational irreducibility, for we can't tell what a particular vertical band in the system will do at a point in the future, or whether it will still exist.

Other rules that you might like to consider for this system are as follows. In each case, try them out with a random start positon, a single-cell initiator and a multiple-cell initiator.

$$1321313223_4 \quad 0133132130_4 \quad 0102030202_4 \quad 0012001200_4$$

Again, experiment with these. You might like to think about modifying Listing 10-1 so that when a random rule is generated, it is also displayed on the screen so that you can make a record of it if the results are interesting.

Non-totalistic rule sets

So far, the rule set used for deriving these automata has been totalistic, in that the state of a progeny cell is derived by adding up the values of the seed cells and using the result to reference the rule set to generate the state of the progeny cell. There are other ways of producing rule sets that do not rely on this simple arithmetical approach. For example:

1. Count the number of cells in a non-zero state around the seed cell. If this is 1, then the new cell has the state 1, otherwise it has a state 0.

2. If the sum of the states of the seed cell and the cells around it is an odd number, let the progeny state be equal to 1, otherwise let it be equal to 0.

There are other techniques that you can include. For example, in two-dimensional systems (see below) we introduce the concept of cells *dying* in subsequent generations. This isn't really possible in one-dimensional systems, but we'll look at this idea again later.

Mutation

We've seen that a single cell or row of cells used as the initial condition changes with time to produce more intricate patterns. If we stretch the biological analogy, we could say that the initial population of cells is evolving; some *strains*, i.e. patterns of cells, will become extinct as time passes, whereas others will proliferate, depending upon the rule set used. Now, in a biological system, the evolution of species from generation to generation is controlled by genes, and occasionally mutations occur. These are caused by damage to the genes of an organism which are passed on to the progeny of that organism and show a new trait. This can be good, or, more usually, bad news for the recipient. Mutations in genes are caused by cosmic radiation, nuclear fallout, background radiation, chemicals, etc. In our cellular automata systems, we could envisage mutation as being a change in one of the rules that controls the production of the next generation of cells. Listing 10-3 shows a version of Listing 10-2 which has a mutation factor built in. It also includes a *noise* factor which will be explained below. The higher the value of *mutate*, the greater the probability that a rule will be changed after each generation is created. In this listing, the rule to be changed is selected randomly, but there is nothing to prevent you including a bias in the program so that certain rules are more prone to change than others. Setting *mutate* to 999 will turn off mutation. Any mutation is a permanent affair until the same rule is mutated again. It is, of course, quite possible that a particular rule can be mutated away from its start value and be returned to the start value by another mutation. In this program, a random number between 0 and 999 is generated and compared with the value of the *mutate* variable to see if a mutation is to be applied. Start off your experiments with a value of *mutate* of about 500, and vary it. You will soon see the effect on the image, as changes to the rule set soon cause changes to the generated image. There are a couple of things you might like to try when modelling mutation in these one-dimensional cellular automata:

1. Print the rule set on the screen when mutation occurs. This will allow you to see what changes have occurred.

2. Add a controlled mutation, i.e. specify the rule that is being affected and even the value to which it is to change if a mutation takes place.

3. Rather than totally altering the value of a rule, as is done in Listing 10-3, you might arrange that mutations only vary the value of a rule by plus or minus 1.

Noise

Real-world systems always experience some noise; this will perturb the system to a greater or lesser degree, and tend to hide the behaviour of the system. We can try simulating noise in cellular automata, a useful trick if we're trying to model real-world physical systems with cellular automata. Listing 10-3 shows how we can do this, and the effects of noise on such a system are quite large; a small amount of noise will cause a large change in the final image in many cases. Note that the effects of noise are not permanent, as is the case when dealing with mutations. Again, experiment with the effects of noise on different systems, and try different ways of introducing noise into the system.

Thermodynamics, reversibility and entropy

Much work on cellular automata has been done by Stephen Wolfram of the Institute of Advanced Study at Princeton University in the United States. He has drawn some rather wide-ranging conclusions from his work. Before moving on to look at two-dimensional cellular automata, it will be interesting to look at Wolfram's work.

He noticed the order in chaos and the long-term regular structures that we have observed in our simple systems, and termed this behaviour *self-organisation*. This is a rather interesting ability of cellular automata that is *not* exhibited by systems in the real world, as in the real world the level of entropy in a system (the level of disorder) does not fall, it increases unless we put a considerable amount of work into keeping things organised. Such systems like this, that develop order from apparent chaos, are known as *thermodynamically irreversible* systems. The real world exhibits reversibility; after all, a pile of bricks doesn't build itself into a wall unless we help out! Wolfram claims that the work being done on cellular automata may have a bearing on developing laws of thermodynamics for these irreversible systems.

Simple two-dimensional automata

Let's now move on to look at two-dimensional cellular automata. These exhibit much more complex behaviour than do one-dimensional cellular automata, but are still quite simple to set up. In two-dimensional cellular automata, the grid of cells is viewed as being infinite; in practical terms we allow cells that 'fall off' the right hand edge to come on again at the left edge, and cells that fall off the bottom come on again at the top of the grid (or vice versa). The simplest two-dimensional cellular automata for us to examine would be the two-dimensional equivalent of the line automata that we've just been studying. Listing 10-4 shows an example of this.

In terms of programming, the only points to watch out for are as follows. These can be applied to all programs for investigating cellular automata, especially two-dimensional ones.

1. Arrays take up a lot of space, so when you use them try and use the data type supported by your language that occupies the least space. In Turbo Pascal, for example, this is the *byte* type, which occupies just one byte per stored number, but in BASIC it is likely to be *integer*, which may occupy two or four bytes depending upon the implementation. This has an impact on the size of the grid that can be modelled if you use arrays, for many PC languages will not allow you to define data structures larger than 64k in size. The way around this, of course, is to define your own structures and address them directly using direct memory access. Alternatively, if you want a large-screen display but are willing to put up with a smaller playing area in which the cellular automata can develop, why not use the 80×25 text mode of the PC screen? You could use different characters for cell values.

2. When checking the edges of the grid, or when using a rule array to get the state of a cell, funny things happen if you try and access a

non-existent array element in certain languages. Many languages will give an error, but others will not and a result will be returned that is taken from a piece of the computer's memory adjacent to but not part of that used for the array storage. This can cause irritating problems; I spent several days trying to work out why, in a prototype cellular automata program, I suddenly ended up with fifteen-colour displays when I was only supposed to have three colours and black! This was traced back to an array-bound error of this sort.

3. Large grids take large amounts of processing time. So, when testing programs you may like to start with small grids, watch for a few generations, then use larger grids if anything interesting develops.

The program in Listing 10-4 generates two-dimensional cellular automata based on totalistic rules like those used in the one-dimensional automata. The rule set used in this program again consists of a counting part, which sums up the values of the cell of generation n under examination and of all cells adjacent to the one being examined, and a rule set array, which is then referenced using the sum just calculated. The result then gives the value of the corresponding cell in generation $n + 1$. The rule set array is larger in this case, as there are more possibilities that we have to consider. Here, our cells can assume any value between 0 and 3 inclusive, so we have to have 27 elements in the rule set (the cell and its surrounding eight cells could all have a value of 3), and each element can have a value of between 0 and 3. The demonstration program generates the rule set randomly, but there is nothing to stop you modifying this program to accept a rule set that is defined. This would then allow you to totally control the evolution of the resulting automata.

Although you can have a random initiator for the start conditions, I've found that much prettier results are obtained by simply placing a few or even just one cell in the middle of the array. The results obtained from a random rule set are fascinating to watch. Like many other dynamic systems, much information is gained from watching how the system develops rather than by looking at the final state it assumes. When this program is running, you will see shapes develop and expand across the whole grid. In some cases, the initial cell(s) may simply disappear, but in most cases symmetrical patterns will be generated. You will also find that the background colour may change in certain cases. The best way to experiment with this system is to define a rule set, rather than use a random rule set. A good starting point would be to modify this program to print up the rule set generated. This will give you a series of starting rule sets for further work.

Selection in two-dimensional systems

We don't have to use totalistic rule sets. We can count the cells around the seed cell and decide whether a cell will appear in the corresponding position in the next generation. Listing 10-5 shows how this can be done. Here all the cells on the grid are either value 0 (background colour) or value 1 (single foreground colour). The cells thus have two states, present or absent. The rule set for this system is as follows:

1. Add up the values of the four cells that are orthogonally adjacent to the cell being examined in generation n. Because cells can only have a value of 1 or 0, the total will be between 0 and 4.

2. If the total is 2 (no more, no less) then the corresponding cell in generation $n + 1$

will be set to 1. Otherwise, it will be set to 0.

The term *orthoganally adjacent* simply means the cells that are to the N, S, E and W of the cell of interest. Cells to the NE, NW, SE and SW are said to be *diagonally adjacent* to this cell, and these are *not* considered in this rule set. If you run this program with a random initiator, then a rather interesting phenomenon can be observed—that of *selection*. From the random collection of cells some degree of order will develop, resulting, after a few generations, in the system settling down to a steady state populated by clumps of cells. These clumps can be quickly categorised into two types:

Fixed clumps can be seen; these are shapes that do not appear to alter with each passing generation.

Oscillators are clumps that flick between discrete states. In this two-state system, where cells can have either 0 or 1 as a value, the oscillators have two distinct states, and flick between these two states as the generations pass. Some oscillators are shown in Figure 10-1.

This is a very interesting result, as from a system that appears to be computationally irreducible when we start, a simple rule set has produced a system that has beome computable. If we know the states of the oscillators at generation n, we can work out their states at any subsequent generation, assuming the rule set doesn't change and no more cells appear on the scene. The rule set has effectively selected a set of shapes that can survive, if we wish to use biological analogies here. Later in this chapter, I'll go on to look at a program that uses similar but slightly more complex rules than this to create what is probably the best known cellular automaton—John Conway's game of *Life*.

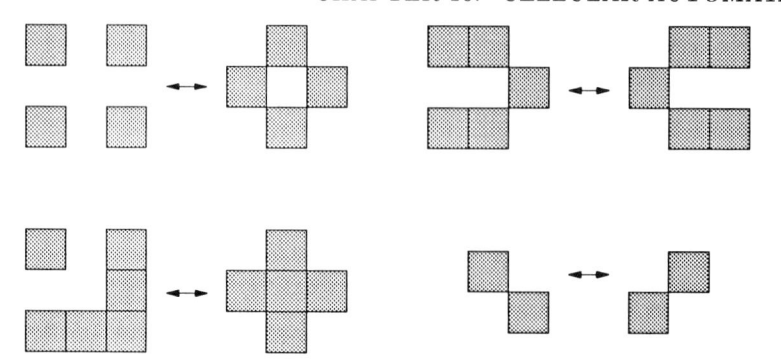

Figure 10-1. A selection of oscillators.

Cell eat cell

A different type of rule set was developed by David Griffeth of the University of Wisconsin and was documented in *Scientific American*. In this system, a cell can actually 'eat' adjacent cells if they have a value 1 less than their own. The fact that a cell has been eaten is indicated by the eaten cell taking on the colour of the eater for the next generation. A cell of value 0 can eat a cell of value n, where n is the largest cell value allowed in the system. Listing 10-6 shows a simple example of this system. The system will again settle down to a final, steady state, but the time taken and the way in which it happens varies enormously for different runs and with different numbers of possible states. You may like to experiment with different numbers of states, but you should take the following into consideration.

Numbers of states

Too few states (below about eight) will take an inordinately long time to settle down and the image will tend to remain a general mess of random colours. This makes this program unsuitable for use on CGA machines or BBC graphics modes with less than eight colours. A good number is fifteen states—this fits conveniently into EGA/VGA modes and quite quickly evolves into a steady state of regular, square, spiral structures.

Grid size

Too small a grid size will cause problems in that the spiral structures that this system usually settles down into will not have room to evolve. Larger grids take longer to settle down.

Time dependent rules

You can have rules for defining cellular automata that are time dependent, generating a cell at a particular position in such a way that is dependent upon the number of generations that have been generated as well as the state of neighbouring cells. Listing 10-7 shows such an example, using a rule described in Clifford Pickover's excellent book *Computers, Pattern, Chaos and Beauty*. A different rule is used on odd and even generations, but you could easily modify this rule to carry out different tasks on every third generation, or every fourth generation, and so on.

That's all I want to say about general purpose cellular automata. It's a field that still needs much work, and I suggest that these programs can provide excellent starting points for further work. To complete this chapter, I'll describe a special case of cellular automata called *Life*.

The game of Life

John Conway, a Cambridge mathematician, developed this game in the late 1960s using, it is said, a square tiled floor and some plates. It is a two-dimensional cellular automaton whose complexity belies the apparently simple rules that control it. Conway called the game 'Life' because the cells in it exist in one of two states—alive or dead. The grid on which Life is played should be infinite, but, as we have already seen with two-dimensional cellular automata, this can be simulated by having a wrap-round grid where cells disappear off one edge only to reappear on the other edge.

The rules of Life are quite straightforward.

1. The grid of squares on which the game takes place must be considered to be infinite.

2. A cell in the grid can have one of two states, either alive, in which a square of the grid is occupied by a cell, or dead, in which case no cell occupies the square of the grid.

3. For a cell in generation n, count the cells that are diagonally and orthogonally adjacent to the cell. This can give a maximum of eight cells.

4. If a cell in generation n has less than two neighbours, then it is deemed to have died of loneliness (these cells are rather sensitive beasties!), and so the corresponding position in generation $n + 1$ will *not* have a live cell in it.

5. If a cell in generation n has more than three neighbours, then the cell will be deemed to have died of overcrowding. The corresponding position in generation $n + 1$ will *not* have a live cell in it.

6. If the cell in generation n has just two or three neighbours, then conditions are just right, and the cell will appear in the same position in the grid at the next generation.

7. These cells reproduce as well. If an empty position on the grid (a dead cell) is surrounded by three living cells to act as parents (my, these creatures must have strange biology!) then a new cell is born in that position in the next generation.

Listing 10-8 shows a program to play the Life game. Unlike the other cellular automata we've examined in this chapter, I've used a text mode for this program. There are two reasons for this:

1. A full screen display on an 80×25 text screen is quite large enough to show all the behaviour of Life, and by using text we can get a very rapid update of the screen. A full screen image using graphics would be very fiddly and slow.

2. It is very useful to be able to set up initial conditions. This is easy to do using text screens but would be harder for graphics screens.

The use of text screens could be considered for some of the automata already described in this chapter; I leave that experiment to you. Within Listing 10-8 there's nothing really fancy in terms of programming. Again, care is needed to make sure that array bounds are not exceeded.

On running the program, leave the default values in to start with. The parameters are as follows:

Setup or random

Entering R here will generate a random initial condition for the grid, populating it with cells at random. This is useful when doing general explorations of the system. Entering S will allow you to actually set up an initial set of cells in the grid. This is done by moving the cursor with the arrow keys to where you wish a cell to be located, then pressing the *Return* key. Pressing the *Spacebar* will delete any cell at the location addressed by the cursor. Once you've set the grid up, press the *Escape* key (PC version) or X (BBC version). This is very useful in Life, as there are many configurations of cells that exhibit interesting behaviour. If you wish to study these, then it's clearly easier to set them up than to hope that they'll occur in the random setting.

Birth parameter

The *birth parameter* is the number of cells around a dead cell on the grid for a live cell to be born in the next generation. For the standard game of

THE GAME OF LIFE

Life, this is 3. Some versions of life allow a second value, which is the most neighbours that a dead cell can have to be born, and in that case this value would indicate the least number of cells needed for birth.

Lowest value for survival

The lowest value for survival is the lowest number of cells that can be adjacent to a cell in the current generation for the cell to continue living in the next generation. The default for normal life is 2.

Highest value for survival

The highest value for survival is the largest number of neighbours that a cell can have and still create a cell in the corresponding position on the grid in the next generation. This is 3 in the standard Life system.

Life nomenclature

A simple shorthand has been devised to describe Life systems. This consists of four numbers, as follows:

1st digit: Fewest living neighbours needed for a cell to be born into the corresponding position of the next generation.

2nd digit: Most living neighbours for a cell to be born in to the corresponding position of the next generation.

3rd digit: Fewest cells needed around a dead cell for a new cell to be born in the position of the dead cell in the next generation.

4th digit: Most cells needed around a dead cell for a new cell to be born in the position of the dead cell in the next generation.

In this notation, normal Life would be described as 2333. There are other versions of Life, using different rules, and these can all be described using this notation.

On running the program, you will see several different types of behaviour exhibited. Some clusters of cells will simply die out, whilst others attain a steady state. Some will oscillate between different forms, apparently forever, whilst other clusters of cells will actually move across the screen. The shapes that you'll encounter in 2333 life have been documented over the years, and some are shown in Figure 10-2. The names are quite descriptive of the behaviour of these cell groupings.

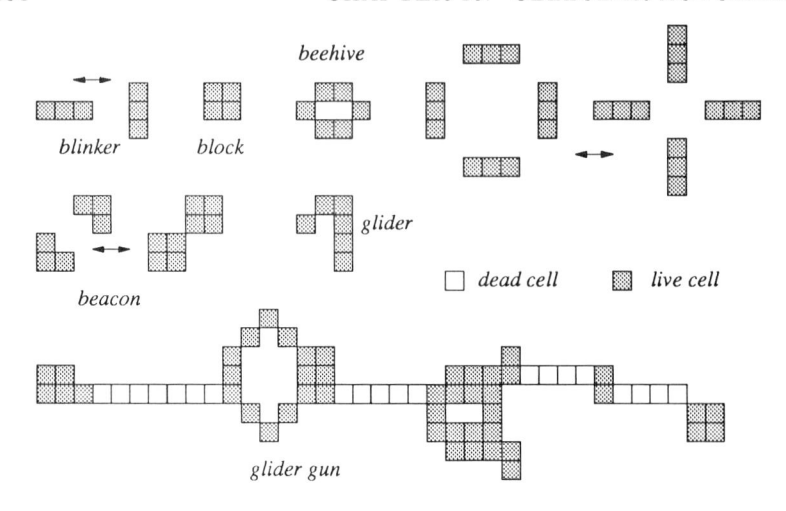

Figure 10-2. Shapes from Life.

Stable structures

The structures that give rise to stable, non-changing states are rather boring but, on the whole, many random initial setups of life will settle down to a cluster of blocks, beehives and other stable entities. In some cases, these are disrupted by the activities of such things as gliders, and in this case a stable structure may be broken apart to give rise to other structures that may or may not be stable.

Oscillating structures

These structures flick between two or more distinct states from generation to generation, but show periodicity. The beacon and blinker are two examples of these shapes. They are similar to the oscillating structures we produced using Listing 10-5. Again, gliders or other dynamic structures may disrupt oscillating structures into other forms.

Generators

These are structures which throw off other structures with no loss to themselves. This is not the same as a structure that evolves into something, then attains a steady state—generators carry on throwing off things for good. The classic example in 2333 life is the *Gosper glider gun*, named after its discoverer. This structure fires off gliders quite regularly; after all the times I've seen it in action, it never ceases to amaze me that it works!

Dynamic structures

Although in one sense of the word the whole cellular automaton that constitutes a run of the life program is a dynamic process as it changes from generation to generation, I use the term here to mean a structure that wanders through the life landscape under its own steam, so to say. The *glider* is a dynamic structure, which moves diagonally across the screen.

Suggestions for further experiments

Running Life will generate a fascinating set of structures that can be documented and analysed. However, here are a few ideas that you might like to try, and a few pointers for playing with the life program.

1. Sometimes the wrap-round grid gets in the way, and you will need a larger grid for some experiments. For this you may need to directly address the memory of your computer using the $PEEK$ and $POKE$ commands or equivalent operations. This is particularly so when you are throwing gliders around and using glider guns, etc., as the glider gun quite often gets hit and knocked out by its own gliders on a wrap-round grid!

2. Examine the effects of changing the rules to generate other versions of Life.

3. In 2333 Life, try the effects of adding some mutation in to the system, by allowing a random event to alter the rules in some way. This will require some changes to the program given, but is well worth trying out.

4. How about shooting gliders at other structures? What sort of debris do you get from these collisions?

Applications of cellular automata

There are, believe it or not, practical applications for cellular automata. For example, one-dimensional cellular automata can be used in cryptography to encrypt a message. If we have a one-dimensional automaton with a large number of possible states and a defined rule set, then after a specified number of generations we could read off that line of the automaton as a sequence of random numbers. That sequence can then be used to encrypt a string of text. Deciphering the message is easy, providing the user knows the initial conditions, the rule set and the generation number that was used to encrypt the original message.

Modelling

We've already seen that cellular automata can be used to model various systems, such as crystal growth, thermodynamically irreversible systems and so on. They are a little like the Hénon map and the logistic equation in that they provide an easily manageable tool for investigating complex mathematical and physical systems.

Pattern design

Finally, if all else fails, you can always use cellular automata as a source of inspiration for fabric design. Some of the patterns obtained are very bold and colourful, whilst others are quite delicate. If you are interested in design at all, a study of cellular automata might easily provide some inspiration.

BBC BASIC listings

Listing 10-1

```
 10 REM cellular automata
 20 :
 30 DIM seed(300)
 40 DIM progeny(300)
 50 DIM rule_set(9)
 60 :
 70 MODE 5
 80 y_limit=1000
 90 max_colours=4
100 REM Initialise a random sseding of values for start case
110 FOR seed_index=0 TO 300
120   seed(seed_index)=RND(4)-1
130   progeny(seed_index)=0
140 NEXT
150 REM Initialise a random rule set.
160 FOR j=0 TO 9
170   rule_set(j)=RND(max_colours)-1
180 NEXT
190 REM Print out the seed line
200 FOR k=0 TO 300
205   GCOL0,rule_set(seed(k))
210   PLOT69,k*4,1
220 NEXT
230 y=4
240 REPEAT
250   FOR seed_index=0 TO 300
260     REM If already at left, then use right hand edge for calculation
270     IF seed_index=0 THEN left_one=seed(300) ELSE left_one=seed(seed_index-1)
280     REM If at right hand edge, use left hand edge in calculation
290     IF seed_index=300 THEN right_one=seed(0) ELSE right_one=seed(seed_index+1)
300     centre_one=seed(seed_index)
310     progeny(seed_index)=rule_set((left_one+right_one+centre_one))
320     GCOL 0,progeny(seed_index)
325     PLOT 69,seed_index*4,y
330   NEXT
335   REM Transfer the progeny back to seed
340   FOR i=0 TO 300
345     seed(i)=progeny(i)
350   NEXT
360   REM   Increment line and check for limit of viewing screen
370   y=y+4
380   IF y>y_limit THEN y=1
390 UNTIL INKEY(2)<>-1
400 END
```

Listing 10-2

```
10 DIM seed(300)
20 DIM progeny(300)
30 DIM rule_set(9)
40 DIM cols(4)
50 MODE 5
60 y_limit=1000
70 max_colours=4
80 cols(0)=0
90 cols(1)=1
```

308   CHAPTER 10. CELLULAR AUTOMATA

```
100 cols(2)=2
110 cols(3)=3
120 FOR seed_index=0 TO 300
130   REM for a random initiator, replace following line with
      seed(seed_index)=random(4)-1
140   REM to set a random number up.
150   seed(seed_index)=0
160   progeny(seed_index)=0
170 NEXT
180 REM Now put one or more non-zero values in the seed array, if you're
    setting up a single
190 REM or multiple cell initiator.
200 seed(150)=1
210 REM Initialise a rule set.
220 rule_set(0)=0
230 rule_set(1)=1
240 rule_set(2)=2
250 rule_set(3)=3
260 rule_set(4)=3
270 rule_set(5)=2
280 rule_set(6)=1
290 rule_set(7)=0
300 rule_set(8)=0
310 rule_set(9)=0
320 REM    Print out the seed line
330 FOR k=0 TO 300
340   GCOL 0,rule_set(seed(k))
350   PLOT 69,k*4,1
355 NEXT
360 REM    Now calculate and print each line of the automata
370 y=4
380 REPEAT
390   FOR seed_index=0 TO 300
400     REM If already at left, then use right hand edge for calculation
410     IF seed_index=0 THEN left_one=seed(300) ELSE
        left_one=seed(seed_index-1)
420     REM If at right hand edge, use left hand edge in calculation
430     IF seed_index=300 THEN right_one=seed(0) ELSE
        right_one=seed(seed_index+1)
440     centre_one=seed(seed_index)
450     progeny(seed_index)=rule_set((left_one+right_one+centre_one))
455     GCOL0,cols(progeny(seed_index))
460     PLOT69,seed_index*4,y
470   NEXT
480   FOR i=0 TO 300
490     seed(i)=progeny(i)
495   NEXT
500   y=y+4
510   IF y>y_limit THEN y=1
520
530
540 UNTIL INKEY(2)<>-1
550 END
```

Listing 10-3

```
10 REM Program OneDCellularAutomataWithMutation
20 DIM seed(300),progeny(300),rule_set(300)
30 MODE 5
40 y_limit=1000
50 max_colours=4
60 FOR seed_index=0 TO 250
70   REM for a random initiator, replace following line with
```

BBC BASIC LISTINGS

```
        seed(seed_index)=random(4)
 80     REM to set a random number up.
 90     seed(seed_index)=0
100     progeny(seed_index)=0
110 NEXT
120 REM Now put one or more non-zero values in the seed array, if you're
    setting up a single
130 REM or multiple cell initiator.
140 seed(125)=1
150 REM Initialise a rule set.
160 rule_set(0)=0
170 rule_set(1)=1
180 rule_set(2)=2
190 rule_set(3)=3
200 rule_set(4)=3
210 rule_set(5)=2
220 rule_set(6)=1
230 rule_set(7)=0
240 rule_set(8)=0
250 rule_set(9)=0
260 REM Initialise probabilities of noise and mutation affecting the system.
    Set
270 REM mutate or noise to 999 to turn off the effect.  Try values of between
    995 and 999 for noise
280 REM and 500 to 998 for mutation
290 mutate=999
300 noise=999
310 REM Print out the seed line
320 FOR k=0 TO 250
330   GCOL0,rule_set(seed(k))
340   PLOT69,k*4,1
350 NEXT
360 REM Now calculate and print each line of the automata
370 y=4
380 REPEAT
390   FOR seed_index=0 TO 250
400     REM If already at left, then use right hand edge for calculation
410     IF (seed_index=0) THEN left_one=seed(250) ELSE
        left_one=seed(seed_index-1)
420     REM If at right hand edge, use left hand edge in calculation
430     IF (seed_index=250) THEN right_one=seed(0) ELSE
        right_one=seed(seed_index+1)
440     centre_one=seed(seed_index)
450     IF RND(1000)-1>noise THEN noise_val=1 ELSE noise_val=0
460     progeny(seed_index)=rule_set(left_one+right_one+centre_one+noise_val
        MOD 9)
470     GCOL0,progeny(seed_index)
480     PLOT69,seed_index*4,y
490   NEXT
500   FOR i=0 TO 250
510     seed(i)=progeny(i)
520   NEXT
530   y=y+4
540   IF (y>y_limit) THEN y=1
550   REM Now check to see if a mutation has occurred
560   IF RND(1000)-1>mutate THEN
      change_rule=RND(10)-1:rule_set(change_rule)=RND(4)-1
570 UNTIL INKEY(10)<>-1
580 END
```

Listing 10-4

```
 10 REM Program Gen2DAutomata
 70 DIM new%(30,30),old%(30,30),rules(30)
 80 I=RND(-TIME)
100 MODE 5
110 PROCInitialise
120 PROCDisplayNewGeneration
130 REPEAT
140   PROCCalculateNextGeneration
150   PROCDisplayNewGeneration
160 UNTIL (INKEY(1)<>-1)
200 END
260 DEFPROCInitialise
280 REM    Initialise size of grid
300 size_of_grid=30
320 REM    Now set up a random rule set for the program. Again, you could define
330 REM    sets of rules. The number 27 is arrived at by assuming that the cell
340 REM    being examined, and the 8 cells around it might all have a value of 3
350 REM    in the current generation, and so we need 9*3 rules = 27 rules
380 FOR i=0 TO 27
390   rules(i)=RND(4)-1
400 NEXT
410 FOR i=1 TO size_of_grid
420   FOR j=1 TO size_of_grid
440     REM   For random seeding - not advised - use new%(i,j)=random(4) here if you
450     REM want it
460     REM   Deliberate seeding of one or two cells in array gives better results.
480     new%(i,j)=RND(4)-1
490     old%(i,j)=new%(i,j)
520   NEXT
530 NEXT
550 REM    If you want to put a known start condition in, set up here - only a few
560 REM    cells need be initiated to give good results. Seed new because it will
570 REM    be copied over to old after the first call to DisplayNewGeneration
590 new%((size_of_grid/2),(size_of_grid/2))=1
610 ENDPROC
650 DEFPROCDisplayNewGeneration
670 REM Display contents of arrays as 2 by 2 pixel blocks for easier viewing
690 FOR i=1 TO size_of_grid
700   m=8*i
710   FOR j=1 TO size_of_grid
720     n=8*j
730     GCOL0,new%(i,j)
740     PLOT69,m,n
750     PLOT69,m+4,n
760     PLOT69,m,n+4
770     PLOT69,m+4,n+4
790     REM transfer current generation to array 'old' so that they can become the
800     REM new seed generation
820     old%(i,j)=new%(i,j)
860   NEXT
870 NEXT
890 ENDPROC
930 DEFPROCCountOldCells(i1,j1)
970 REM First of all, check that the limits of the grid will not be exceeded
980 REM and set up variables to allow addressing of the 8 cells around the
```

BBC BASIC LISTINGS

```
 990 REM one being examined.
1010 IF j1=size_of_grid THEN l=1 ELSE l=j1+1
1020 IF (j1-1)=0 THEN m=size_of_grid ELSE m=j1-1
1030 IF i1=size_of_grid THEN n=1 ELSE n=i1+1
1040 IF (i1-1)=0 THEN o=size_of_grid ELSE o=i1-1
1070 k=old%(m,o)+old%(m,j1)+old%(m,n)+old%(i1,o)
1080 k=k+old%(i1,j1)+old%(i1,n)+old%(l,o)+old%(l,j1)+old%(l,n)
1100 CountOldCells=k
1130 ENDPROC
1170 DEFPROCCalculateNextGeneration
1190 REM Scan the grid of this generation of cells and place the values of
1200 REM the new generation in the array 'new'. Use rules array to define the
1210 REM cell value.
1230 FOR i=1 TO size_of_grid
1240   FOR j=1 TO size_of_grid
1250     PROCCountOldCells(i,j)
1260     LiveOld=CountOldCells
1270     new%(i,j)=rules(LiveOld)
1290   NEXT
1300 NEXT
1320 ENDPROC
```

Listing 10-5

```
  10 REM Program Gen2DAutomataMark2
  80 MODE 5
  90 PROCInitialise
 100 REPEAT
 110   PROCCalculateNextGeneration
 120   PROCDisplayNewGeneration
 130 UNTIL (INKEY(1)<>-1)
 150 END
 190 DEFPROCInitialise
 210 size_of_grid=20
 220 DIM old(20,20),new(20,20)
 230 FOR i=1 TO size_of_grid
 240   FOR j=1 TO size_of_grid
 260     REM Initialise the grid. Stick a 1 in old randomly.
 280     new(i,j)=0
 290     old(i,j)=new(i,j)
 300     IF (RND(10)-1>7) THEN old(i,j)=1
 330   NEXT
 340 NEXT
 350 ENDPROC
 380 DEFPROCDisplayNewGeneration
 410 FOR i=1 TO size_of_grid
 420   tx=8*i
 430   FOR j=1 TO size_of_grid
 440     ty=8*j
 450     GCOL0,new(i,j)
 460     PLOT69,tx,ty
 470     PLOT69,tx+4,ty
 480     PLOT69,tx,ty+4
 490     PLOT69,tx+4,ty+4
 520     old(i,j)=new(i,j)
 530     new(i,j)=0
 560   NEXT
 570 NEXT
 590 ENDPROC
 630 DEFPROCCountOldCells(i1,j1)
 650 REM Count cells orthogannally adjacent to the one of interest - again,
 660 REM check the edges of the grid to see if we need to wrap around.
 680 IF j1=size_of_grid THEN l=1 ELSE l=j1+1
```

```
690 IF (j1-1)=0 THEN m=size_of_grid ELSE m=j1-1
700 IF i1=size_of_grid THEN n=1 ELSE n=i1+1
710 IF (i1-1)=0 THEN o=size_of_grid ELSE o=i1-1
730 k=old(i1,l)+old(i1,m)+old(n,j1)+old(o,j1)
750 CountOldCells=k
770 ENDPROC
810 DEFPROCCalculateNextGeneration
830 FOR i=1 TO size_of_grid
840   FOR j=1 TO size_of_grid
860     REM Here's where the rules get implemented only create a new cell if the
870     REM original cell
880     REM was surrounded by two others - no more, no less. This can, of course, be
890     REM easily
900     REM changed to see what happens.
920     PROCCountOldCells(i,j) : LiveOld=CountOldCells
930     IF LiveOld=2 THEN new(i,j)=1 ELSE new(i,j)=0
950   NEXT
960 NEXT
980 ENDPROC
```

Listing 10-6

```
10 REM Program CellEatCell
30 REM Based on Scientific American Algorithm, August 1989
35 DIM old(20,20),new(20,20)
90 MODE 5
100 PROCInitialise
110 PROCDisplayNewGeneration
120 REPEAT
130   PROCCalculateNextGeneration
140   PROCDisplayNewGeneration
150 UNTIL INKEY(1)<>-1
190 END
230 DEFPROCInitialise
260 REM   Initialise size of grid
290 REM   Default size of 120 is good. Try varying sizes from 20 upwards.
310 size_of_grid=20
330 REM   This program really only works properly with the number of states
340 REM available to
350 REM   a cell of 10 or more. Less than this, and the system takes a long time
370 REM to settle down into mopre interesting states. With 4 states, as you might
390 REM get using mode 5 BBC displays, you never really get there!
410 t_states=4
450 FOR i=1 TO size_of_grid
460   FOR j=1 TO size_of_grid
480     new(i,j)=RND(t_states)-1
490     old(i,j)=new(i,j)
510   NEXT
520 NEXT
540 ENDPROC
580 DEFPROCDisplayNewGeneration
600 REM Display contents of arrays as 2 by 2 pixel blocks for easier viewing
620 FOR i=1 TO size_of_grid
630   m=8*i
640   FOR j=1 TO size_of_grid
650     n=8*j
660     GCOL0,new(i,j)
670     PLOT69,m,n
680     PLOT69,m+4,n
```

BBC BASIC LISTINGS   313

```
 690     PLOT69,m,n+4
 700     PLOT69,m+4,n+4
 720     REM    transfer current generation to array 'old' so that they can
         become the
 730     REM    new seed generation
 750     old(i,j)=new(i,j)
 790   NEXT
 800 NEXT
 820 ENDPROC
 870 DEFPROCCalculateNextGeneration
 910 REM   Scan the grid of this generation of cells and place the values of
 920 REM     the new generation in the array 'new'. Examine all neighbouring
         cells
 930 REM     but do not include the cell being studied.
 960 FOR i=1 TO size_of_grid
 970   FOR j=1 TO size_of_grid
1000     REM  t_old is the current value of the 'old' cell. If it's 0 then
         assume
1010     REM    it's t_states for the rest of the procedure.
1030     t_old=old(i,j)
1040     IF (t_old=0) THEN t_old=t_states
1060     REM   Now define the new cell
1080     IF (i-1)=0 THEN m=size_of_grid ELSE m=i-1
1090     IF i=size_of_grid THEN l=1 ELSE l=i+1
1100     IF (j-1)=0 THEN o=size_of_grid ELSE o=j-1
1110     IF j=size_of_grid THEN n=1 ELSE n=j+1
1140     IF old(m,o)=((t_old+1) MOD t_states) THEN new(i,j)=old(m,o)
1160     IF old(m,n)=((t_old+1) MOD t_states) THEN new(i,j)=old(m,n)
1170     IF old(l,o)=((t_old+1) MOD t_states) THEN new(i,j)=old(l,o)
1190     IF old(l,n)=((t_old+1) MOD t_states) THEN new(i,j)=old(l,n)
1220     IF old(i,o)=((t_old+1) MOD t_states) THEN new(i,j)=old(i,o)
1240     IF old(m,j)=((t_old+1) MOD t_states) THEN new(i,j)=old(m,j)
1250     IF old(l,j)=((t_old+1) MOD t_states) THEN new(i,j)=old(l,j)
1270     IF old(i,n)=((t_old+1) MOD t_states) THEN new(i,j)=old(i,n)
1300   NEXT
1310 NEXT
1330 ENDPROC
```

Listing 10-7

```
  10 REM Program TimeDependantRules
  30 REM Uses a time dependant rule set.
  40 MODE 5
  42 DIM new(20,20), old(20,20)
  60 PROCInitialise
  70 generation=1
  80 REPEAT
 100   PROCCalculateNextGeneration
 110   PROCDisplayNewGeneration
 120   generation=generation+1
 140 UNTIL INKEY(1)<>-1
 180 ENDPROC
 240 DEFPROCInitialise
 260 size_of_grid=20
 280 FOR i=1 TO size_of_grid
 290   FOR j=1 TO size_of_grid
 310     REM   Initialise the grid. Stick a 1 in old randomly.
 330     new(i,j)=0
 340     old(i,j)=new(i,j)
 370   NEXT
 380 NEXT
 390 old(10,10)=1
 410 ENDPROC
```

314   CHAPTER 10. CELLULAR AUTOMATA

```
440 DEFPROCDisplayNewGeneration
470 FOR i=1 TO size_of_grid
480   tx=8*i
490   FOR j=1 TO size_of_grid
500     ty=8*j : GCOL0,new(i,j)
520     PLOT69,tx,ty
530     PLOT69,tx+4,ty
540     PLOT69,tx,ty+4
550     PLOT69,tx+4,ty+4
580     old(i,j)=new(i,j)
590     new(i,j)=0
620   NEXT
630 NEXT
650 ENDPROC
690 DEFPROCCountOldCells(i1,j1)
730 REM   Count cells orthogannally adjacent to the one of interest - again,
740 REM   check the edges of the grid to see if we need to wrap around.
760 IF j1=size_of_grid THEN l=1 ELSE l=j1+1
770 IF (j1-1)=0 THEN m=size_of_grid ELSE m=j1-1
780 IF i1=size_of_grid THEN n=1 ELSE n=i1+1
790 IF (i1-1)=0 THEN o=size_of_grid ELSE o=i1-1
820 orthoganal=old(i1,l)+old(i1,m)+old(n,j1)+old(o,j1)
830 diagonal=old(m,o)+old(m,n)+old(l,o)+old(l,n)
860 CountOldCells=orthoganal+diagonal
880 ENDPROC
920 DEFPROCCalculateNextGeneration
960 FOR i=1 TO size_of_grid
970   FOR j=1 TO size_of_grid
990     REM   Here's where the rules get implemented only create a new cell
              on even
1000    REM generations
1010    REM with orthoganal neighbours=1 OR odd geneerations with 1 diagonal
              and
1030    REM orthog. neighbour - a time dependant rule.
1050    PROCCountOldCells(i,j) : LiveOld=CountOldCells
1060    IF ((orthoganal=1) AND ((generation MOD 2)=0)) THEN new(i,j)=1
1080    IF ((LiveOld=1) AND ((generation MOD 2)=1)) THEN new(i,j)=1
1130  NEXT
1140 NEXT
1160 ENDPROC
```

Listing 10-8

```
10 REM Game of Life for BBC
20 :
25 DIM new%(80,25), old%(80,25)
30 *FX4,1
40 MODE 3
50 INPUT "Set up or Random (S/R) ",pos_or_rand$
60 INPUT "Birth Parameter (3) ",birth
70 INPUT "Lowest value to survive (2) ",lowest
80 INPUT "Highest value to survive (3) ",highest
90 MODE 3
100 IF pos_or_rand$="R" OR pos_or_rand$="r" THEN PROCrandom ELSE PROCsetup
110 REPEAT
120   PROCCalculateNextGeneration
130   PROCDisplayNewGeneration
140 UNTIL FALSE
150 :
160 DEFPROCrandom
170 FOR I=1 TO 80
180   FOR J=1 TO 23
190     IF RND(10)>7 THEN new%(I,J)=1 ELSE new%(I,J)=0
```

BBC BASIC LISTINGS

```
200     old%(I,J)=new%(I,J)
210   NEXT J
220 NEXT I
230 ENDPROC
240 :
250 DEFPROCsetup
260 FOR I=1 TO 80
270   FOR J=1 TO 23
280     new%(I,J)=0 : old%(I,J)=0
290   NEXT J
300 NEXT I
310 I=1 : J=1
320 PRINTTAB(I,J);
330 REPEAT
340   CH=GET
350   IF CH=138 THEN PROCdown
360   IF CH=139 THEN PROCup
370   IF CH=136 THEN PROCleft
380   IF CH=137 THEN PROCright
390   IF CH=32 THEN PROCspace
400   IF CH=13 THEN PROCbug
410 UNTIL CH=ASC("X") OR CH=ASC("x") : PRINT CHR$(7);
420 ENDPROC
430 :
440 DEFPROCdown
450 IF J>23 THEN J=1 ELSE J=J+1
460 PRINTTAB(I,J);
470 ENDPROC
480 :
490 DEFPROCup
500 IF J<1 THEN J=23 ELSE J=J-1
510 PRINTTAB(I,J);
520 ENDPROC
530 :
540 DEFPROCleft
550 IF I<1 THEN I=80 ELSE I=I-1
560 PRINTTAB(I,J);
570 ENDPROC
580 :
590 DEFPROCright
600 IF I>80 THEN I=1 ELSE I=I+1
610 PRINTTAB(I,J);
620 ENDPROC
630 :
640 DEFPROCspace
650 PRINT" ";
660 new%(I,J)=0 : old%(I,J)=0
670 ENDPROC
680 :
690 DEFPROCbug
700 PRINT "#";
710 new%(I,J)=1 : old%(I,J)=1
720 ENDPROC
740 :
750 DEFPROCDisplayNewGeneration
760 FOR I=1 TO 80
770   FOR J=1 TO 23
780     PRINTTAB(I,J);
790     IF new%(I,J)=1 THEN PRINT"#"; ELSE PRINT" ";
800     old%(I,J)=new%(I,J)
810   NEXT
820 NEXT
830 ENDPROC
840 :
850 DEFPROCCountOldCells(I1,J1)
```

```
 860 IF (J1+1)=24 THEN L=1 ELSE L=J1+1
 870 IF (J1-1)=0 THEN M=23 ELSE M=J1-1
 880 IF (I1+1)=81 THEN N=1 ELSE N=I1+1
 890 IF (I1-1)=0 THEN O=80 ELSE O=I1-1
 900 X=old%(I1,L)+old%(I1,M)+old%(N,J1)+old%(O,J1)
 910 CountOldCells=X+old%(N,L)+old%(O,L)+old%(N,M)+old%(O,M)
 920 ENDPROC
 930 :
 940 DEFPROCCalculateNextGeneration
 950 FOR I=1 TO 80
 960   FOR J=1 TO 23
 970     PROCCountOldCells(I,J)
 980     LiveOld=CountOldCells
 990     IF (old%(I,J)=0) AND (LiveOld=birth) THEN new%(I,J)=1
1000     IF (old%(I,J)=1) AND ( (LiveOld<lowest) OR (LiveOld>highest)) THEN
         new%(I,J)=0
1010   NEXT
1020 NEXT
1030 ENDPROC
```

Turbo Pascal listings

Listing 10-1

```pascal
Program OneDCellularAutomata;

Uses crt,graph,extend;

Var

seed:                         array[0..700] of byte;
progeny:                      array[0..700] of byte;
rule_set:                     array[0..9] of byte;
x,y,i,j,k:                    integer;
seed_index:                   integer;
left_one,right_one,centre_one: integer;
max_colours:                  integer;
y_limit:                      integer;

Begin

ChoosePalette;
y_limit:=GetMaxY;
max_colours:=4;

randomize;

{ Initialise a random sseding of values for start case }

for seed_index:=0 to 500 do begin
    seed[seed_index]:=random(4);
    progeny[seed_index]:=0;
end;

{   Initialise a random rule set.        }

for j:=0 to 9 do begin
    rule_set[j]:=random(max_colours);
end;

{   Print out the seed line   }

for k:=0 to 500 do
    putpixel(k,1,rule_set[seed[k]]);

{   Now calculate and print each line of the automata   }

y:=2;
repeat begin
    for seed_index:=0 to 500 do begin

{   If already at left, then use right hand edge for calculation   }

        if (seed_index=0) then
           left_one:=seed[500]
        else
           left_one:=seed[seed_index-1];

{   If at right hand edge, use left hand edge in calculation   }

        if (seed_index=500) then
           right_one:=seed[0]
        else
           right_one:=seed[seed_index+1];
```

318   CHAPTER 10. CELLULAR AUTOMATA

```
            centre_one:=seed[seed_index];
            progeny[seed_index]:=rule_set[(left_one+right_one+centre_one)];
            putpixel(seed_index,y,progeny[seed_index]);

    end;

{ Transfer the progeny back to seed }

    for i:=0 to 500 do
        seed[i]:=progeny[i];
{   Increment line and check for limit of viewing screen }
    y:=y+1;
    if (y>y_limit) then
        y:=1;
    end;

{ Repeat until key pressed on keyboard }

    until (KeyPressed);

end.
```

Listing 10-2

```
Program OneDCellularAutomataWithChoice;

Uses crt,graph,extend;

Var

seed:                            array[0..700] of byte;
progeny:                         array[0..700] of byte;
rule_set:                        array[0..9] of byte;
cols:                            array[0..4] of byte;
x,y,i,j,k:                       integer;
seed_index:                      integer;
left_one,right_one,centre_one:   integer;
max_colours:                     integer;
y_limit:                         integer;

Begin

ChoosePalette;
y_limit:=GetMaxY;
max_colours:=4;

{   Next lines define a palette for VGA / EGA screens.  Standard
    CGA Palette is OK for this program           }

cols[0]:=Black;
cols[1]:=Red;
cols[2]:=Green;
cols[3]:=Yellow;

randomize;

{ Initialise a 0 sseding of values for start case }

for seed_index:=0 to 500 do begin

{   for a random initiator, replace following line with
```

TURBO PASCAL LISTINGS

```pascal
      seed[seed_index]:=random(4)
      to set a random number up.   }

      seed[seed_index]:=0;
      progeny[seed_index]:=0;
end;

{ Now put one or more non-zero values in the seed array, if you're setting up
  a single
  or multiple cell initiator. }

   seed[200]:=1;

{  Initialise a rule set.          }

   rule_set[0]:=0;
   rule_set[1]:=1;
   rule_set[2]:=2;
   rule_set[3]:=3;
   rule_set[4]:=3;
   rule_set[5]:=2;
   rule_set[6]:=1;
   rule_set[7]:=0;
   rule_set[8]:=0;
   rule_set[9]:=0;

{  Print out the seed line    }

for k:=0 to 500 do
   putpixel(k,1,rule_set[seed[k]]);

{  Now calculate and print each line of the automata  }

y:=2;
repeat begin
   for seed_index:=0 to 500 do begin

{  If already at left, then use right hand edge for calculation   }

         if (seed_index=0) then
            left_one:=seed[500]
         else
            left_one:=seed[seed_index-1];

{  If at right hand edge, use left hand edge in calculation    }

         if (seed_index=500) then
            right_one:=seed[0]
         else
            right_one:=seed[seed_index+1];

         centre_one:=seed[seed_index];
         progeny[seed_index]:=rule_set[(left_one+right_one+centre_one)];
         putpixel(seed_index,y,cols[progeny[seed_index]]);

   end;
   for i:=0 to 500 do
      seed[i]:=progeny[i];
   y:=y+1;
   if (y>y_limit) then
      y:=1;
   end;
   until (KeyPressed);

end.
```

Listing 10-3

```
Program OneDCellularAutomataWithMutation;

Uses crt,graph,extend;

Var

seed:                              array[0..700] of byte;
progeny:                           array[0..700] of byte;
rule_set:                          array[0..9] of byte;
cols:                              array[0..4] of byte;
x,y,i,j,k:                         integer;
seed_index:                        integer;
left_one,right_one,centre_one:     integer;
max_colours:                       integer;
y_limit:                           integer;
mutate:                            integer;
noise,noise_val:                   integer;
change_rule:                       integer;

Begin

ChoosePalette;
y_limit:=GetMaxY;
max_colours:=4;

{   Next lines define a palette for VGA / EGA screens.  Standard
    CGA Palette is OK for this program            }

cols[0]:=Black;
cols[1]:=Red;
cols[2]:=Green;
cols[3]:=Yellow;

randomize;

{ Initialise a 0 sseding of values for start case }

for seed_index:=0 to 500 do begin

{    for a random initiator, replace following line with
     seed[seed_index]:=random(4)
     to set a random number up.   }

     seed[seed_index]:=0;
     progeny[seed_index]:=0;
end;

{ Now put one or more non-zero values in the seed array, if you're setting up
  a single
  or multiple cell initiator. }

   seed[200]:=1;

{   Initialise a rule set.          }

   rule_set[0]:=0;
   rule_set[1]:=1;
   rule_set[2]:=2;
   rule_set[3]:=3;
   rule_set[4]:=3;
   rule_set[5]:=2;
   rule_set[6]:=1;
   rule_set[7]:=0;
```

TURBO PASCAL LISTINGS   321

```pascal
   rule_set[8]:=0;
   rule_set[9]:=0;

{  Initialise probabilities of noise and mutation affecting the system.   Set
   mutate or noise to 999 to turn off the effect.  Try values of between 995
   and 999 for noise
   and 500 to 998 for mutation  }

mutate:=999; noise:=999;

{  Print out the seed line   }

for k:=0 to 500 do
    putpixel(k,1,rule_set[seed[k]]);

{  Now calculate and print each line of the automata   }

y:=2;
repeat begin
    for seed_index:=0 to 500 do begin
{   If already at left, then use right hand edge for calculation     }

          if (seed_index=0) then
             left_one:=seed[500]
          else
             left_one:=seed[seed_index-1];

{   If at right hand edge, use left hand edge in calculation    }

          if (seed_index=500) then
             right_one:=seed[0]
          else
             right_one:=seed[seed_index+1];

          centre_one:=seed[seed_index];

          if random(1000)>noise then
             noise_val:=1
          else
             noise_val:=0;

          progeny[seed_index]:=rule_set[((left_one+right_one+centre_one+noise_val mod 9))];
          putpixel(seed_index,y,cols[progeny[seed_index]]);

    end;
    for i:=0 to 500 do
        seed[i]:=progeny[i];
    y:=y+1;
    if (y>y_limit) then
       y:=1;
    end;

{   Now check to see if a mutation has occurred              }

    if random(1000)>mutate then begin
       change_rule:=random(10);
       rule_set[change_rule]:=random(4);
    end;

    until (KeyPressed);

end.
```

Listing 10-4

```
Program Gen2DAutomata;

Uses Crt,Graph,Extend;

Var
    old:            array[1..160,1..160] of byte;
    new:            array[1..160,1..160] of byte;
    i,j,n,o:        integer;
    i1,j1,k,l,m:    integer;
    LiveOld:        integer;
    gd,gm:          integer;
    rules:          array[0..27] of integer;
    size_of_grid:   integer;

Procedure Initialise;
Begin

{   Initialise size of grid }

size_of_grid:=100;

{   Now set up a random rule set for the program.  Again, you could define
    sets of rules.  The number 27 is arrived at by assuming that the cell
    being examined, and the 8 cells around it might all have a value of 3
    in the current generation, and so we need 9*3 rules = 27 rules          }

for i:=0 to 27 do
    rules[i]:=random(4);

    for i:=1 to size_of_grid do begin
        for j:=1 to size_of_grid do begin

{   For random seeding - not advised - use new[i,j]:=random(4) here if you want
    it.
    Deliberate seeding of one or two cells in array gives better results. }

            new[i,j]:=0;
            old[i,j]:=new[i,j];

        end;
    end;

{   If you want to put a known start condition in, set up here - only a few
    cells need be initiated to give good results.  Seed new because it will
    be copied over to old after the first call to DisplayNewGeneration  }

new[trunc(size_of_grid/2),trunc(size_of_grid/2)]:=2;

end;

Procedure DisplayNewGeneration;
Begin

{   Display contents of arrays as 2 by 2 pixel blocks for easier viewing }

    for i:=1 to size_of_grid do begin
        m:=2*i;
        for j:=1 to size_of_grid do begin
            n:=2*j;
            PutPixel(m,n,new[i,j]);
            PutPixel(m+1,n,new[i,j]);
            PutPixel(m,n+1,new[i,j]);
```

TURBO PASCAL LISTINGS

```pascal
              PutPixel(m+1,n+1,new[i,j]);
{   transfer current generation to array 'old' so that they can become the
    new seed generation      }

              old[i,j]:=new[i,j];

        end;
    end;
end;

Function CountOldCells(i1,j1: integer): integer;
Begin

{   First of all, check that the limits of the grid will not be exceeded
    and set up variables to allow addressing of the 8 cells around the
    one being examined.    }

    if j1=size_of_grid then l:=1
        else
    l:=j1+1;
    if (j1-1)=0 then m:=size_of_grid
        else
    m:=j1-1;
    if i1=size_of_grid then n:=1
        else
    n:=i1+1;
    if (i1-1)=0 then o:=size_of_grid
        else
    o:=i1-1;

    k:=old[m,o]+old[m,j1]+old[m,n]+old[i1,o];
    k:=k+old[i1,j1]+old[i1,n]+old[l,o]+old[l,j1]+old[l,n];

    CountOldCells:=k;
end;

Procedure CalculateNextGeneration;
Begin

{   Scan the grid of this generation of cells and place the values of
    the new generation in the array 'new'. Use rules array to define the
    cell value.     }

    for i:=1 to size_of_grid do begin
        for j:=1 to size_of_grid do begin

            LiveOld:=CountOldCells(i,j);
            new[i,j]:=rules[LiveOld];

        end;
    end;
end;

Begin

Randomize;

ChoosePalette;

Initialise;
DisplayNewGeneration;
Repeat
    CalculateNextGeneration;
    DisplayNewGeneration;
```

324   CHAPTER 10. CELLULAR AUTOMATA

```
until keypressed;

end.
```

Listing 10-5

```
Program Gen2DAutomataMark2;

{   Uses a different rule set.   }

Uses Crt,Graph,Extend;

Var
    old:                array[1..170,1..170] of byte;
    new:                array[1..170,1..170] of byte;
    i,j,n,o:            integer;
    i1,j1,k,l,m:        integer;
    LiveOld:            integer;
    tx,ty:              integer;
    size_of_grid:       integer;

Procedure Initialise;
Begin
size_of_grid:=100;

    for i:=1 to size_of_grid do begin
       for j:=1 to size_of_grid do begin

    {   Initialise the grid. Stick a 1 in old randomly.         }

            new[i,j]:=0;
            old[i,j]:=new[i,j];
            if (random(10)>7) then
                old[i,j]:=1;

       end;
     end;
end;

Procedure DisplayNewGeneration;
Begin
   for i:=1 to size_of_grid do begin
      tx:=2*i;
      for j:=1 to size_of_grid do begin
         ty:=2*j;

         PutPixel(tx,ty,new[i,j]);
         PutPixel(tx+1,ty,new[i,j]);
         PutPixel(tx,ty+1,new[i,j]);
         PutPixel(tx+1,ty+1,new[i,j]);

         old[i,j]:=new[i,j];
         new[i,j]:=0;

      end;
    end;
end;

Function CountOldCells(i1,j1: integer): integer;
Begin

    {   Count cells orthogannally adjacent to the one of interest - again,
```

TURBO PASCAL LISTINGS   325

```
        check the edges of the grid to see if we need to wrap around.     }

    if j1=size_of_grid then l:=1
        else
    l:=j1+1;
    if (j1-1)=0 then m:=size_of_grid
        else
    m:=j1-1;
    if i1=size_of_grid then n:=1
        else
    n:=i1+1;
    if (i1-1)=0 then o:=size_of_grid
        else
    o:=i1-1;

    k:=old[i1,l]+old[i1,m]+old[n,j1]+old[o,j1];

    CountOldCells:=k;
end;

Procedure CalculateNextGeneration;
Begin

    for i:=1 to size_of_grid do begin
        for j:=1 to size_of_grid do begin
{   Here's where the rules get implemented; only create a new cell if the
    original cell
    was surrounded by two others - no more, no less. This can, of course, be
    easily
    changed to see what happens.     }

            LiveOld:=CountOldCells(i,j);
            if liveold=2 then
                new[i,j]:=1
            else
                new[i,j]:=0;

        end;
    end;
end;

Begin

ChoosePalette;

Initialise;
Repeat

    CalculateNextGeneration;
    DisplayNewGeneration;

until keypressed;

end.
```

Listing 10-6

```
Program CellEatCell;

{   Based on Scientific American Algorithm, August 1989     }

Uses Crt,Graph,Extend;
```

```
Var

    old:               array[1..160,1..160] of byte;
    new:               array[1..160,1..160] of byte;
    i,j,n,o:           integer;
    i1,j1,k,l,m:       integer;
    LiveOld:           integer;
    size_of_grid:      integer;
    t_old,t_states:    integer;

Procedure Initialise;
Begin

{   Initialise size of grid }

{   Default size of 120 is good.  Try varying sizes from 20 upwards.    }

size_of_grid:=120;

{   This program really only works properly with the number of states available
    to
    a cell of 10 or more.  Less than this, and the system takes a long time to
    settle down into mopre interesting states.  With 4 states, as you might get
    using CGA displays, you never really get there!      }

t_states:=15;

    for i:=1 to size_of_grid do begin
        for j:=1 to size_of_grid do begin

            new[i,j]:=random(t_states);
            old[i,j]:=new[i,j];

        end;
    end;

end;

Procedure DisplayNewGeneration;

Begin

{   Display contents of arrays as 2 by 2 pixel blocks for easier viewing }

    for i:=1 to size_of_grid do begin
        m:=2*i;
        for j:=1 to size_of_grid do begin
            n:=2*j;
            PutPixel(m,n,new[i,j]);
            PutPixel(m+1,n,new[i,j]);
            PutPixel(m,n+1,new[i,j]);
            PutPixel(m+1,n+1,new[i,j]);

{   transfer current generation to array 'old' so that they can become the
    new seed generation    }

            old[i,j]:=new[i,j];

        end;
    end;

end;

Procedure CalculateNextGeneration;
```

TURBO PASCAL LISTINGS 327

```
Begin
{   Scan the grid of this generation of cells and place the values of
    the new generation in the array 'new'.  Examine all neighbouring cells
    but do not include the cell being studied.               }

    for i:=1 to size_of_grid do begin
      for j:=1 to size_of_grid do begin

{   t_old is the current value of the 'old' cell.  If it's 0 then assume
    it's t_states for the rest of the procedure.    }

            t_old:=old[i,j];
            if (t_old=0) then
               t_old:=t_states;

{   Now define the new cell }

            if (i-1)=0 then
               m:=size_of_grid
            else
               m:=i-1;
            if i=size_of_grid then
               l:=1
            else
               l:=i+1;
            if (j-1)=0 then
               o:=size_of_grid
            else
               o:=j-1;
            if j=size_of_grid then
               n:=1
            else
               n:=j+1;

            if old[m,o]=((t_old+1) mod t_states) then
               new[i,j]:=old[m,o];

            if old[m,n]=((t_old+1) mod t_states) then
               new[i,j]:=old[m,n];
            if old[l,o]=((t_old+1) mod t_states) then
               new[i,j]:=old[l,o];

            if old[l,n]=((t_old+1) mod t_states) then
               new[i,j]:=old[l,n];

            if old[i,o]=((t_old+1) mod t_states) then
               new[i,j]:=old[i,o];

            if old[m,j]=((t_old+1) mod t_states) then
               new[i,j]:=old[m,j];
            if old[l,j]=((t_old+1) mod t_states) then
               new[i,j]:=old[l,j];

            if old[i,n]=((t_old+1) mod t_states) then
               new[i,j]:=old[i,n];
      end;
   end;

end;

Begin
```

328   CHAPTER 10.   CELLULAR AUTOMATA

```
Randomize;

ChoosePalette;

Initialise;
DisplayNewGeneration;
Repeat
   CalculateNextGeneration;
   DisplayNewGeneration;
until keypressed;

end.
```

Listing 10-7

```
Program TimeDependantRules;

{   Uses a time dependant rule set.   }

Uses Crt,Graph,Extend;

Var

    old:              array[1..170,1..170] of byte;
    new:              array[1..170,1..170] of byte;
    i,j,n,o:          integer;
    i1,j1,k,l,m:      integer;
    LiveOld:          integer;
    tx,ty:            integer;
    size_of_grid:     integer;
    orthoganal,diagonal,generation : integer;

Procedure Initialise;
Begin
size_of_grid:=100;

    for i:=1 to size_of_grid do begin
       for j:=1 to size_of_grid do begin

{   Initialise the grid.  Stick a 1 in old randomly.       }

          new[i,j]:=0;
          old[i,j]:=new[i,j];

       end;
    end;
    old[50,50]:=1;

end;

Procedure DisplayNewGeneration;

Begin
    for i:=1 to size_of_grid do begin
       tx:=2*i;
       for j:=1 to size_of_grid do begin
          ty:=2*j;

          PutPixel(tx,ty,new[i,j]);
          PutPixel(tx+1,ty,new[i,j]);
          PutPixel(tx,ty+1,new[i,j]);
          PutPixel(tx+1,ty+1,new[i,j]);
```

TURBO PASCAL LISTINGS

```pascal
            old[i,j]:=new[i,j];
            new[i,j]:=0;

      end;
   end;

end;

Function CountOldCells(i1,j1: integer): integer;

Begin
{  Count cells orthogannally adjacent to the one of interest - again,
   check the edges of the grid to see if we need to wrap around.     }

   if j1=size_of_grid then l:=1
      else
   l:=j1+1;
   if (j1-1)=0 then m:=size_of_grid
      else
   m:=j1-1;
   if i1=size_of_grid then n:=1
      else
   n:=i1+1;
   if (i1-1)=0 then o:=size_of_grid
      else
   o:=i1-1;

   orthoganal:=old[i1,l]+old[i1,m]+old[n,j1]+old[o,j1];
   diagonal:=old[m,o]+old[m,n]+old[l,o]+old[l,n];

   CountOldCells:=orthoganal+diagonal;

end;

Procedure CalculateNextGeneration;

Begin

   for i:=1 to size_of_grid do begin
      for j:=1 to size_of_grid do begin

{  Here's where the rules get implemented; only create a new cell on even
   generations
   with orthoganal neighbours=1 OR odd geneerations with 1 diagonal and orthog.
   neighbour - a time dependant rule.    }

            LiveOld:=CountOldCells(i,j);
            if ((orthoganal=1) and ((generation mod 2)=0)) then
               new[i,j]:=1;
            if ((liveold=1) and ((generation mod 2)=1)) then
               new[i,j]:=1;

      end;
   end;

end;

Begin

ChoosePalette;

Initialise;
generation:=1;
```

```
Repeat

  CalculateNextGeneration;
  DisplayNewGeneration;
  generation:=generation+1;

until keypressed;

end.
```

Listing 10-8

```
Program Life;

{   Play the life game      }

Uses Crt,Graph,Extend;

Var
    old:               array[1..80,1..25] of byte;
    new:               array[1..80,1..25] of byte;
    i,j,n,o:           integer;
    i1,j1,k,l,m:       integer;
    LiveOld:           integer;
    ch:                char;
    FuncKey:           boolean;
    Lowest,Highest:    integer;
    Birth:             integer;
    birth_s,low_surv_s: string;
    high_surv_s,pos_or_rand: string;

Procedure Initialise;
Begin

TextBackGround(Blue);
TextColor(Yellow);

ClrScr;
pos_or_rand:='R';
birth_s:='3';
low_surv_s:='2';
high_surv_s:='3';

PrintAt(1,2,'Set up position or random? (S/R) ',Green);
StringEdit(pos_or_rand,1,35,2,Yellow);
PrintAt(1,3,'Birth parameter: (3)',Green);
StringEdit(birth_s,1,35,3,Yellow);
PrintAt(1,4,'Lowest value for survival: (2)',Green);
StringEdit(low_surv_s,1,35,4,Yellow);
PrintAt(1,5,'Highest value for survival: (3)',Green);
StringEdit(high_surv_s,1,35,5,Yellow);

val(birth_s,birth,i);
val(low_surv_s,lowest,i);
val(high_surv_s,highest,i);

ClrScr;

if (pos_or_rand='R') or (pos_or_rand='r') then begin

{ This section of code sets up a random seeding of the arrays for a random
  start position.                                                        }
```

```
   for i:=1 to 80 do begin
      for j:=1 to 24 do begin

         if random(10)>7 then
            new[i,j]:=1
         else
            new[i,j]:=0;

         old[i,j]:=new[i,j];

      end;
   end;
 end

else
begin

   for i:=1 to 80 do begin
      for j:=1 to 24 do begin
         new[i,j]:=0;
         old[i,j]:=0;
      end;
   end;

i:=1;
j:=1;

GoToXY(i,j);

repeat
   ch:=ReadKey;
   if ch<>#0 then Funckey:=False else
   begin
      Funckey:=True;
      ch:=Readkey;
   end;
   if Funckey then begin

      if ch=#72 then begin
         if (j<1) then
            j:=24
         else
            j:=j-1;
      end;
      if ch=#80 then begin
         if (j>24) then
            j:=1
         else
            j:=j+1;
      end;
      if ch=#75 then begin
         if (i<1) then
            i:=80
         else
            i:=i-1;
      end;
      if ch=#77 then begin
         if (i>80) then
            i:=1
         else
            i:=i+1;
      end;

      GoToXY(i,j);
```

```
          end
        else begin

          if ch=#32 then begin
            write(' ');
            new[i,j]:=0;
            old[i,j]:=0;
          end;
          if ch=#13 then begin
            write(chr(1));
            new[i,j]:=1;
            old[i,j]:=1;
          end;

          GoToXY(i,j);

        end;

      until (ch=#27);

    end;

  end;

  Procedure DisplayNewGeneration;

  Begin
    for i:=1 to 80 do begin
      for j:=1 to 24 do begin

        GoToXY(i,j);
        if new[i,j]=1 then
          write(chr(1))
        else
          write(' ');

        old[i,j]:=new[i,j];

      end;
    end;

  end;

  Function CountOldCells(i1,j1: integer): integer;

  Begin
    if (j1+1)=25 then l:=1
               else
    l:=j1+1;
    if (j1-1)=0 then m:=24
               else
    m:=j1-1;
    if (i1+1)=81 then n:=1
               else
    n:=i1+1;
    if (i1-1)=0 then o:=80
               else
    o:=i1-1;

    k:=old[i1,l]+old[i1,m]+old[n,j1]+old[o,j1];
    k:=k+old[n,l]+old[o,l]+old[n,m]+old[o,m];

    CountOldCells:=k;
```

```
end;

Procedure CalculateNextGeneration;

Begin

    for i:=1 to 80 do begin
       for j:=1 to 24 do begin

            LiveOld:=CountOldCells(i,j);
            if (old[i,j]=0) and (LiveOld=Birth) then
               new[i,j]:=1;

            if (old[i,j]=1) and ((LiveOld<Lowest) or (LiveOld>Highest)) then
               new[i,j]:=0

       end;
    end;

end;

Begin

Randomize;

Initialise;
CursorControl(0);
Repeat
   CalculateNextGeneration;

   DisplayNewGeneration;

until keypressed;

CursorControl(2);

end.
```

Chapter 11

Practical chaos

In this final chapter I want to make a brief foray into the real world of fractals and chaos, just to point out that the subjects we've been examining aren't just mathematical abstracts. You can walk out of your house and find an object or situation that is exhibiting chaotic or fractal behaviour; I guarantee it! Of course, finding such a system is the start of our problems; in order to analyse it on a PC using the techniques we've explored in this book, we have to get some measurable aspect of the behaviour system that we can use to get numbers into the computer.

The easiest way to see how we can do this is to look at a few chaotic systems that you might like to try setting up yourself. If you're not in to messing about with real systems, then the first project is probably for you! These descriptions are by necessity brief; to detail experimental set-ups for each of these, and full software listings, would probably take a chapter for each system, but I hope you'll feel encouraged to try out some of these ideas.

The Stock Market

An examination of the financial pages, or the last couple of seconds of many news bulletins, will reveal the FTSE index being quoted. This is a measure of the performance of the top 100 shares on the London Stock Exchange, and is often viewed as a measure of the performance of the economy. Whether you hold shares or not, the economic system has a great impact on our lives, and there have been various efforts made to predict the behaviour of the stock exchange as a whole or of the stocks of an individual company. However, the behaviour of the stock market defies description by computational means, so playing the stock market remains a game for the wealthy. Because of the unpredictability of this system, various efforts have

Data acquisition

Legend has it that Mrs Beeton's recipe for jugged hare starts off with something along the lines of 'first catch your hare'. (For all you cooks out there, I checked, and it doesn't—another legend bites the dust!) The same caveat applies to us here; we first need a large amount of data in a machine-readable form before we can actually get on with the job of analysing the data. This particular data is widely available, from newspapers or yearbooks, but the data will have to be typed in to the computer. This is a problem that besets all users of real-world data—the method of storing it. The data is presented in the form $nnnn.nn$, with two decimal places. If you need to store a lot of data it's often a good idea to try and store numbers as integers. This can be done by simply removing the decimal point, bearing in mind how many decimal places the number had. For example, we could store the number 2004.34 as 200434, and remember that if we ever want the real value we have to divide the value 200434 by 100. However, storing the number as an integer rather than as a real number will save considerable amounts of space. It is a straightforward task to write a program in any language that will allow data to be typed in and stored in an array variable. Once the data is in the computer, it can be processed, but it's a good idea to write the data out to a disc or tape file before you start.

The data can be stored in a file in a variety ways, but the simplest is to use the sequential file operations that the language possesses. For example, in some versions of BASIC we might use code like:

```
open 'fred' as #1
  for i=1 to 200
    print #1,data_item(i)
  next i
close #1
```

to write the contents of the array *data_item* to file. Similar code can be written to read such data in from a file and place it in an array.

As to processing the data, we here have a time series of data points, where the value of the FTSE index varies each day. There are, therefore, certain types of analysis that aren't really available to us. For example, phase diagrams are out to start with. The simplest thing to do with a time series is, of course, to plot a line graph and see what you get. The resultant line is frequently very similar to the sort of graph we get from a fractal Brownian motion program.

If we wish to try and find a strange attractor in this set of data, then we will certainly need a large amount of data, assuming there is one to find! You will need to look at the data using the method of creating false observables that was mentioned earlier in this book. For example, we could take pairs of data values and plot them on a graph. We should expect a random sprinkling within the area that corresponds to the range of values plotted on the graph, assuming that there is no attractor present. For example, for a sequence of numbers like:

$$123, 453, 442, 765, 431, 987$$

we could plot values as follows:

$$(123, 453) \quad (442, 765) \quad (431, 987)$$

Of course, you may have to scale these values to bring them into the range of your graphics screen. By plotting these false observables you may see some order in the apparent random changes noted in the FTSE index. You might also like to consider constructing a simple equation system to model price rises and falls in an economic system; in general, increasing availability of a commodity causes price falls, decreasing availability causes a price rise. A similar relationship occurs with demand; high demand can generate high prices, low demand can cause low prices. Such relationships can give rise to equations with similar characteristics to the logistic equation explored in Chapter 2.

Weather

We've already learnt that the model weather explored in Chapter 4 is subject to the whims of a strange attractor; what about the real weather? Well, we can't measure the same parameters that are represented in the Lorenz equations but what about other observables? It's possible to measure temperature and pressure fairly easily through a thermometer and a barometer, so you might like to see if you can create a phase portrait from these parameters.

The dripping tap

The dripping tap is an almost classical, real-world, chaotic system. It's also a fairly simple system to set up yourself to investigate, although you'll have problems if you try to do it with the kitchen tap! The reason for this is purely one of control, so if you're feeling adventurous you *could* always try it and see what happens. The observable parameter in this system is the time between subsequent drops. In practical terms, if you start a tap dripping you'll go through the following phases:

Irregular drips This is effectively a settling down stage which soon dies out as the drips from the tap settle down into a regular pattern of drips.

Regular drips These drips will be found to be a regular period of time apart, and the number of drips noticed in a set period of time will increase as you adjust the flow rate. The control of flow rate is quite important, and this is why a kitchen tap isn't the best experimental set-up for this type of work; you just can't get the fine control needed.

Period doubling Continuing to increase the rate of flow (slowly) will lead to period doubling, in which the drip sequence moves from the same time between subsequent drops to an alternating sequence of two different times, then four times; it's unlikely you'll measure other transitions as the next stage is the onset of chaos.

Chaotic dripping At this point, the drips come with apparently random times between them.

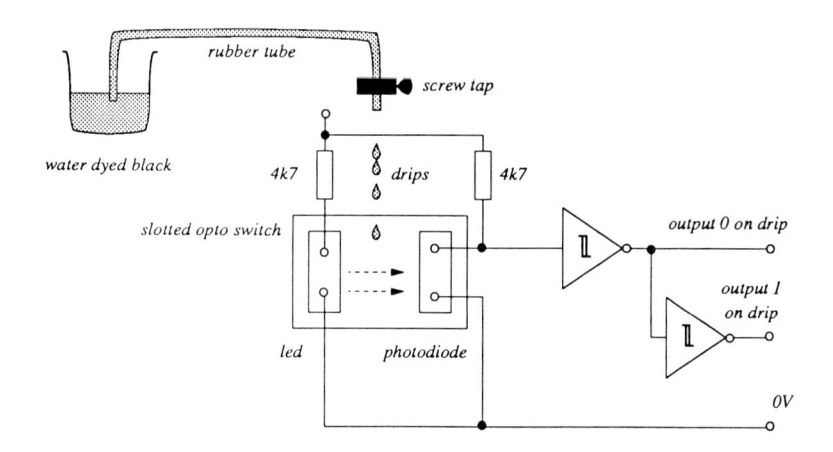

Figure 11-1. Monitoring a dripping tap.

Figure 11-1 shows a simple experimental set-up for timing drips. I leave the actual business of interfacing the TTL signal from this circuit to your computer to you, but it is easily done via the BBC Micro User Port or via a digital I/O card in the PC. The fine control over drip rate is carried out through the clamp on the rubber tube; you only need a very slow rate of flow to explore chaotic behaviour in this system. By measuring and logging the times between drips (i.e. between pulses from the circuit in Figure 11-1) you can plot a bifurcation diagram, work out the Feigenbaum constant, and plot an attractor by the process of creating false observables.

The main requirement of software for this investigation is that it needs to be able to measure the elapsed time between two drops. This can be done in BASIC or Pascal; faster languages, such as C or Assembler, aren't really required as the times being estimated are fairly long in computing terms. It doesn't really matter if you can't get the recorded times in terms of real time units, like seconds, because to see the period doubling, etc. develop we only need to study comparative times. An algorithm for this sort of time recording might be as follows, assuming the passage of a drip through the sensor creates a logic '1' input to the computer:

```
Start:
    Pointer=1
    Time=0
    Repeat
    Until Input = 1        ;Wait for an input from the drip
                           ;counter.
Loop:
    Repeat
    Until Input = 0        ;Now wait for the drip to pass
                           ;all the way through the sensor

    Repeat
        Time = Time + 1    ;Increment Time until a new drip
    Until Input = 1
    Times[pointer]=Time    ;Store time in an array of times
    pointer=pointer+1      ;and increment the pointer
    Time=0                 ;zero the time
    Go To Loop             ;Round we go again.
```

This process will give an array of numbers representing times between subsequent drips. The array can be processed in the following ways.

Draw a graph

For a fixed rate of flow, you can draw a simple graph of time between drips against drip. So, for the second drip, we'd plot 2 on the x-axis, and the time between drip 1 and drip 2 on the y-axis. For the third drip, we'd plot 3 on x (and time3) time2 on y, and so on.

Bifurcation plot

You will need to get arrays of times for different flow rates. Then, plot flow rate on the x-axis and the time between drips on the y-axis. For a single flow rate, therefore, we'd end up with a single vertical line of dots if we were observing chaotic drips or a single dot or pair of dots for the ordered dripping and the period 2 dripping.

A chaotic pendulum

One of the more predictable things in life is a pendulum; set it gently swinging and it will execute a regular path with a fixed period until friction causes the pendulum to stop. However, if you introduce some non-linearities into this simple system we can end up with a chaotic pendulum, where the course and period of the pendulum are very sensitive to the initial position of the pendulum. The easiest way to introduce some chaos into a pendulum system is to use a handful of small magnets, as shown in Figure 11-2. The bob is a small magnet, and it swings over a surface containing four other magnets, all repelling the bob. The path taken by the bob is irregular and unpredictable as the bob gets different amounts of 'push' from each of the magnets.

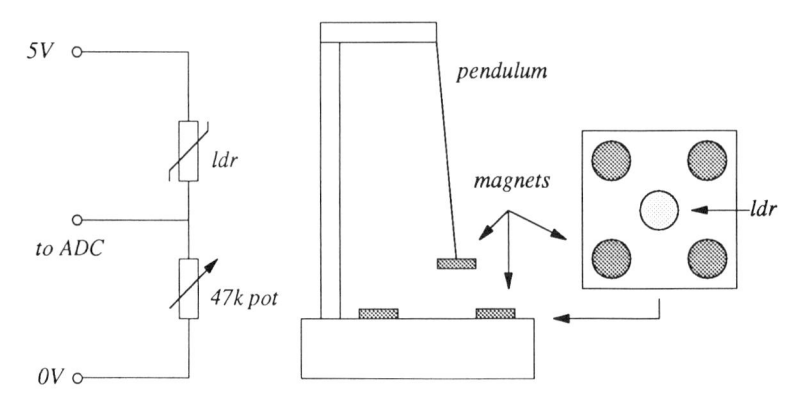

Figure 11-2. A chaotic pendulum.

A practical model

I made my chaotic pendulum out of wood, and used pieces from a magnetic travelling chess set for the magnets. Four of these were glued on the baseboard, all with the same face, either N or S, pointing upwards. The magnet acting as the bob was glued to the cotton with the same face facing down towards the other magnets, and was suspended so that it was a few millimetres above the four repelling magnets.

To see chaos in action, pull the pendulum to one side and then release it. As the pendulum bob enters the magnetic fields of the other magnets it will be repelled into an elaborate, curved path. Varying the point of release will cause a difference in the path taken—the chaos hallmark of a small change in initial conditions (the start position) causing a large change in the final result (the path taken).

Recording this behaviour isn't as difficult as it looks at first sight. I used a simple light dependent resistor and fixed resistor to form a potential divider which will give a varying output voltage dependant upon the light reaching the LDR. If the LDR is positioned so that the shadow of the swinging pendulum falls on it, then the output voltage can be fed to an analogue to digital converter and recorded on a computer. This set up will allow you to demonstrate quite effectively the fact that for virtually identical initial pendulum positions the path travelled by the pendulum will be quite different, as indicated by the voltage readings from the potential divider. Of course, a more accurate representation of the path of the pendulum could be made if we uses more than one light sensor. There are other ways of monitoring the motion of the pendulum; for example, a magnetic field sensor, called a *Hall effect device*, could be used to follow the changes in the local magnetic field caused by the moving magnet on the pendulum.

A driven pendulum

Another chaotic pendulum is obtained by driving a pendulum at a frequency near its natural resonant frequency. This is basically a driven oscillator, as was mentioned in Chapter 5. The actual setting up of such an arrangement is not exactly simple, but Figure 11-3 should give you some ideas to try. It's again rather awkward to make qualitative measurements from this model, due to the difficulty of logging the movement of the pendulum bob, but you will see a variety of effects as the frequency of the driver oscillator is varied around the natural frequency of swing of the pendulum. The natural frequency of the pendulum is measured by timing how long it takes to make a number of full swings. Say it did 10 swings in 5 seconds. This is a period of 0.5 seconds, or a frequency of $1/0.5\text{Hz} = 2\text{Hz}$. Thus, driving the pendulum at frequencies in the range 1 to 3 Hz will give different effects.

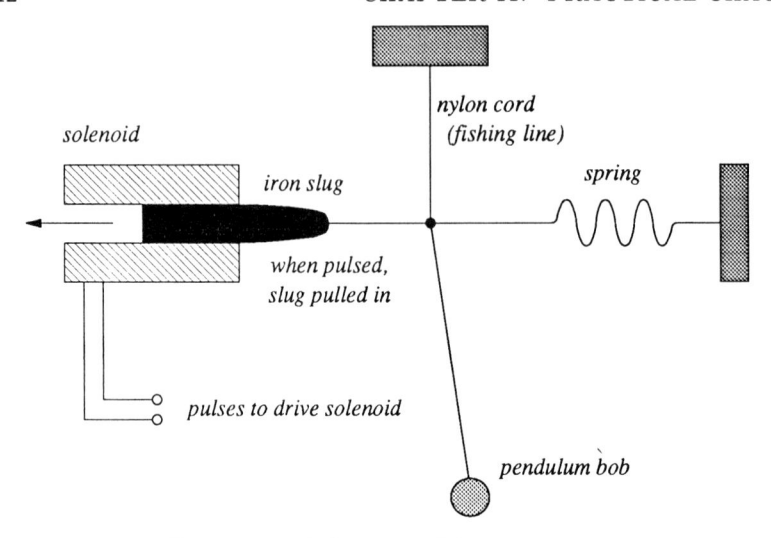

Figure 11-3. A driven pendulum experiment.

Electronic chaos

For the experimenter with a little knowledge of electronics and access to some test equipment, particularly an oscilloscope, many chaotic effects can be explored using simple electronic equipment. These experiments have some considerable advantages as far as we're concerned. The first is that they're more easily built and played around with than the mechanical models detailed above. In addition, the voltage and currents in the circuits provide us with easily measured observables against time; we don't have to mess about timing drips, etc. From my point of view, this section of the book offers some complications as I've not got the space for a crash course in electronics! However, any standard 'practical electronics' manual will be helpful here, and I've listed a couple in the Bibliography.

Driven LCR network

There are a variety of systems that can be examined; the simplest are those based around semiconductor diodes, and Figure 11-4 shows one possible circuit to try out. This is known as a *driven LCR network*. This indicates that the circuit consists of electrical inductance L, capacitance C and resistance R. Circuits with these elements exhibit a behaviour that is frequency dependent, so driving this circuit with an electrical signal of varying frequency will probably show some interesting effects. However, the capacitance element in this circuit is not provided by a capacitor; instead, it is provided by the semiconductor diode, which is a special component called a *varactor*, or variable capacitance diode. This component exhibits electrical capacitance whose value varies when the voltage across the component varies. This can lead to interesting effects when a signal is applied around the resonant frequency of the LCR network, as the capacitance of the diode will vary as the signal voltage varies!

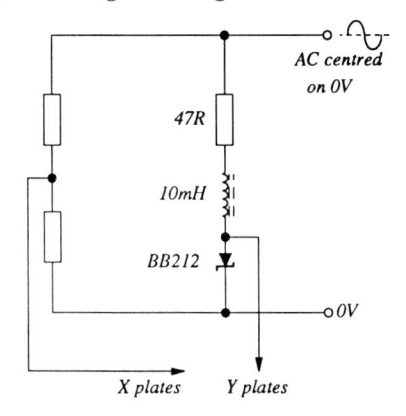

Figure 11-4. A driven non-linear LCR circuit.

The actual theoretical behaviour of this circuit is quite complex; apart from the capacitance offered by the diode, which combines with the resistance and inductance in the circuit to give a frequency-dependent behaviour, the diode also only allows current to pass through it when the anode is positive with respect to the cathode. This non-linearity is at the heart of the chaotic behaviour offered by this circuit. When the driving voltage is such that the anode is positive with respect to the cathode, the diode conducts current and the varactor effect is not present; however, when the diode is subject to a signal that leaves the cathode more positive than the anode, the capacitance effect becomes very marked and this modifies the frequency response of the circuit.

The behaviour of the circuit is monitored using an oscilloscope, as shown. Varying the frequency and amplitude of the applied sine wave will give a variety of different effects on the waveform. Some of these can be seen in Figure 11-4. A paper on this topic by Testa, Perez and Jeffries was published in *Physical Review Letters*, (1982), 714-17. In their work, an 1N953 silicon varactor was used, but I've had similar results from other components as well, such as the BB212. If you're really into experimenting, then you might like to try a standard small signal diode in the circuit rather than a varactor. Whatever you use, the driving voltage is the equivalent to k in the logistic equation and the voltage measured across the varactor is the observable. I used a signal generator to produce the driving voltage, but any circuit could be used that can produce a sine wave output with variable frequency and a variable output amplitude. There have been several designs published in recent years using the 8038 function generator integrated circuit, and any of these would be fine. In practical terms, you may need to mess around with the frequency and amplitude of the driving signal to get the best results, but with care this system generates some rather interesting phase portraits and shows the onset of period doubling in both the phase portrait and as a straight plot of varactor voltage against time. In the circuit shown, the best example of period doubling was seen with a driving voltage at about 120kHz and sweeping the level of the voltage between 3 and 4V (as measured at the diode). The driving voltage was a sine wave that was symmetrical around 0V, though experimenting with the DC offset of the signal, if you can do this, will often give differing results.

The period doubling is clearly seen in the straight plot of varactor voltage and the phase portrait, and suddenly commences as the drive voltage level is increased through a threshold level. At frequencies away from the resonant frequency, other effects are noticed, giving quite attractive phase portraits. Although I've not done this, I feel that it should be possible to run this circuit at audio frequencies and monitor the output on a loudspeaker; this could be interesting to see what the different states of the system sound like as well as what they look like!

Plotting phase portraits with an oscilloscope

To see the phase portrait using an oscilloscope, it is essential that the 'scope allows you to drive the X plates from an external signal source. Normally, the X plates are driven from an internal timebase circuit, but for a phase portrait we need to drive the X plates with a signal representing one of the parameters of the system being measured. In this system, the X plates can be driven by the voltage from the signal generator. You may need to apply this voltage to the X input of the scope through a potential divider of some sort if your 'scope doesn't have a separate X gain control. Applying the drive signal without an input to the Y input of the 'scope will generate a horizontal line on the screen. Application of the Y signal will create an image that represents the phase portrait of the system. Again, you may need to adjust the Y gain of the 'scope to get a suitable image. The rules for photographing such displays are similar to those for photographing computer displays, though more experimentation is needed. I find that black and white film, processed yourself, is the most economical way of getting permanent records of 'scope displays. I've used Ilford HP5 for this work, and quite acceptable prints can be made. Any decent photography book will show you what to do to process films and make simple prints; it's not that difficult.

Driven Oscillator

A slightly more complex case is given by the electronic version of the driven oscillator mentioned in Chapter 5. This is very much a classic system in the study of non-linear differential equations, and is often described under the heading *Van der Pohl oscillator*. The usual arrangement for this system that is described in theoretical text books is to use an oscillating circuit based on a thermionic valve. I prefer to use something a little more up to date, and in Figure 11-5a an oscillator using a *field effect transistor*. The FET exhibits similar electrical characteristics to the valve in many respects, and so is well suited to this application. Alternatively, you might like to try the other circuit shown in Figure 11-5b using a BC108 transistor. Again, the principal effects are noted around the resonant frequency of the circuit, but both of these oscillators are well worth investigating as they give rise to some very attractive phase portraits, even if they're not always chaotic!

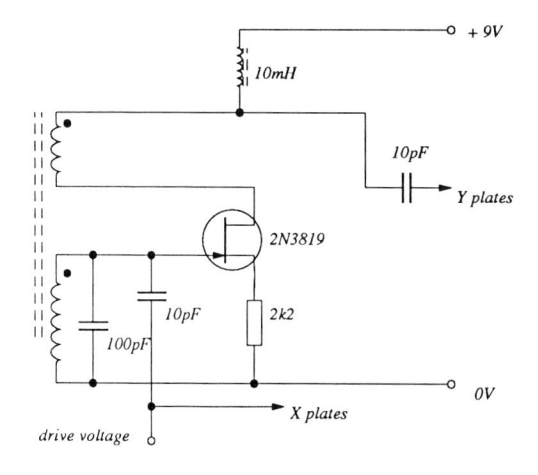

Figure 11-5a. A driven oscillator using an FET.

One point to note in this experiment, and whenever using a driven oscillator, is that the driving signal should be linked in to the driven oscillator as loosely as possible, as otherwise the driving oscillator can swamp the effects that you want to see. In some cases, particularly with the bipolar oscillator above, a too tight connection of the driving signal can prevent the oscillator working by itself.

346   CHAPTER 11. PRACTICAL CHAOS

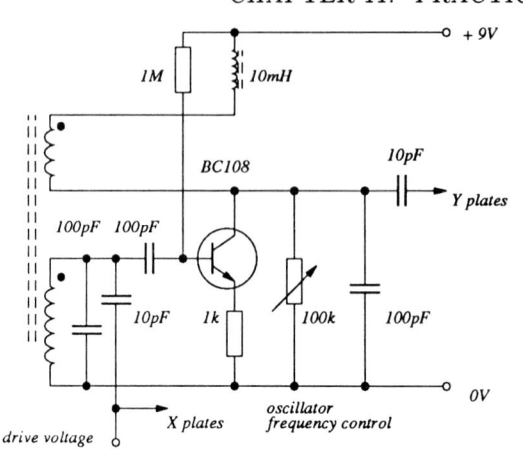

Figure 11-5b. A driven oscillator using a bipolar transistor.

Analogue computers

Long before digital computers, certain problems, particularly those involving differential equations, were solved using electronic computers that represented values in the equations by voltage levels. Computers like this were called analogue computers, and are now an almost extinct breed. However, because they allow us to directly model differential equations, they allow us an alternative way to explore chaotic behaviour by setting up systems of equations and letting them run, monitoring the results on an oscilloscope screen or voltmeter. The use of oscilloscopes in this work allows phase portraits to be drawn on the screen, and the parameters of the equation can be varied by modifying component values in the circuits modelling the system. Don't forget, this is how the original work on the Lorenz equations was done, so there's clearly a precedent for using them!

The *Analogue Computer* in ETI (see Bibliography) is a rather nice introduction to the field of analogue computing, and this practical design for a simple analogue computer to build is well worth a look. The simple computer described is perfectly adequate for running some simple models, and one of my current projects is to use these circuits to build an electronic version of the Lorenz equations; if successful, I will have gone full circle, from Lorenz's simple analogue computer in the early 1960s, through a 386SX personal computer, back to a 1990s version of an analogue computer.

That's really all I've got space for in this book. There are many other systems that can be investigated, and an examination of the literature listed in the Bibliography will no doubt give you more food for thought. Chaos is still young, and there is scope for experimentation with simple electronic and mechanical systems, as well as with mathematical models. I hope that I've stimulated a few ideas for you in this book; who knows, you may come up with a whole new chunk of chaos for yourself!

Appendix

Many of the Turbo Pascal listings make use of a unit called *Extend*. This unit gathers together a number of useful routines for control of the screen, printer and serial port.

Listing for the Extend unit

```
Unit EXTEND;

Interface
Uses Crt,Dos,Graph;

Type

FileArray=       Array[0..10] of Real;

Var temp : string;
    nwid : integer;
    ok   : boolean;
    Xorg,Yorg : byte;
    Regs: Registers;
    input_string: String;
    ScreenType : integer;
    GraphDriver,GraphMode : integer;
    Tcolor,MaxColor:        Byte;
    PixelColor:             byte;
    x1,x2,y1,y2:            Integer;
    FracParams:             FileArray;
    Mode:                   Byte;
    title:                  String;
    TurtleX,TurtleY:        Integer;
    TurtleTheta:            Integer;
    distance,Theta:         Integer;

Procedure CursorControl(Size: byte);
Function Max(a,b: integer): integer;
Function Min(a,b: integer): integer;
Procedure StringEdit(Var input_string: String; Width,Xorg,Yorg,Tcolor: Byte);
Function Spaces(num: word): string;
Function Trim(s: String): string;
Procedure PrintScreen;
Function Sgn(z: real): integer;
Function PrinterStatus: integer;
Function InitComPort(port,config: byte): integer;
Function WriteComPort(port,value: byte): integer;
Function ReadComPort(port: byte): integer;
Function ReadComStatus(port: byte): integer;
```

350   APPENDIX

```
Procedure ChoosePalette;
Procedure PrintAt(a,b:integer; input_string: String; z:integer);
Function PickPixelColor(r,n: Integer) : integer;
Procedure SaveScreen(input_string: String);
Procedure LoadScreen(input_string: String);
Procedure SaveFractalScreen(input_string,title: String; FracParams: FileArray;
                            mode: byte);
Procedure LoadFractalScreen(input_string: String; Var title: string;
                            Var FracParams: FileArray; Var mode: byte);
Procedure Turn(theta : Integer);
Procedure Front(distance : integer);

Implementation

Procedure CursorControl(Size: byte);
{ 0 - Turn Cursor OFF
  1 - Small Cursor
  2 - Big Cursor                            }
Begin
   With Regs Do Begin
      AX:=$100;
      Case Size of
         0 : CX:=$3030;
         1 : CX:=$0F;
         2 : CX:=$607;
      end;
      Intr($10,Regs);
   End;
End;

Function Max(a,b: integer): integer;

{   Returns the bigger of a and b }

Begin
If a>b then max:=a else max:=b;
end;

Function Min(a,b: integer): integer;

{   Returns the smallest of a and b }

Begin
If a<b then min:=a else min:=b;
end;

Function Spaces(Num: Word): String;
Const bl=' '
Begin
   Spaces:=Copy(Bl,1,Num);
end;

Function Trim(s: string): string;
Var start,finish : Integer;
Begin
   start:=1;

{   work along string until first non-space character   }

   While s[start]= ' ' do start:=start+1;
   finish:=Length(s);

{   Now work from end to last character of string      }
```

351

```
        While s[finish]=' ' do finish:=finish-1;
        Trim:=Copy(s,start,finish-start+1);
end;

Procedure StringEdit(Var input_string: String; Width,Xorg,Yorg,Tcolor: Byte);
Var       InChr: Char;
          cursor,Endstr: Byte;
          Inserting,FirstChar:     Boolean;
          Old: String;
Begin
   TextColor(Tcolor);
   old:= input_string;
   cursor:=1;
   FirstChar:=True;
   GotoXY(Xorg,Yorg);
   Repeat
      EndStr:=Length(input_string);
      GoToXY(Xorg,Yorg);
      If EndStr <= Width then
         Write(input_string+spaces(width-Endstr));

      GotoXY(Xorg+cursor-1,Yorg);
      InChr:=Readkey;
      Case InChr of
         #27 : input_string:=old;
         #0  : Begin

{ Now look after the control keys to move the cursor     }

                  InChr:=Readkey;
                  Case Inchr of

{ Move to the right using right arrow key}

                     #77 : cursor:=Min(cursor+1,EndStr+1);

{ Move to the left using the left arrow key  }

                     #75 : cursor:=Max(cursor-1,1);

{ Delete the character at the cursor   }

                     #83 : input_string:=Copy(input_string,1,cursor-1)+
                           Copy(input_string,cursor+1,Endstr-cursor);
                  end;
               end;

{ Now look after the backspace key    }

         #8 : Begin
                 input_string:=Copy(input_string,1,cursor-2)+
                 Copy(input_string,cursor,Endstr-cursor+1);
                 cursor:=Max(cursor-1,1);
              end;

   #32..#254 : Begin

{ Now look after all the other key presses.....  }

                  If endstr < Width then begin
                     input_string:=Copy(input_string,1,cursor-1)+ InChr;
                     cursor:=Min(cursor+1,Width);
                  end;
               end;
      end;
```

```
            Until (Inchr=#13) or (inchr=#27);
            input_string:=Trim(input_string);
end;

Procedure PrintScreen;
Begin
  FillChar(regs,SizeOf(regs),0);
  Regs.AH:=$05;
  Intr($5,regs);
end;

Function PrinterStatus: integer;
{ Bit 7 set - Printer not busy.
  Bit 6 set - Acknowledge from printer.
  Bit 5 set - Out of paper.
  Bit 4 set - Printer selected.
  Bit 3 set - I/O error.
  Bit 2 set - Not used.
  Bit 1 set - Not Used.
  Bit 0 set - Time Out.                              }

Begin
  FillChar(regs,SizeOf(regs),0);
  Regs.AH:=$02;
  Intr($17,regs);
  PrinterStatus:=regs.AH;
end;

Function Sgn(z: real) : integer;
Begin
if z=0 then
   Sgn:=0;
if z>0 then
   Sgn:=1;
if z<0 then
   Sgn:=-1;
end;

Function InitComPort(port,config:byte): integer;
{ Initialise a given COM port

    port = 0 uses COM1
         = 1 uses COM2

    config    Bits 0 to 1 set word length
                     10 = 7 bits
                     11 = 8 bits

              Bit 2 Stop Bits
                     0 = 1 stop bit
                     1 = 2 stop bit

              Bits 3 to 4 parity
                     00 = None
                     10 = None
                     01 = Odd
                     11 = Even

              Bits 5 to 7 parity
                     000 = 110 baud
                     001 = 150 baud
                     010 = 300 baud
                     011 = 600 baud
                     100 = 1200 baud
                     101 = 2400 baud
```

```
                        110 = 4800 baud
                        111 = 9600 baud    }

Begin
   FillChar(regs,SizeOf(regs),0);
   Regs.AH:=$00;
   Regs.AL:=config;
   Regs.DX:=port;
   Intr($14,regs);
   InitComPort:=regs.AX;
end;

Function WriteComPort(port,value: byte): integer;
{ port is the com port number

         0 = COM1
         1 = COM2

  value is the byte to be written           }

Begin
   FillChar(regs,SizeOf(regs),0);
   Regs.AH:=$01;
   Regs.AL:=value;
   Regs.DX:=port;
   Intr($14,regs);
   WriteComPort:=regs.AH;
end;

Function ReadComPort(port: byte): integer;
{ Read the byte from the com port

              port = 0     COM1
              port = 1     COM2
                                          }
Begin
   FillChar(regs,SizeOf(regs),0);
   Regs.AH:=$02;
   Regs.DX:=port;
   Intr($14,regs);
   ReadComPort:=regs.AL;
end;

Function ReadComStatus(port: byte): integer;
{ Get status byte from serial port in port value:

         0    =   COM1
         1    =   COM2

  Function returns the Port Status

         Bit 7    =    Timed Out
         Bit 6    =    Transmit shift register empty
         Bit 5    =    transmission hold register empty
         Bit 4    =    break detected
         Bit 3    =    framing error
         Bit 2    =    parity error
         Bit 1    =    over-run error
         Bit 0    =    data ready
                                                       }
Begin
   FillChar(regs,SizeOf(regs),0);
   Regs.AH:=$03;
   Regs.DX:=port;
```

353

```
    Intr($14,regs);
    ReadComStatus:=regs.AH;
end;

Procedure ChoosePalette;
{ sets palette and mode according to whether a CGA/ VGA or EGA
  card has been detected by the system                            }

Begin
    GraphDriver:=Detect;
    GraphMode:=0;
    InitGraph(GraphDriver,GraphMode,' ');
    if GraphDriver=CGA then
        SetGraphMode(0);
    if (GraphDriver=EGA) or (Graphdriver=VGA) then
        SetGraphMode(1);
    MaxColor:=GetMaxColor;
end;

Procedure PrintAt(a,b: integer;input_string: string; z:integer);
{ Prints a text string at position a,b on a Text Screen. String
  is in the variable input_string and the colour is in z          }

Begin
    GoToXY(a,b);
    TextColor(z);
    Write(input_string);
end;

Function PickPixelColor(r,n: Integer) : integer;

{ Simple function to alter Pixel Colour every 'n' iterations }
Begin
    if r mod n=0 then
        begin
            PixelColor:=PixelColor+1;
        end;
    if PixelColor>MaxColor then
        pixelcolor:=1;
    PickPixelColor:=PixelColor;
end;

Procedure SavePart(x1,y1,x2,y2: word; Var F : File);
{   Saves a block of screen to a previously opened file. }

    Var        block_start:      Pointer;
               Block_size:       Word;

Begin
    Block_size:=ImageSize(x1,y1,x2,y2);
    GetMem(block_start,Block_size);
    Getimage(x1,y1,x2,y2,block_start^);
    BlockWrite(f,block_start^,Block_size);
    FreeMem(block_start,Block_Size);
End;

Procedure SaveScreen(input_string: string);
{   Save a whole screen; split into 4 chunks as in some screen modes
    the resultant block would be larger than can be addressed in a single
    block of memory by Turbo Pascal Vertsion 4   }

    Var    F:            File;
           X:            Word;
           I:            Byte;
```

```
Begin
    X:=GetMaxX Div 4;
    Assign(F,input_string);
    Rewrite(f,1);
    for i:=0 to 3 do
        SavePart((X*i),0,(x*(i+1)),GetMaxY,F);
    Close(f);
End;

Procedure PutPart(x1,y1,x2,y2: word; Var F: File);
Var     block_start:        Pointer;
        Block_Size:         Word;

Begin
    Block_size:=ImageSize(x1,y1,x2,y2);
    GetMem(block_start,Block_Size);
    BlockRead(F,block_start^,Block_size);
    PutImage(x1,y1,block_start^,0);
    FreeMem(block_start,Block_Size);
End;

Procedure LoadScreen(input_string: String);
Var     F:                  File;
        X:                  Word;
        I:                  Byte;

Begin
    X:=GetMaxX Div 4;
    Assign(f,input_string);
    Reset(f,1);
    For i:=0 to 3 Do
        PutPart((i*x),0,(x*(i+1)),GetMaxY,F);
    Close(f);

end;

Procedure SaveFractalScreen(input_string,title: string; FracParams: FileArray;
                            Mode: Byte);
{ Saves a screen in BGI format for loading back into Turbo Pascal Applications
and also saves an array of 11 elements which may be used to hold a variety
of parameters - e.g. Real and Imaginary parts of Mandelbrot Set Computations,
Parameters of Strange Attractors, and so on. }

Var     F:                  File;
        X:                  Word;
        I:                  Byte;

Begin

    X:=GetMaxX Div 4;
    Assign(F,input_string);
    Rewrite(f,1);
    BlockWrite(f,FracParams,SizeOf(FracParams));
    BlockWrite(f,title,SizeOf(Title));
    BlockWrite(f,mode,SizeOf(Mode));
    for i:=0 to 3 do
        SavePart((X*i),0,(x*(i+1)),GetMaxY,F);
    Close(f);
End;

Procedure LoadFractalScreen(input_string: String; Var title: String;
                            Var FracParams: FileArray; Var Mode: Byte);
{ Loads a screen to video display after being saved with SaveFractalScreen.
Also loads the parameters in FracParams into the array that was passed
to the procedure. }
```

```
Var     F:              File;
        I:              Byte;
        X:              Word;
Begin
   X:=GetMaxX Div 4;
   Assign(f,input_string);
   Reset(f,1);
   BlockRead(f,FracParams,SizeOf(FracParams));
   BlockRead(f,title,SizeOf(Title));
   BlockRead(f,mode,SizeOf(Mode));
   For i:=0 to 3 Do
      PutPart((i*x),0,(x*(i+1)),GetMaxY,F);
   Close(f);
end;

Procedure Front(distance : Integer);
Begin
   TurtleX:=trunc( distance*cos(TurtleTheta*3.14/180)) + TurtleX;
   TurtleY:=trunc( distance*sin(TurtleTheta*3.14/180)) + TurtleY;
   LineTo(TurtleX,TurtleY);
end;

Procedure Turn(theta : integer);
Begin
   TurtleTheta:=(TurtleTheta + theta) mod 360;
end;

end.
```

Bibliography

Adams, D. *The Hitch-hiker's guide to the galaxy.* London: Pan Books, 1979.

Becker, K.-H. and Doerfler, M. *Dynamical systems and fractals: Computer graphics experiments in Pascal.* Cambridge: Cambridge University Press, 1989.

Borland International. *Turbo Pascal: Owner's handbook, version 4.0.* Scotts Valley: Borland International, 1987.

Bruce, J.W., Giblin, P.J. and Rippon, P.J. *Microcomputers and mathematics.* Cambridge: Cambridge University Press, 1990.

Coll, J. *The BBC microcomputer: User guide.* London: British Broadcasting Corporation, 1982.

Cvitanovic, P., ed. *Universality in chaos.* 2nd ed. Bristol: Adam Hilger, 1989.

Davies, P. *The cosmic blueprint.* London: Unwin Hyman, 1989.

Devaney, R.L. *An introduction to chaotic dynamical systems.* 2nd ed. Redwood City: Addison–Wesley, 1989.

— *Chaos, fractals and dynamics.* Addison–Wesley, 1991.

Dewdney, A.K. *The armchair universe: An exploration of computer worlds.* New York: W.H. Freeman, 1988.

Falconer, K. *Fractal geometry: Mathematical foundations and applications.* Chichester: John Wiley, 1990.

Gleick, J. *Chaos: Making a new science.* London: Sphere Books, 1988.

Hickman, I. *Analog electronics.* Oxford: Heinemann Newnes, 1990.

— *Oscilloscopes: How to use them, how they work.* 3rd ed. Oxford: Heinemann Newnes, 1990.

Holden, A.V., ed. *Chaos.* Manchester: University of Manchester, 1986.

Kaye, B.H. *A random walk through fractal dimensions.* New York: VCH, 1989.

McGregor, J.J. and Watt, A.H. *Pascal for science and engineering.* London: Pitman, 1983.

McGuire, Michael. *An eye for fractals.* Addison–Wesley, 1991.

Mandelbrot, B.B. *The fractal geometry of nature.* (Updated and augmented.) New York: W.H. Freeman, 1983.

Peitgen, H.-O. and Richter, P.H. *The beauty of fractals: images of complex dynamical systems.* Berlin: Springer–Verlag, 1986.

Peitgen, H.-O. and Saupe, D., eds. *The science of fractal images.* New York: Springer–Verlag, 1988.

Pickover, C.A. *Computers, pattern, chaos and beauty: graphics from an unseen world.* Stroud: Alan Sutton, 1990.

Pitt, V.H., ed. *The Penguin dictionary of physics.* Harmondsworth: Penguin, 1977.

Press, Flannery, Teukolsky and Vetterling. *Numerical recipes in Pascal.* Cambridge: Cambridge University Press, 1989.

Rucker, R. *Mind tools: The mathematics of information.* Harmondsworth: Penguin, 1988.

Stevens, R.T. *Fractal programming in C.* Redwood City: M&T Books, 1989.

Stewart, I. *Does God play dice? The mathematics of chaos.* Harmondsworth: Penguin, 1990.

Tooley, M. *Electronic circuits handbook: design, testing and construction.* Oxford: Heinemann Newnes, 1988.

Wagner, T. and Peterson, M. *Fractal creations* (includes copy of Fractint) Waite Group, 1991.

Periodicals

Chaitin, G. 'A random walk in arithmetic.' *New Scientist* (24 March 1990): 44–6.

Cuthbertson, P. 'Analogue computer.' *ETI.* (July, August 1988).

Davies, P. 'Chaos frees the universe.' *New Scientist* (6 October 1990): 48–51.

Dewdney, A.K. 'Computer recreations.' *Scientific American* (August 1989): 88–91.

Juergens, H., Peitgen, H.-O. and Saupe, D. 'The language of fractals.' *Scientific American* (August 1990): 40–7.

Lesurf, J. 'Chaos on the circuit board.' *New Scientist*. (30 June 1990): 63–6.

McRobie, A. and Thompson, M. 'Chaos, catastrophes and engineering.' *New Scientist* (9 June 1990): 41–5.

Martin, B. and Mudge, M. 'From chaos to beauty.' *Personal Computer World* (November 1987): 118–22.

Mullin, T. 'Turbulent times for fluids.' *New Scientist* (11 November 1989): 52–4.

Ojha, J. 'Turn of the skew.' *The Guardian* (22 June 1989): 31.

Palmer, T. 'A weather eye on unpredictability.' *New Scientist*. (11 November 1989): 56–9.

Percival, I. 'Chaos: a science for the real world.' *New Scientist*. (21 October 1989): 42–7.

Savit, R. 'Chaos on the trading floor.' *New Scientist* (11 August 1990): 48–51.

Stewart, I. 'Portraits of chaos.' *New Scientist* (4 November 1989): 42–7.

Tritton, D. 'Chaos in the swing of a pendulum.' *New Scientist* (24 July 1986): 37–40.

Vivaldi, F. 'An experiment with mathematics.' *New Scientist* (28 October 1989): 46–9.

Wolfram, S. 'Computer software in science and mathematics.' *Scientific American* (September 1984): 140–51.

Wood, K. 'Designing fractals to order.' *Electronics World and Wireless World* (August 1990): pp.703–8.

Other reading

Fractal Report. Newsletter with a fractal bias. Very useful, subscribe and support! Contact John de Rivaz, Reeves Telecommunications Laboratories, West Towan House, Porthtowan, Cornwall, TR4 8AX. SAE for details.

Chaos and Complex Cartography. Another newsletter, but with a Chaos bias. Again, useful and interesting. Contact Andy Lunness, 36 Linton Avenue, Bury, Lancashire, BL9 6NL for details. Please enclose an SAE.

Fractal Digest. New quarterly newsletter covering chaos, cellular automata, etc. I've not yet seen a copy of this. Information from Le Mont Ardaine, Rue Des Ardaines, St Peters, Guernsey, Channel Islands.

The above periodicals are published by individuals on a tight budget and are available by subscription only. They provide a valuable source of information, and are well worth a look.

Equinox. Channel 4's science magazine program produced an interesting program on chaos; worth a viewing if you can find a repeat showing!

New Scientist. This weekly science magazine carries articles and news items on chaos and associated subjects. In 1990 a very useful series was run on the subject.

Scientific American. Again, readable articles on chaos and associated topics appear, along with a 'Mathematical Recreations' column which often gives good inspiration.

Software

There is a variety of software packages available that contain programs that generate the images shown in this book. Some of these programs are very sophisticated, others are more simple. Some, so-called public domain programs are free of charge, you simply pay a small fee for the copying of the software. Shareware software is 'try before you buy', that is, if you like the software you're expected to pay a fee. Finally, commercial software is software that you simply buy like a book or a record. In this section, I'll briefly discuss a few packages that you might like to look at.

Chaos: The software. This is a software package that covers a variety of chaotic systems, from Mandelbrot sets to a 'funny pendulum' system. PC or compatibles with EGA graphics or better, Commercial, costs around £50, published by Autodesk. Contact on 0483–303322.

SOFTWARE

Rudy Rucker's CA Lab. This package, again available from Autodesk, concentrates on cellular automata. Commercial software for PCs with CGA or better graphics. Costs around £50. Contact on 0483-303322.

The Desktop Fractal Design System. Concentrates on iterated function systems. PC compatibles, available for about £30 from Academic Press, Harcourt Brace Jovanvitch Limited on 071-4857074.

Dragons. This is a really nice package from Larry Cobb, c/o Bay House, Dean Down Drove, Littleton, Winchester, Hants, SO22 6PP. Concentrates on the Mandelbrot/Julia set type images.

Fractint. As far as I'm concerned, *the* public domain fractal generator. It's almost worth getting a PC with decent graphics for this package alone! Make sure you get the latest version (v16). It covers literally dozens of different types of fractal, and the documentation files provide masses of valuable info. I got my copy from the Public Domain Software Library, Winscombe House, Beacon Road, Crowborough, Sussex, TN6 1UL.

There are many other fractal generator programs around in the public domain and shareware libraries, and it's worth keeping an eye on the various computer magazines and bulletin boards for news of new releases or reviews.

The author can be contacted via the publisher, as *jpritchard* on CIX or on CompuServe [100010,2243].

Index

absolute value, 188
Argand plane, 187
artifacts, 11
attractor, 21, 68, 231
    Hénon, 119–122
    point, 32
    Rossler, 126–127
    strange, 38, 103, 119–133

backward iteration, 237
basin, 231
Beetle, 193
bifurcation, 24
    diagram, 26, 38, 205
biology, 161
boundary, 186, 232, 236
buds, 195, 235
Butterfly Effect, 28, 97–98

c plane, 229
Calculus, 19
Cantor
    dust, 151
    set, 151–153
cardiac arrhythmias, 129
cardioid, 194
catalysts, 162
cellular
    automata, 289
    structures, 289
chaos game, 265
chaotic pendulum, 340–341
clouds, 159
coastline, 160
Collage Theorem, 271

comma separated value, 17
compiled language, 9
complex number, 186
computational irreducibility, 294
connectedness, 232, 234
contour, 194
critical points, 67

data export, 17
degrees of freedom, 21
dendrites, 234
deterministic, 20
diagonally adjacent, 299
differential equations, 59, 97
diffusion limited aggregation, 162
dimension
    fractal, 122
    Hausdorff-Besicovitch, 123
dimensions in automata, 290
disc
    floppy, 6, 7
    hard, 7
discontinuity, 152
dissipative system, 21
dragons, 155
dripping tap, 337
driven
    LCR network, 342
    oscillator, 345
    pendulum, 341
Duffing
    equations, 127
    oscillator, 127
dust, 152, 233
dynamic systems, 60

electronic chaos, 342
entropy, 296
enzymes, 162
equation
    logistic, 25, 32
equations
    differential, 59, 97
    Duffing, 127
    Lorenz, 91–104
    Rossler, 126–127
    Volterra's, 65
equilibrium points, 67
equipotential, 194
Euler's method, 70

false observables, 132, 339
fault line modelling, 262
feedback, 30, 97
Feigenbaum
    constant, 39
    diagram, 26
filaments, 195, 236
finite attractor basin, 194
fixed
    clumps, 299
    field, 17
forced non-linear oscillators, 72
fractal, 123
    Brownian motion, 150
    dimension, 122
    geometry, 143
    landscape, 261
frequency spectrum, 24
function
    iterative, 11
    square, 31
    square root, 31
    two-dimensional, 41

Gaussian plane, 187
generator, 148, 304
geography, 160
GIF file, 13
Gingerbread Man, 193

Gosper glider gun, 304
graftals, 264
graphics
    adapters, 3
    modes, 2

Hénon attractor, 119–122
Hausdorff-Besicovitch dimension, 123
Hele-Shaw cell, 162
high-resolution graphics, 2

image
    processing, 12
    storing, 13
imaginary number, 186
initial states, 294
initiator, 148
integration, 60
intermittency, 152
interpreted language, 8
inversion, 241
inverted Julia set, 241
iterated function systems, 155, 264
iteration, 26
iterative
    array, 289
    functions, 11, 29

Julia sets, 229–242

KAM Theorem, 123
Koch snowflake, 144, 159

L-language, 156
landscape, 261
Laplace, 20
Libchaber, 23
Life, 301–305
limit cycle, 68
logistic equation, 25, 32
Lorenz
    equations, 27, 91–104

Mandelbroids, 202, 206

INDEX

Mandelbrot set, 185–206
Martin's mappings, 41
maths coprocessor, 5, 195
memory, 2, 4
microprocessor, 4, 5
mid-point displacement, 150, 262
modelling, 306
mutation, 295

Newton's method, 237
Newtonian system, 19
nodes, 68
noise, 132, 152, 296
numerical
    accuracy, 11
    overflow, 11
    resolution, 10

observable parameters, 131
orbit
    stable, 31
    unstable, 31
orthogonally adjacent, 299
oscillators, 299, 304

pacemaker, 129
PCX file, 13
Peano curve, 153
pendulum, 21
    chaotic, 340–341
    driven, 341
period, 234
    attractor, 68
    doubling, 24
    doubling cascade, 37, 97
periodicity, 36, 204
phase
    diagram, 64
    plane, 64
    portrait, 64, 101
    space, 21, 101
pinch point, 232
plane-filling curve, 153
Poincaré section, 23

point attractor, 32
Prandtl number, 92
predator–prey, 64, 65

quadric curve, 155
Quaternions, 240

rain, 160
random fractal curves, 150
Rayleigh number, 92
real number, 187
reciprocal, 241
recursion, 145
reversibility, 296
Rossler
    attractor, 126–127
    equations, 126–127
    funnel, 126
rule sets, 291
    non-totalistic, 295
run length coding, 14
Runge–Kutta method, 70

saddle, 68
satellites, 195
screen
    dump, 15
    grabber, 14
    photography, 15
    resolution, 2
self-organisation, 297
self-similarity, 38, 185
Sierpinski carpet, 155, 266
SIR model, 130
snowflakes, 160
space-filling curves, 149
stability, 71
stable
    focus, 69
    node, 68
    structures, 304
statistical self-similarity, 144
Stock Market, 335
strange attractor, 38, 103, 119–133

string rewriting systems, 156

tessellation automata, 289
thermodynamically irreversible, 297
thermodynamics, 296
TIFF file, 13
tiling, 272
time
    dependent, 301
    history, 64
    progression, 64, 93, 94
    series, 64
trajectory, 64, 67, 153, 231
transformation
    affine, 268
    contractive, 269
triadic curve, 155

unstable
    focus, 69
    node, 68

Van der Pohl oscillator, 129, 345
Vast Intellect, 20
Volterra's equations, 65

weather, 97–98, 159, 160, 337
wrap-round, 289

z plane, 229